Emergence, Complexity and Computation

Volume 44

The Emergence, Complexity and Computation (ECC) series publishes new developments, advancements and selected topics in the fields of complexity, computation and emergence. The series focuses on all aspects of reality-based computation approaches from an interdisciplinary point of view especially from applied sciences, biology, physics, or chemistry. It presents new ideas and interdisciplinary insight on the mutual intersection of subareas of computation, complexity and emergence and its impact and limits to any computing based on physical limits (thermodynamic and quantum limits, Bremermann's limit, Seth Lloyd limits…) as well as algorithmic limits (Gödel's proof and its impact on calculation, algorithmic complexity, the Chaitin's Omega number and Kolmogorov complexity, non-traditional calculations like Turing machine process and its consequences,…) and limitations arising in artificial intelligence. The topics are (but not limited to) membrane computing, DNA computing, immune computing, quantum computing, swarm computing, analogic computing, chaos computing and computing on the edge of chaos, computational aspects of dynamics of complex systems (systems with self-organization, multiagent systems, cellular automata, artificial life,…), emergence of complex systems and its computational aspects, and agent based computation. The main aim of this series is to discuss the above mentioned topics from an interdisciplinary point of view and present new ideas coming from mutual intersection of classical as well as modern methods of computation. Within the scope of the series are monographs, lecture notes, selected contributions from specialized conferences and workshops, special contribution from international experts.

Indexed by zbMATH.

More information about this series at https://link.springer.com/bookseries/10624

Hector Zenil · Fernando Soler Toscano ·
Nicolas Gauvrit

Methods and Applications of Algorithmic Complexity

Beyond Statistical Lossless Compression

Springer

Hector Zenil
Algorithmic Nature Group
Paris, UK

Oxford Immune Algorithmics
Reading, UK

The Alan Turing Institute
The University of Cambridge
London, UK

Algorithmic Dynamics Lab
Karolinska Institute
Stockholm, Sweden

Nicolas Gauvrit
Universite de Paris VII
Paris, France

Fernando Soler Toscano
Departamento de, Lógica y Filosofía de la Ciencia
Universidad de Sevilla
Sevilla, Spain

ISSN 2194-7287 ISSN 2194-7295 (electronic)
Emergence, Complexity and Computation
ISBN 978-3-662-64983-1 ISBN 978-3-662-64985-5 (eBook)
https://doi.org/10.1007/978-3-662-64985-5

This Springer imprint is published by the registered company Springer-Verlag GmbH, DE part of Springer Nature.
The registered company address is: Heidelberger Platz 3, 14197 Berlin, Germany

Contents

Part I Theory and Methods

1 Preliminaries ... 3
1.1 Computability and the Behavior of Computing Programs ... 3
1.1.1 Deterministic Turing Machines ... 3
1.1.2 The Theory of Cellular Automata ... 4
1.1.3 Elementary Cellular Automata ... 4
1.1.4 Wolfram's Classification ... 5
1.2 Chance and Classical Probability Theory ... 7
1.2.1 Conditional Probability ... 8
1.2.2 Shannon's Information Theory ... 8
1.2.3 Noisy-Channel Coding Theorem and Redundancy ... 10
1.2.4 Bayes' Theorem ... 10
1.2.5 Data Compression ... 10
1.2.6 Compressibility of Cellular Automata ... 12
1.3 Kolmogorov Complexity and Algorithmic Probability ... 13
References ... 14

2 Enumerating and Simulating Turing Machines ... 17
2.1 The Complete Enumeration of (s, k) ... 18
2.2 Graphical Representation of Turing Machines ... 20
2.3 The Reduced Enumeration ... 20
2.4 From Reduced to Complete Enumeration ... 25
2.5 Simulating Turing Machines ... 27
2.6 Decoding the Enumeration ... 27
2.7 Detecting Non-halting Machines ... 27
2.7.1 Machines Without Transitions to the Halting State ... 27
2.7.2 Detecting Escapees ... 28
2.7.3 Detecting Cycles ... 29
2.8 Running Machines and Storing the Output Strings ... 29
2.8.1 Producing Random Machines ... 30
2.9 Halting and Runtime Distributions ... 30

2.9.1 Halting History of (2, 2) and (3, 2) Turing Machines 31
2.9.2 Returning the Runtime 31
2.9.3 Two-dimensional Turing Machines 34
References 38

3 The Coding Theorem Method 39
3.1 The Limits of Statistical Compression Algorithms 40
3.1.1 The Problem of Short Strings 42
3.2 Approximating the Universal Distribution 42
3.3 The Empirical Distribution D 44
3.4 Methodology 46
3.4.1 Numerical Calculation of $D(4)$ 46
3.4.2 Algorithmic Probability Tables 47
3.4.3 Derivation and Calculation of Algorithmic Complexity 55
3.4.4 Runtimes Investigation 57
3.5 Calculating $D(5)$ 62
3.5.1 Setting the Runtime 62
3.6 A Glance at $D(5)$ 62
3.7 Reliability of the Approximation of $D(5)$ 63
3.7.1 Exponential Regression 65
3.7.2 The Exponential Fit 67
3.8 Features of $D(5)$ 67
3.8.1 Lengths 67
3.8.2 Global Simplicity 68
3.8.3 Binomial Behavior 69
3.8.4 A Bayesian Approach 70
3.9 Comparing $D(4)$ and $D(5)$ 73
3.9.1 Agreement in Probability 74
3.9.2 Agreement in Rank 74
3.10 Kolmogorov Complexity Calculation 76
3.10.1 Randomness in $D(5)$ 78
3.10.2 Robustness of $K_{D(n)}$ 79
3.11 Validation by the Number of Instructions Used 80
3.11.1 Logical Depth and $K_{D(5)}$ 82
3.12 Summary 86
References 88

4 Theoretical Aspects of Finite Approximations to Levin's Semi-measure 89
4.1 Algorithmic Information Measures 89
4.2 Turing Machines as a Prefix-Free Set of Programs 90
4.3 A Levin-Style Algorithmic Measure 92
4.3.1 Finite Approximations to m 93
4.3.2 Properties of m and m_k 94
4.4 Computing m_5 96
References 100

5 Validation and Generalization of CTM 101
5.1 Multidimensional Complexity 101
5.2 Deterministic Two-Dimensional Turing Machines 101
5.3 An Approximation to the Universal Distribution 102
5.4 Evaluating 2-Dimensional Kolmogorov Complexity 105
5.5 Cross-Validation of the Coding Theorem Method by Compressibility 110
5.6 Limitations of Lossless Compression Approaches 110
5.7 CTM and Compression on Strings of Length 10 to 15 111
5.8 Comparing 1-Dimensional Strings in $(4, 2)_{2D}$ to $(4, 2)$ 114
5.9 Comparison of CTM and Compression of Cellular Automata 117
References 123

6 The Block Decomposition Method 125
6.1 The Use and Misuse of Lossless Compression 125
6.1.1 Building Upon Entropy Rate 126
6.2 The Block Decomposition Method 127
6.2.1 l-Overlapping String Block Decomposition 128
6.2.2 2- and w-Dimensional Complexity 129
6.2.3 BDM Upper and Lower Absolute Bounds 131
6.3 Dealing with Object Boundaries 132
6.3.1 Recursive BDM 133
6.3.2 Periodic Boundary Conditions 133
6.4 Error Estimation 134
6.4.1 BDM Convergence Towards Shannon Entropy 136
6.5 Normalized BDM 137
6.6 Behaviour of CTM to BDM Transition 139
6.6.1 Smooth BDM (and 'Add Col') 141
6.6.2 Weighted Smooth BDM with Mutual Information 142
6.7 A Logical Depth-Based Measure 143
6.8 BDM Asymptotic Time Complexity and Scalability 145
6.9 Algorithmic Complexity of Integer Sequences 146
6.10 Summary 148
References 149

7 Conditional BDM 151
7.1 Introduction 151
7.1.1 On the Second Term 152
7.2 Joint and Mutual BDM 152
7.3 The Relationship Between Conditional, Joint and Mutual Information 153
7.4 Properties and Relationship with Entropy 154
7.4.1 The Impact of Partition Strategy 155
7.4.2 Conditional BDM as an Extension of Conditional Entropy 155
7.5 Numerical Results 157

7.5.1 Conditional BDM Compared to Conditional Entropy over Biased Distributions ... 157
7.5.2 Conditional BDM over Different Partition Sizes ... 158
7.6 Distance to Conditional Kolmogorov Complexity ... 159
7.7 Hybrid Classical and Algorithmic Conditional BDM ... 160
References ... 161

Part II Applications

8 Applications to Graph and Network Complexity ... 165
8.1 Introduction ... 165
8.2 Preliminaries ... 166
8.3 Applying BDM to Graphs ... 167
8.3.1 Complexity and Graph Vertex Order ... 168
8.3.2 Complexity and Edge Density ... 169
8.3.3 Graph Duality ... 171
8.4 Testing BDM and Boundary Condition Strategies ... 172
8.5 Graph Automorphisms and Kolmogorov Complexity ... 176
8.6 Applying BDM to Real-World Natural and Social Networks ... 179
8.7 The Algorithmic Randomness of Synthetic Complex Networks ... 180
8.7.1 Network Connectedness and Complexity ... 181
8.7.2 Topological Characterization of Artificial Complex Networks ... 184
References ... 188

9 Algorithmic Complexity in Cognition ... 191
9.1 Working Memory and Intelligence ... 191
9.1.1 Participants ... 193
9.1.2 Task ... 193
9.1.3 Results ... 194
9.1.4 Discussion ... 196
9.2 Randomness ... 197
9.2.1 The Usual Measures of Randomness ... 198
9.2.2 Understanding Random "Biases" ... 202
9.2.3 Applications ... 205
9.2.4 Comments ... 212
9.3 Development ... 213
9.3.1 Methods ... 215
9.3.2 Modulating Factors ... 218
9.3.3 Experiment ... 219
9.3.4 Results and Discussion ... 220
9.4 Aesthetic Preferences ... 223
9.4.1 General Method ... 225
9.4.2 Results ... 228
9.4.3 Conclusive Discussion ... 229

9.5 Visual Cognition ... 230
9.5.1 Method ... 232
9.5.2 Results and Discussion ... 233
9.6 Conspiracy Theories ... 235
9.6.1 General Method ... 236
9.6.2 Experiment 1 ... 237
9.6.3 Experiment 2 ... 240
9.6.4 Experiment 3 ... 241
9.6.5 General Discussion ... 244
9.7 The Emergence of Grammar ... 245
9.7.1 Method ... 247
9.7.2 Results ... 247
9.7.3 Discussion ... 251
References ... 253

Appendix: Source Code of the Turing Machine Simulator ... 257

Index ... 265

Part I
Theory and Methods

Chapter 1
Preliminaries

1.1 Computability and the Behavior of Computing Programs

An abstract computer program (or machine) can be defined in terms of input, output, and a set of allowable operations used to turn the input into the output. Examples of popular models of computation are Turing machines (sometimes denoted by TM), cellular automata (denoted by CA) and Post Tag systems, to mention but three examples of models of computation (there is a very long list). What make these models of computation is that they are capable of computation universality which is a concept mostly, intuitively and expressively grasped, that is they can simulate any machine of the same class and in fact they can simulate each other.

The Turing machine model represents the basic framework underlying many concepts in computer science, including the definition of algorithmic complexity. The cellular automaton is a well-known model which, together with the Post Tag system model, has been studied since the foundation of the field of abstract computation by some of its first pioneers.

1.1.1 Deterministic Turing Machines

A Turing machine is a theoretical computing machine invented by Alan Turing (1937) to serve as an idealized model for mathematical calculation. The Turing machine description consists of a list of rules (a finite program) capable of manipulating a linear list of cells, called the *tape*, using an access pointer called the *head*. The finite program can be in any one of a finite set of states Q numbered from 1 to n, with 1 the state at which the machine starts its computation. There are many Turing machine conventions, the one used in this book will be provided in Chap. 2. As an example, consider a Turing machine with the binary alphabet $\Sigma = \{0, 1\}$ and n states

H. Zenil et al., *Methods and Applications of Algorithmic Complexity*,
Emergence, Complexity and Computation 44,
https://doi.org/10.1007/978-3-662-64985-5_1

$\{1, 2, \ldots n\}$ and an additional halting state denoted by 0 (as defined by Rado in his original Busy Beaver paper [1]). The machine runs on a 2-way unbounded tape. At each step:

1. the machine's current "state" (instruction); and
2. the tape symbol the machine's head is scanning

define each of the following:

1. a unique symbol to write (the machine can overwrite a 1 on a 0, a 0 on a 1, a 1 on a 1, and a 0 on a 0);
2. a direction to move in: -1 (left), 1 (right) or 0 (none, when halting); and
3. a state to transition into (which may be the same as the one it was in).

The machine halts if and when it reaches the special halt state 0. There are $(4n + 2)^{2n}$ Turing machines with n states and 2 symbols according to this formalism. The output string is taken from the number of contiguous cells on the tape the head of the halting n-state machine has gone through. A Turing machine is considered to produce an output string only if it halts. The output is what the machine has written on the tape.

It is worth noting that the Busy Beaver problem is defined for Turing machines with initial empty tapes, and Turing machines studied in this book are all provided with an initially empty tapes too. However, for every Turing machine with an arbitrary input there is another Turing machine with empty input producing the same output, hence Turing machines with empty tapes produce all possible output (the translation may only result in some extra states).

1.1.2 The Theory of Cellular Automata

Another type of computer program that is called a cellular automaton is a collection of cells on a grid of specified shape, that evolves through a number of discrete time steps according to a set of rules based on the states of neighboring cells. The rules are applied iteratively for as many time steps as desired. The number of states (or colors) (k) of a cellular automaton is a non-negative integer. In addition to the grid on which a cellular automaton evolves and the colors its cells may assume, the neighborhood over which cells affect one another must also be specified and can vary in many forms. Cellular automata can also be defined in n different dimensions, evolving in space-time diagrams that can be visualized in $n + 1$ dimensions.

1.1.3 Elementary Cellular Automata

The simplest type of cellular automaton is the one-dimensional automaton, $k = 2$ (binary), nearest-neighbor, (named *Elementary* by Wolfram [2]). There are 256 such automata, each of which can be indexed by a unique binary number whose decimal representation is known as the rule for the particular automaton.

A one-dimensional CA can be represented by an array of *cells* x_i where $i \in \mathbb{Z}$ (integer set) and each x takes a value from a finite alphabet Σ. Thus, a sequence of cells $\{x_i\}$ of finite length n describes a string or *global configuration* c on Σ. This way, the set of finite configurations will be expressed as Σ^n. An evolution comprises a sequence of configurations $\{c_i\}$ produced by the mapping $\Phi : \Sigma^n \rightarrow \Sigma^n$; thus the global relation is symbolized as:

$$\Phi(c^t) \rightarrow c^{t+1} \tag{1.1}$$

Where t represents time and every global state of c is defined by a sequence of cell states. The global relation is determined over the cell states in configuration c^t updated simultaneously at the next configuration c^{t+1} by a local function φ as follows:

$$\varphi(x^t_{i-r}, \ldots, x^t_i, \ldots, x^t_{i+r}) \rightarrow x^{t+1}_i. \tag{1.2}$$

Wolfram [2] represents 1-dimensional cellular automata (CA) with two parameters (k, r) where $k = |\Sigma|$ is the number of states, and r is the neighborhood radius. Hence this type of CA is defined by the parameters (2, 1). There are Σ^n different neighborhoods (where $n = 2r + 1$) and k^{k^n} distinct evolution rules. The evolutions of these cellular automata usually have periodic boundary conditions. Wolfram calls this type of CA Elementary Cellular Automata (denoted simply by ECA) and there are exactly $k^{k^n} = 256$ rules of this type. They are considered the most simple cellular automata (and among the simplest computing programs) capable of great behavioral richness.

1.1.4 Wolfram's Classification

Regardless of the apparent simplicity of their formal description, cellular automata are capable of displaying a wide range of interesting and different dynamical properties as thoroughly investigated by Wolfram in [2]. The problem of classification is a central topic in cellular automata theory.

Wolfram identifies and classifies cellular automata (and other discrete systems) as displaying these four different classes of behavior. Wolfram's classes can be characterised as follows:

- Class I. CA evolving to a homogeneous state
- Class II. CA evolving periodically
- Class III. CA evolving chaotically
- Class IV. Includes all previous cases, known as a class of *complex rules*

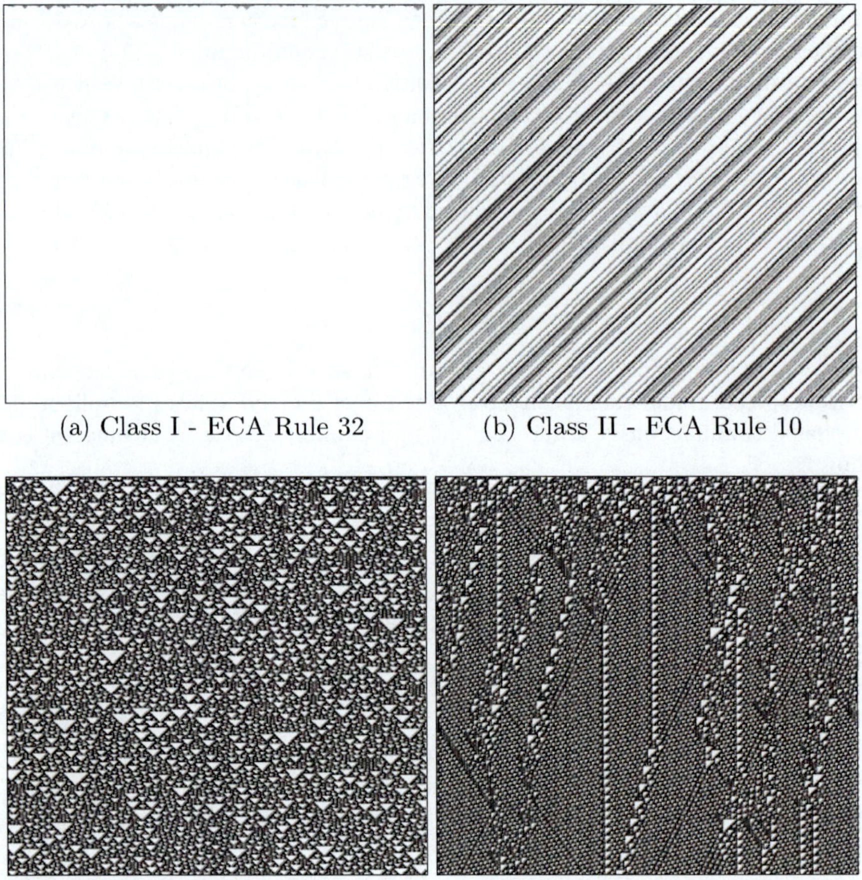

(a) Class I - ECA Rule 32
(b) Class II - ECA Rule 10
(c) Class III - ECA Rule 126
(d) Class IV - ECA Rule 110

Fig. 1.1 Wolfram's classes represented by ECA rules. We have the same initial condition in all these cases, with a density of 50% for state 0 (white dots) and state 1 (black dots). The evolution space begins with a ring of 358 cells and lasts 344 generations

Figure 1.1 illustrates Wolfram's classes, focusing on a specific ECA evolution rule (following Wolfram's notation for ECA [2]). All evolutions begin with the same random initial condition. Thus, Fig. 1.1a displays ECA Rule 32 converging quickly to a homogeneous state, Class I. Figure 1.1b displays blocks of cells in state one which evolve periodically showing a leftward shift, Class II. Figure 1.1c displays a typical chaotic evolution, where no pattern can be recognised or any limit point identified, Class III. Finally, Fig. 1.1d displays the so called complex class or Class IV. Here we see non-trivial patterns emerging in the evolution space. Such patterns possess a defined form and travel along the evolution space. They interact (collide), giving rise to interesting reactions such as annihilations, fusions, solitons, and reflections, or they produce new structures. These patterns are referred to as *gliders* in the CA literature ('glider' is a widely accepted concept popularised by John Conway through his well-known additive 2D CA, the *Game of Life* (GoL) [3]). In Class IV CA we see regions with periodic evolutions and chaos, and most frequently in complex rules the background is dominated by stable states, such as in GoL.

1.2 Chance and Classical Probability Theory

It is a common observation that some events are less predictable than others. When is is justified to regard a given phenomenon as random?

The most common approach in classical probability theory is to frame events in a set-theoretic framework. Events have an outcome that belongs to a set of possible outcomes. Traditionally events are assumed to be repeatable, and such a repetition is called a trial. Outcomes are traditionally direct observables. Let us denote the probability P of an event A_n as $P(A_n)$. For example, in the roll of a die, events $A_n = \{1, 2, \ldots, 6\}$ are elements of a space of possible outcomes of A_n. Two events A and B are called mutually exclusive, or disjoint, if the occurrence of A rules out the occurrence of B. In the dice example, if it turns up "three dots", this event rules out the event "six dots". Two non-disjoint events would be "a roll of less than 3" and "an even number of dots", to mention but one example.

Associated with each possible event A is its "probability" of occurrence $P(A)$. This numbers obeys the following axioms:

1. $P(A) \geq 0$. That is, a probability is never negative.
2. $P(\textit{all events}) = 1$.
3. If events A and B are disjoint, $P(A \textit{ or } B) = P(A) + P(B)$ where *or* is the traditional logical disjunction.

These three axioms, together with another version of axiom 3 including an infinity of disjoint events, imply all the rules of traditional probability theory, and hence many statistical methods.

With 2 and 3 we have that $\sum_{i=1}^{n} P(A_i) = 1$. That is, the sum of the probabilities of all the events is 1.

The word "probability" is but a mathematical abstraction for the intuitively more meaningful term "frequency of occurrence" that will be often used throughout this book (together with probability). The use of one or another term is mathematically justified because of the "law of large numbers" connecting frequency of an event A, that is $f(A)$ and its probability in the context of the limit of an infinity of independent occurrences. Formally, it says that $f(A)$ asymptotically approaches $P(A)$ as $n \to \infty$:

$$P(A) = \lim_{n\to\infty} \frac{m}{n} \tag{1.3}$$

1.2.1 Conditional Probability

Conditional probability answers the question of what is the probability of an event if another event has already occurred. This is formally defined as:

$$P(B|A) = \frac{P(A \cap B)}{P(A)} \tag{1.4}$$

where $P(B|A)$ denotes the conditional probability of event "B if A" and $P(A \cap B)$ is the probability of the simultaneous occurrence of A and B.

1.2.2 Shannon's Information Theory

The central paradigm of classical information theory is the engineering problem of the transmission of information over a noisy channel. Claude Shannon [4] came up with an extraordinary idea today at the center of information theory establishing the bit as the unit of information, given that one can encode any amount of digital information in bits.

Bits can be very powerful as they can encode a lot of information. For example with only 20 bits one can identify any town in United States. With a simple strategy consisting of asking twenty questions which answer can be yes or no, hence corresponding to 20 bits of information. There are $2^{20} = 1\,048\,576$ equally probable alternatives to accommodate 20 bits in a string, hence equal number of towns that each configuration can uniquely encode. In general, to encode data with n entries one will need only about $\log n$ questions (bits). However what bit encodes what answer is a different matter. The limit of this approach is most visible at the level of semantics because no meaning is captured.

Shannon's measure also called information entropy has been mistakenly taken as a general measure of complexity or information content but was proposed in the context of communication theory and heavily draws upon classical probability theory. The measure is intended to quantify how much information of a message is received in an imperfect communication channel which introduces some uncertainty, but has nothing o do with the message itself. Formally, the information in bits $I(A, B)$ in a message A about an event B is:

$$I(A, B) = \log_2 \frac{P(B|A)}{P(B)} \tag{1.5}$$

By this definition, I measures simply the probability $P(B|A)$ that event B occurred if message A is received, relative to the probability $P(B)$ that B occurred regardless of any message. If the message A contains significant information about the occurrence B, once A is known surely the occurrence of B should be more probable than if no message were received, but it is by no means a measure of information content, randomness or structure about the message itself. It heavily depends upon the concept of conditional probability and it has also been related to a measure of surprise, given that the more unexpected is an event B (i.e. the smaller $P(B)$), the higher I, with a perfectly reliable channel when $I(A, B) = -\log_2 P(B)$. It can grow, however, unlimitedly with the degree to which $P(B|A)$ exceeds $P(B)$ with the logarithm slowing down the growth, and more importantly providing additivity of independent messages.

For example, specifying the outcome of a fair coin flip (two equally likely outcomes) requires one bit at a time, because the results are independent and therefore each result conveys maximum entropy. Things begin to get interesting when the coin is not fair. If one considers a two-headed coin, then the tossing experiment always results in heads, and the message will always be 1. One caveat regarding Shannon's entropy is that one is forced to make an arbitrary choice when the basic units are ambiguous. Take for example the bit string 01010101010101. The Shannon entropy of the string at the level of single bits is maximal, as there are the same number of 1s and 0s, but the string is clearly regular when two-bit blocks are taken as basic units, in which instance the string has minimal complexity because it contains only 1 symbol (01) from among the 4 possible ones (00, 01, 10, 11). One way to overcome this problem is by taking into consideration all possible granularities (we call this *Block Entropy*), from length 1 to n where n is the length of the sequence. To proceed this way is computationally expensive as compared to proceeding in a linear fashion for fixed n, as it entails producing all possible overlapping $\binom{i}{n}$ substrings for all $i \in \{1, \ldots, n\}$. One can limit the growth by picking a small n, say $n = 4$, as we did. This means that Block Entropy will only identify short range regularities of up to 4 bits.

The capacity C of a channel is then the maximum number of bits (binary digits) per second that can be transmitted. Suppose that a source transmits messages selected from a finite set, and the relative frequencies of each message define a discrete probability distribution. Let p_A be the relative frequency that message A is transmitted from a source. The entropy of an ensemble of messages X, in which events s_i occur with probabilities p_i is calculated as

$$H(X) = \sum_{i=1}^{n} P(A_i) \log_2(P(A_i)). \tag{1.6}$$

Where the sum goes over all different genotypes i in X. If $P(A_i) = 0$ for some i it is defined $H = 0$.

Shannon's entropy H is zero when probability for $P(A_i) = 0$ or $P(A_i) = 1$ because if the probability for any one of the variables is 1 then the entropy is minimal, that is why this measure has been associated to uncertainty. On the other hand, H is maximum when $P(A_i) = 1/n$ or in other words, a uniform distribution. H has its largest value only when the symbols are equally probable.

1.2.3 Noisy-Channel Coding Theorem and Redundancy

Shannon's noisy-channel coding theorem [5], establishes that for any given degree of noise in a communication channel, it is possible to communicate a message nearly error-free up to some maximum rate through the channel. More formally, for every digital memoryless (no storing capabilities) channel, the channel capacity given by $C = \sup_{pX} I(X; Y)$. For $\epsilon > 0$ and $R < C$, there exists a code of length N and rate $\geq R$ and a decoding algorithm, such that the maximal probability of block error is $\leq \epsilon$ for N sufficiently large. In other words, arbitrarily low error probability communicating across the noisy channel is achievable up to a computable maximum rate associated to the noise in the channel that asymptotically guarantees that a block of data can be communicated nearly error-free (at a cost of increasing computational time).

1.2.4 Bayes' Theorem

In its simplest form, Bayes' theorem establishes a relationship between the probabilities of 2 events A and B, $P(A)$ and $P(B) \neq 0$, and the conditional probabilities $P(A|B)$ and $P(B|A)$, then Bayes' theorem establishes that:

$$P(A|B) = \frac{P(B|A)P(A)}{P(B)} \tag{1.7}$$

Bayes' theorem as a rule provides some kind of "backwards probability" where $P(A)$ is traditionally called the prior (or initial degree of belief in A) and $P(A|B)$ is the constructed *posterior*.

1.2.5 Data Compression

There[1] are two basic forms of compression, lossy and lossless. A good introduction can be found in [6]. Compression is a powerful tool for pattern recognition and has often been used for classification and clustering. Lossless compression reduces bits by identifying and eliminating statistical redundancy and constructing a dictionary where shorter code words are assigned to the most frequent segments. No information is lost in lossless compression. Lossy compression, however, reduces bits by identifying marginally important information and removing it.

---

[1] Some material in this section was adapted from **On the Algorithmic Nature of the World**, Authors: Hector Zenil and Jean-Paul Delahaye, Published in *World Scientific Series in Information Studies: Volume 2, Information and Computation Essays on Scientific and Philosophical Understanding of Foundations of Information and Computation*, pp. 477–496, Copyright 2011, with permission from World Scientific.

Lempel-Ziv (LZ)-like lossless data compressors have been proven to be universally optimal and are therefore good candidates as approximations of the Program-size (Kolmogorov) complexity of strings. LZ compression methods are among the most popular algorithms for lossless storage. DEFLATE is a variation on LZ optimized for decompression speed and compression ratio, but compression rate can be determined at execution. DEFLATE is used in programs and formats such as PKZIP, GZIP and PNG. LZW (Lempel–Ziv–Welch) was also used in GIF images. LZ methods use a table-based compression model where table entries are substituted for repeated strings of data. For most LZ methods, this table is generated dynamically from earlier data in the input.

Lossless compression-based mathematical characterizations and techniques for classifying and clustering have been suggested and successfully developed in areas as diverse as languages, literature, genomics, music, and astronomy, but also to classify the evolution of abstract symbolic systems such as cellular automata (see e.g. [7]).

1.2.5.1 Frequency Distributions

The distribution of a variable is a description of the relative number of times each possible outcome occurs in a number of trials. One of the most common probability distributions describing physical events is the normal distribution, also known as the Gaussian or Bell curve distribution, with values more likely to occur due to small random variations around a mean.

There is also a particular scientific interest in power-law distributions, partly from the ease with which certain general classes of mechanisms generate them. The demonstration of a power-law relation in some data can point to specific kinds of mechanisms that might underlie the natural phenomenon in question, and can indicate a connection with other, seemingly unrelated systems.

As explained however, when no information is available, the simplest distribution is the uniform distribution, in which values are equally likely to occur. In a macroscopic system at least, it must be assumed that the physical laws which govern the system are not known well enough to predict the outcome. If one does not have any reason to choose a specific distribution and no prior information is available, the uniform distribution is the one making no assumptions according to the principle of indifference. This is supposed to be the distribution of a balanced coin, an unbiased die or a casino roulette where the probability of an outcome k_i is $1/n$ if k_i can take one of n possible different outcomes.

1.2.5.2 Correlation Coefficients

In statistics, correlation refers to any statistical relationship between two random variables or two sets of data. The correlation coefficient, sometimes also called the cross-correlation coefficient, is a quantity that gives the quality of a least squares fitting to the original data. There are several correlation coefficients, often denoted ρ or r, measuring the degree of correlation. The most common of these is the Pearson correlation coefficient, here we use quite often the Spearman correlation coefficient that allows ranking comparisons between classifications.

The Spearman rank correlation coefficient [8] is a non-parametric measure of correlation that makes no assumptions about the frequency distribution of the variables. Spearman's rank correlation coefficient is equivalent to the Pearson correlation on ranks. Spearman's rank correlation coefficient is usually denoted by the Greek letter ρ.

The Spearman rank correlation coefficient is calculated as follows:

$$\rho = 1 - \frac{6 \sum d_i^2}{n(n^2 - 1)}$$

where d_i is the difference between each rank of corresponding values of x and y, and n the number of pairs of values.

Spearman's rank correlation coefficient can take real values from -1 to 1, where -1 is a perfect negative (inverse) correlation, 0 is no correlation and 1 is a perfect positive correlation.

The approach to testing whether an observed ρ value is significantly different from zero, considering the number of elements, is to calculate the probability that it would be greater than or equal to the observed ρ given the null hypothesis using a permutation test [9] to ascertain that the obtained value of ρ obtained is unlikely to occur by chance (the alternative hypothesis). The common convention is that if the value of ρ is between 0.01 and 0.001 the correlation is strong enough, indicating that the probability of having found the correlation is very unlikely to be a matter of chance, since it would occur one time out of hundred (if closer to 0.01) or a thousand (if closer to 0.001), while if it is between 0.10 and 0.01 the correlation is said to be weak, although yet quite unlikely to occur by chance, since it would occur one time out of ten (if closer to 0.10) or a hundred (if closer to 0.01). The lower the significance level, the stronger the evidence in favor of the null hypothesis.

1.2.6 Compressibility of Cellular Automata

1-dimensional ECA can be visualized in 2-dimensional space-time diagrams where every row is an evolution in time of the ECA rule. By their simplicity and because we have a good understanding about them (e.g. at least one ECA is known to be capable of Turing universality [2, 10]) they are excellent candidates to test our work, being just as effective as other methods that approach ECA using compression algorithms [7] that have yielded the results that Wolfram obtained heuristically.

A compression-based classification of CA (and other systems) was proposed in [7], based on concepts from algorithmic complexity. The technique is based on the notion of asymptotic behaviour, and unlike the mean field theory technique, it analyses the statistical properties of CA by looking at the evolution space of individual rules. The method produced the following variation of Wolfram's classification [11].

- Class I: highly compressible evolutions for any number of steps;
- Class II: highly compressible evolutions for any number of steps;
- Class III: the lengths of compressed evolutions asymptotically converge to the uncompressed evolution lengths;
- Class IV: the lengths of compressed evolutions asymptotically converge to the uncompressed evolution lengths.

Wolfram's classes can be reformulated using a compression-based approach [7, 11–13]:

- Class I: insensitivity to initial configurations, inability to transfer information other than isolated bits;
- Class II: sensitivity to initial conditions, ability to transfer some information;
- Class III: insensitivity to initial configurations, inability to transfer information, perhaps due to lack of (evident means of) control;
- Class IV: sensitivity to initial conditions, ability to transfer some information.

In other words, when one changes the initial configuration of a system in either of classes I and III the system's behaviour remains the same (each evolution is equally compressible), and it is therefore considered unable to or inefficient at transferring information or programming a CA to perform (universal) computation. On the other hand, classes II and IV are better at transferring information, even if they may do so in different ways. This classification tells us that classes II and IV are more sensitive to initial configurations (e.g. Wolfram's ECA rule 22, considered to belong to Class II, and Wolfram's ECA Rule 110 belonging to Class IV).

1.3 Kolmogorov Complexity and Algorithmic Probability

Central to this book and to Algorithmic Information Theory is the definition of plain algorithmic (Kolmogorov–Chaitin or program-size) complexity [14–16].

Extensive introductions are given in [6, 17–19] so we will not cover the basics but rather or its vast theoretical side but instead we will focus on the new methods to approximate it beyond popular lossless compression algorithms that has so far been used and abused to approximate K. Abused because they offer little related to the algorithmic side of K and instead are closer to classical Shannon entropy estimators [20].

Formally, the plain algorithmic complexity K of a string s is given by:

$$K_T(s) = \min\{|p| \; : \; T(p) = s\} \tag{1.8}$$

where $|p|$ represents the length in bits of program p and $T(p) = s$ means that the universal Turing machine T with input p produces s and halts. Then, the complexity

$K_T(s)$ of string s is the length of the shortest program producing s in the universal Turing machine T.

$K_T(s)$ is a measure of the information content of s because it is the size of its shortest description. It is also a measure of the randomness of s because random string without repeated patterns cannot have descriptions shorter than themselves.

The measure K_T is said to be universal because the dependence on the chosen machine T is only up to an additive constant. This is guaranteed by the *Invariance Theorem* [6, 21]): If U_1 and U_2 are two universal Turing machines and $K_{U_1}(s)$ and $K_{U_2}(s)$ the algorithmic complexity of s for U_1 and U_2, then there exists a constant c such that:

$$|K_{U_1}(s) - K_{U_2}(s)| < c \tag{1.9}$$

Hence the longer the string, the less important c is (i.e. the choice of programming language or universal Turing machine). Then, we can simply write $K(s)$ instead of $K_T(s)$ when the choice of T is not relevant. However, in practice c can be arbitrarily large, thus having a very great impact on finite short strings.

A major challenge of K is its uncomputability. That is, there is no program which takes a string s as input and produces the integer $K(s)$ as output. But because of the many applications of K in quite different fields, there are some approximations to it. The most frequent is lossless compression, where the compressed file for s is considered as the program that will produce s in T (decompressor program).

In this book we will introduce another method to approximate K different to popular compression algorithms that is based on what we call the Coding Theorem Method (Chap. 3). It uses the relation between $K(s)$ and the algorithmic probability (also known as Levin's semi-measure) of s given by [22]:

$$m_T(s) = \sum_{p\,:\,T(p)=s} \frac{1}{2^{|p|}} \tag{1.10}$$

Measure $m_T(s)$ is equal to the probability that a random program (programmed by coin flips) halts and produces s in the universal Turing machine T. Is is called the Universal Distribution [23]. The relation between $K_T(s)$ and $m_T(s)$ is given by the Coding Theorem [22]:

$$|-\log_2 m_T(s) - K_T(s)| < c \tag{1.11}$$

for some constant c. The choice of T is relevant only up to a multiplicative constant, so we will frequently write $m(s)$ instead of $m_T(s)$.

References

1. Rad, T.: On non-computable functions. Bell Syst. Tech. J. **41**(3), 877–884 (1962)
2. Wolfram, S.: A New Kind of Science. Wolfram Media, Champaign (2002)

3. Gardner, M.: Mathematical games - the fantastic combinations of John Conway's new solitaire game "life". Sci. Amer. **23**, 120–123 (1970)
4. Shannon, C.E.: A mathematical theory of communication. Bell Syst. Tech. J. **27**, 379–423, 623–656 (1948)
5. Shannon, C.E., Weaver, W.: The Mathematical Theory of Communication. University of Illinois Press, Champaign (1949)
6. Li, M., Vitnyi, P.: An Introduction to Kolmogorov Complexity and Its Applications. Springer, Berlin (2008)
7. Zenil, H.: Compression-based investigation of the dynamical properties of cellular automata and other systems. Complex Syst. **19**(1), 1–28 (2010)
8. Snedecor, W., Cochran, W.G.: Statistical Methods, 8th edn. Iowa State University Press, Iowa City (1989)
9. Good, P.: Permutation Tests. Springer, Berlin (2000)
10. Cook, M.: Universality in elementary cellular automata. Complex Syst. **15**, 1–40 (2004)
11. Zenil, H., Villarreal-Zapata, E.: Asymptotic behavior and ratios of complexity in cellular automata. Int. J. Bifurc. Chaos **23**(09), 1350159 (2013)
12. Zenil, H.: On the dynamic qualitative behaviour of universal computation. Complex Syst. **20**(3) (2012)
13. Martinez, G.J., Seck-Tuoh-Mora, J.C., Zenil, H.: Computation and universality: class IV versus class III cellular automata (2013). arXiv:1304.1242
14. Kolmogorov, A.N.: Three approaches to the quantitative definition of information. Probl. Inf. Trans. **1**(1), 1–7 (1965)
15. Solomonoff, R.J.: A formal theory of inductive inference: parts 1 and 2. Inf. Control **7**(1–22), 224–254 (1964)
16. Chaitin, G.J.: On the length of programs for computing finite binary sequences. J. ACM (JACM) **13**(4), 547–569 (1966)
17. Calude, C.: Information and Randomness. An Algorithmic Perspective, 2nd revised and extended. Springer, Berlin (2002)
18. Downey, R., Hirschfeldt, D.R.: Algorithmic Randomness and Complexity. Springer, Heidelberg (2010)
19. Nies, A.: Computability and Randomness, vol. 51. Oxford University Press, Oxford (2009)
20. Zenil, H., et al.: Coding-theorem like behaviour and emergence of the universal distribution from resource-bounded algorithmic probability. Int. J. Parallel Emerg. Distrib. Syst. (2018)
21. Calude, C.S.: Information and Randomness. Springer, Berlin (2002)
22. Levin, L.: Laws of information conservation (non-growth) and aspects of the foundation of probability theory. Probl. Form. Trans. **10**, 206–210 (1974)
23. Kircher, W., Li, M., Vitnyi, P.: The miraculous universal distribution. Math. Intell. **19**(4), 7–15 (1997)

Chapter 2
Enumerating and Simulating Turing Machines

Our method alternative to popular lossless compression algorithms such as LZW consists in mining the space of all possible computer programs (from shorter to longer) to produce an empirical estimation of the so-called universal distribution [1, 2] related to algorithmic probability [1, 3]. To this end, we have chosen the most standard and studied model of computation, that of Turing machines [4] as used in the 'Busy Beaver game' [5].

Turing machines with s states $\{1, 2, \ldots, s\}$ plus the halting state 0 and k symbols $\{0, 1, \ldots, k-1\}$ have sk entries (s_1, k_1) in the transition table, each one with one instruction that determines the behavior of the machines. Such instructions are represented by

$$(s_1, k_1) \rightarrow (s_2, k_2, d)$$

where s_1 and k_1 are respectively the current state and the symbol being read and (s_2, k_2, d) represents the instruction to be executed: s_2 is the new state, k_2 the symbol to write and d the direction. If s_2 is the halting state 0, then $d = 0$, otherwise d is 1 (right) or -1 (left).

With these constraints there are $2sk + k$ different instructions (s_2, k_2, d). These are k instructions when $s_2 = 0$ (given that $d = 0$ is fixed and k_2 can be one of the k different symbols) and $2sk$ instructions if $s_2 \neq 0$ (2 possible moves, s states and k symbols). Let (s, k) be the set of Turing machines following these conventions. Then, considering the sk entries in the transition table,

$$|(s, k)| = (2sk + k)^{sk} \tag{2.1}$$

These machines can be enumerated from 0 to $|(s, k)| - 1$. An online program illustrates the enumeration and is available at the Wolfram Demonstrations project at http://demonstrations.wolfram.com/SmallTuringMachinesWithHaltingStateEnumerationAndRunningOnAB/ [6].

H. Zenil et al., *Methods and Applications of Algorithmic Complexity*, Emergence, Complexity and Computation 44, https://doi.org/10.1007/978-3-662-64985-5_2

Table 2.1 Enumeration of instructions in (3, 2)

$n(0, 0, 0) = 0$	$n(1, 0, -1) = 4$	$n(2, 0, -1) = 8$	$n(3, 1, 1) = 11$
$n(0, 1, 0) = 1$	$n(1, 1, -1) = 5$	$n(2, 1, -1) = 9$	$n(3, 0, -1) = 12$
$n(1, 0, 1) = 2$	$n(2, 0, 1) = 6$	$n(3, 0, 1) = 10$	$n(3, 1, -1) = 13$
$n(1, 1, 1) = 3$	$n(2, 1, 1) = 7$		

2.1 The Complete Enumeration of (s, k)

To enumerate all Turing machines in (s, k) it is possible to use natural numbers from 0 to $|(s, k)| - 1$. First, note that each instruction (s_i, k_i, d_i) can be represented by a natural number $n(s_i, k_i, d_i)$ from 0 to $2sk + k - 1$, given the $2sk + k$ different instructions. The used convention is

$$n(s_i, k_i, d_i) = \begin{cases} k_i & \text{iff } d_i = 0 \\ 2k(s_i - 1) + k_i + k & \text{iff } d_i = 1 \\ 2ks_i + k_i & \text{iff } d_i = -1 \end{cases} \tag{2.2}$$

As an example, Table 2.1 shows the enumeration of all the different instructions that can use Turing machines in (3, 2). Given a natural number X from 0 to $2sk + k - 1$, Algorithm 1 shows the process to obtain the instruction (s_X, k_X, d_X) such that $n(s_X, k_X, d_X) = X$.

Algorithm 1 Calculate (s_X, k_X, d_X), the instruction number X in (s, k)

Require: $0 \leq X < 2sk + k$
Ensure: $n(s_X, k_X, d_X) = X$
  $s \leftarrow$ number of states
  $k \leftarrow$ number of symbols
  if $X < k$ **then**
    $(s_X, k_X, d_X) \leftarrow (0, X, 0)$
  else
    $X \leftarrow X - k$
    $k_X \leftarrow (X \mod k)$
    $X \leftarrow \lfloor X/k \rfloor$
    if $(X \mod 2) = 0$ **then**
      $d_X \leftarrow 1$
    else
      $d_X \leftarrow -1$
    end if
    $X \leftarrow \lfloor X/2 \rfloor$
    $s_X \leftarrow (X \mod s) + 1$
  end if

$$3737616 = \overbrace{6}^{(1,1)\to(2,0,1)} \cdot 14^5 + \overbrace{13}^{(1,0)\to(3,1,-1)} \cdot 14^4 + \overbrace{4}^{(2,1)\to(1,0,-1)} \cdot 14^3 + \overbrace{1}^{(2,0)\to(0,1,0)} \cdot 14^2 + \overbrace{6}^{(3,1)\to(2,0,1)} \cdot 14 + \overbrace{8}^{(3,0)\to(2,0,-1)}$$

Fig. 2.1 Example of the Turing machine enumeration

So there is a one-to-one correspondence between the set of possible instructions of a Turing machine in (s, k) and natural numbers in $\{0, \ldots, 2sk + k - 1\}$. Equation (2.2) and Algorithm 1 show the conversions in both directions.

Now see the correspondence between Turing machines in (s, k) and natural numbers. Attending to (2.2), the sk entries on the transition table of a Turing machine can be represented by a vector with the following structure

$$(n_1^{k-1}, n_1^{k-2}, \ldots, n_1^0, n_2^{k-1}, \ldots, n_2^0, \ldots, n_s^{k-1}, \ldots, n_s^0) \tag{2.3}$$

where $n_{s'}^{k'}$ is the number given by (2.2) representing the instruction at the entry (s', k') in the transition table. Note that the order of the instructions in (2.3) goes from state 1 to s. The instructions for a given state are all together going from symbol $k - 1$ to 0.

To get the number M_n corresponding to a given Turing machine M, let $(d_1^M, d_2^M, \ldots, d_{sk}^M)$ be the vector of natural numbers representing the instructions of M, according to (2.2) and (2.3). Then, M_n is the number given by digits $(d_1^M, d_2^M, \ldots, d_{sk}^M)$ in base $2sk + k$, that is,

$$M_n = \sum_{j=1}^{sk} d_j^M (2sk + k)^{sk-j} \tag{2.4}$$

Figure 2.1 shows an example of the correspondence between natural numbers and the set of instructions of a Turing machine in (3, 2). Look at Table 2.1 for the numbers corresponding to each instruction and Eq. (2.3) for the order of instructions.

Given a decimal number M_n the digits $(i_1^M, \ldots, i_{sk}^M)$ of its representation in base $2sk + k$ provide the set of instructions of the Turing machine corresponding to M_n, according to Algorithm 1.

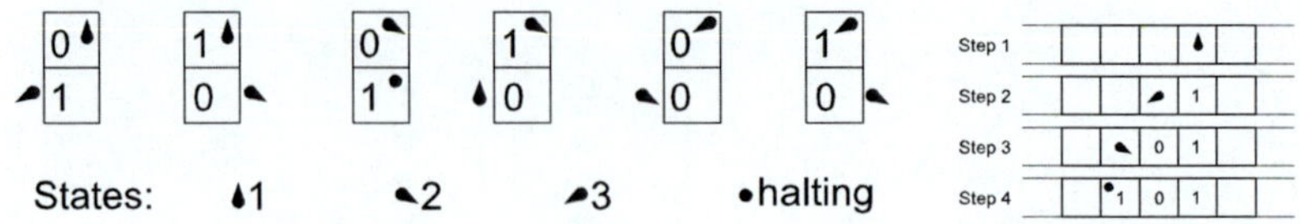

(a) Transition table (b) Execution

Fig. 2.2 Graphical representation of Turing machine 3737616 in (3, 2)

2.2 Graphical Representation of Turing Machines

Look at the Turing machine in Fig. 2.1. Entries in the transition table have been represented following the format $(s_1, k_1) \rightarrow (s_2, k_2, d)$. The same machine is represented in Fig. 2.2a. There are 6 pictures for the different entries in the transition table. States are represented by black drops pointing at different directions, as shown in the legend. The halting state is represented by a black dot. The top part of each transition corresponds to (s_1, k_1), the state and the symbol ('0' or '1') that the machine is reading. The bottom part is the instruction (s_2, k_2, d). The cell is filled with the symbol k_2 ('0' or '1') and the new state s_2 is drawn at the right (if $d = 1$) or the left (if $d = -1$). For transitions to the halting state, as $d = 0$, the black dot is drawn inside the cell.

The execution of a Turing machine can also be represented. Figure 2.2b shows the execution of the example machine over a blank tape (taking '0' as the blank symbol, the reader may check that this machine does not halt when starting on a tape filled with '1'). Each row in the diagram represents a computation step, being the first row the initial state. The symbol representing the state is inside the read cell. It can be seen that the example machine (number 3737616 in (3, 2), as shown in Fig. 2.1) halts after 3 steps. The output tape, from left to right, is "101".

2.3 The Reduced Enumeration

There are many symmetries amongst machines in (s, k). As the experiments require running complete (s, k) spaces or great samples of them and counting the number of occurrences of each output, it is important to use some of these symmetries to reduce the number of machines to run.

A first symmetry is with respect to the blank symbol. The experiments need to consider all k symbols as possible blank symbols, to give the result of running all machines in (s, k) starting in a blank tape, with all possible blank symbols. But it is enough to consider one blank symbol, say '0', because for any Turing machine M that starting in a k'-filled tape ($k' \neq 0$) halts and produces string x, there is another machine $M_{0\leftrightarrow k'}$ ($M_{0\leftrightarrow k'}$ obtained by interchanging the occurrences of '0' and k' in the transition table of M) that starts in a 0-filled tape and halts and produces $x_{0\leftrightarrow k'}$,

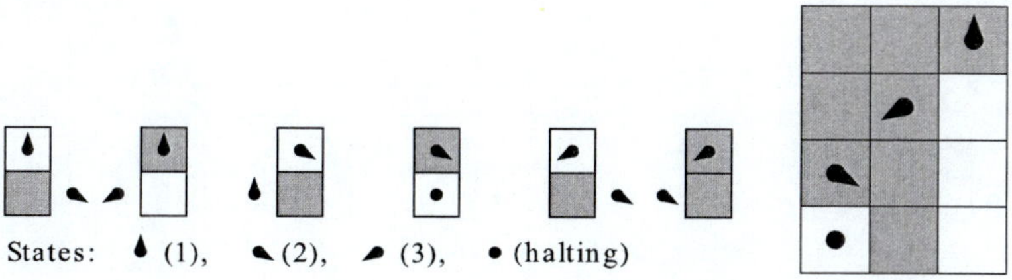

(a) Instructions of the '1'-'0' symmetric machine (b) Execution

Fig. 2.3 Interchanging '0' and '1' in (3, 2) machine 3737616. Symbol '0' is represented by white cells and '1' by gray cells

where $x_{0\leftrightarrow k'}$ is the result of interchanging '0' and k' in x one by the other. So the machines are run just starting in a '0'-filled tape and for any produced string x^* appearing n times there are added to the final result n occurrences of the string $x^*_{0\leftrightarrow k'}$, for every $k' \neq$ '0'.

Figure 2.3 shows an example of this symmetry. The left side shows the result of interchanging the occurrences of '0' and '1' in the transition table of (3, 2) machine 3737616. The resulting machine can be called $3737616_{0\leftrightarrow 1}$. Compare Figs. 2.2a and 2.3a, only the roles of '0' and '1' are interchanged. So when starting $3737616_{0\leftrightarrow 1}$ in a '1'-filled tape (Fig. 2.3b) the execution is symmetric to that of 3737616 (Fig. 2.2b), the apparition of '0' and '1' are interchanged. So the output string is '0'-'1' symmetric to that of machine 3737616, that is, "010". When there are more than 2 symbols the same symmetry happens. It is possible to interchange the role of '0' and any other symbol k'. In the output strings the apparitions of '0' and k' are interchanged accordingly.

The other used symmetry reduces the number of machines to be run. First, consider that machines with the initial transition $(init, blank) \rightarrow (init, k_2, d)$ keeping in the initial state will never halt. With independence of the written symbol k_2 and direction d, the machine, which runs on a blank tape, will find another new blank symbol, so the same transition $(init, blank) \rightarrow (init, k_2, d)$ is applied again, keeping in the initial state, and as it goes again in the same direction d, it finds a new blank symbol, so the machine never halts. Then it is possible to avoid running machines with the initial transition (the entry in the transition table that corresponds to $(init, blank)$) keeping in the initial state. As there are $2k$ possible initial transitions keeping in the initial state (2 directions and k symbols) and the other $sk - 1$ transitions may have any of the $2sk + k$ different instructions, this avoids running $2k(2sk + k)^{sk-1}$ non-halting machines.

Moreover, those machines moving to the halting state at the initial transition can be quantified. Those machines halt in just one step, but it can be saved a considerable time by skipping their generation. There are k possible initial transitions $(init, blank) \rightarrow (0, k', 0)$ going to the halting state, one for each different symbol k'. These machines produce the string "k'". As the other $sk - 1$ transitions may have any of the $2sk + k$ different instructions, the generation of $k(2sk + k)^{sk-1}$ machines is avoided.

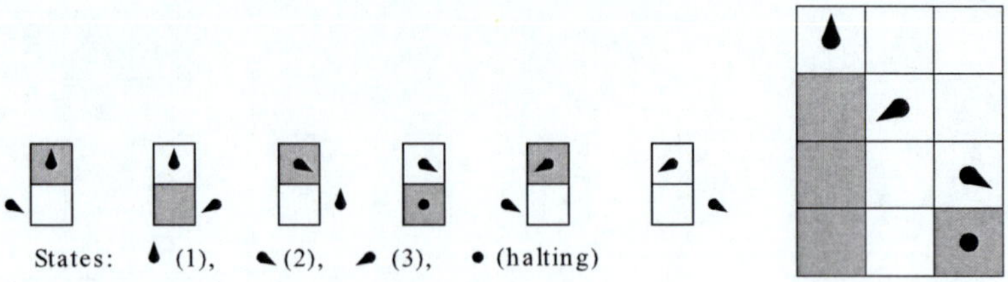

(a) Instructions of the left-right symmetric machine (b) Execution

Fig. 2.4 Left-right symmetric machine to 3737616 in (3, 2)

But the mayor reduction is obtained by exploiting the left-right symmetry. Look at the example of Fig. 2.4. The transition table shown in the left picture has exactly the transitions $(s_1, k_1) \to (s_2, k_2, -d)$ where machine 3737616 (Fig. 2.2a) has $(s_1, k_1) \to (s_2, k_2, d)$. So left and right movements are interchanged. The executions shown in the right part of Figs. 2.2 and 2.4 are thus left-right symmetric. So the resulting strings are each one the reversal of the other. So it is not necessary to run all machines. It is possible to run only those machines moving to the right at the initial transition, and for any produced string count the same number of occurrences of the reversed string. Of course, for cases as in Fig. 2.4 where the symmetric string is equal to the original one, the number of occurrences is counted twice.

Now a brief summary of the reductions. First, machines keeping in the initial state for the initial transition were skipped. Also machines whose initial transition goes to the halting state. And finally, machines moving to the left at the initial transition. Call $Red(s, k)$ to the reduced set of machines in (s, k), that is the result of eliminating those machines. So $Red(s, k)$ contains only machines whose initial transition is like

$$(init, blank) \to (s_2, k_2, right)$$

where s_2 is neither the initial or the halting state (so there are $s - 1$ possible values for s_2) and k_2 is one of the k symbols. Then, there are $k(s - 1)$ possible initial transitions. The other $sk - 1$ transitions may have any of the $2sk + k$ different instructions presented in Sect. 2.1. So,

$$|Red(s, k)| = k(s - 1)\,(2sk + k)^{sk-1} \tag{2.5}$$

By (2.1) and (2.5), $\lim_{s\to\infty} \frac{|Red(s,k)|}{|(s,k)|} = \frac{1}{2}$, so in the limit $Red(s, k)$ is a half of (s, k), but as Table 2.2 shows, for small values of s the advantage is even greater.

Now let see how $Red(s, k)$ can be enumerated. As it was done in Sect. 2.1 the first step is to establish a correspondence between instructions and natural numbers. For non-initial instructions, the same convention (2.2) as before is used. But as the possibilities for the initial transition have been reduced, a special enumeration for it is needed:

$$n_{init}(s_i, k_i, 1) = (s_i - 2)k + k_i \tag{2.6}$$

Table 2.2 The impact of reducing the enumeration in $(s, 2)$

(s, k)	$\|(s, k)\|$	$\|Red(s, k)\|$
(2, 2)	10 000	2 000
(3, 2)	7 529 536	2 151 296
(4, 2)	11 019 960 576	3 673 320 192
(5, 2)	26 559 922 791 424	9 658 153 742 336

Table 2.3 Enumeration of initial instructions in $Red(3, 2)$

$n_{init}(2, 0, 1) = 0$	$n_{init}(2, 1, 1) = 1$
$n_{init}(3, 0, 1) = 2$	$n_{init}(3, 1, 1) = 3$

Algorithm 2 Calculate (s_X, k_X, d_X), the initial instruction number X in $Red(s, k)$

Require: $0 \leq X < k(s-1)$
Ensure: $n_{init}(s_X, k_X, d_X) = X$
  $s \leftarrow$ number of states
  $k \leftarrow$ number of symbols
  $d_X \leftarrow 1$
  $k_X \leftarrow (X \mod k)$
  $X \leftarrow \lfloor X/k \rfloor$
  $s_X \leftarrow (X \mod (s-1)) + 2$

Table 2.3 shows the enumeration of the initial instructions in $Red(3, 2)$ according to (2.6). Algorithm 2 shows the construction of the initial transition from its value.

Non-initial transitions can be enumerated using function n as given in (2.2). Then, for a given Turing machine M in $Red(s, k)$, the vector

$$(n_s^{k-1}, n_s^{k-2}, \ldots, n_s^0, \ldots, n_1^{k-1}, \ldots, n_1^0) \tag{2.7}$$

is constructed, where each $n_{s'}^{k'}$ represents the number for the instruction at the entry (s', k') of the transition table of M, n_1^0 obtained according to (2.6) and the others with (2.2). Note the difference with (2.3). Now the states are sorted from s to 1. This is to consider the initial transition as the least significant digit in the enumeration. Now the number M_r for M in $Red(3, 2)$ is given by the mixed radix number

$$(n_s^{k-1})_{2sk+k} \cdots (n_1^1)_{2sk+k}(n_1^0)_{k(s-1)}$$

equal to

$$M_r = n_1^0 + k(s-1)\left(\sum_{j=1}^{k-1} n_1^j\,(2sk+k)^{j-1} + \sum_{i=2}^{s}\sum_{j=0}^{k-1} n_i^j\,(2sk+k)^{k(i-1)+j-1}\right) \tag{2.8}$$

$$1013011 = \overbrace{\mathbf{6}}^{(3,1)\to(2,0,1)} \cdot 4 \cdot 14^4 + \overbrace{\mathbf{8}}^{(3,0)\to(2,0,-1)} \cdot 4 \cdot 14^3 + \overbrace{\mathbf{4}}^{(2,1)\to(1,0,-1)} \cdot 4 \cdot 14^2 + \overbrace{\mathbf{1}}^{(2,0)\to(0,1,0)} \cdot 4 \cdot 14 + \overbrace{\mathbf{6}}^{(1,1)\to(2,0,1)} \cdot 4 + \overbrace{\mathbf{3}}^{(1,0)\to(3,1,1)}$$

Fig. 2.5 Example of the reduced enumeration

Figure 2.5 shows an example of the enumeration of $Red(3, 2)$. The numbers that correspond to non-initial transitions are given by (2.2) (as depicted in Table 2.1). The number of the initial transition is given by (2.6) (look at Table 2.3).

Algorithm 3 Calculate the numbers n_i^j, $1 \le i \le s, 0 \le j < k$, corresponding to the instructions in the transition table for Turing machine $M \in Red(s, k)$ with number X

Require: $0 \le X < |Red(s, k)|$
Ensure: n_i^j, $1 \le i \le s$, $0 \le j < k$, is the transition for state i and symbol j in machine $M \in Red(s, k)$
  $s \leftarrow$ number of states
  $k \leftarrow$ number of symbols
  $n_1^0 \leftarrow (X \mod k(s-1))$
  $X \leftarrow \lfloor X/(k(s-1)) \rfloor$
  for $j = 1$ **to** $k - 1$ **do**
    $n_1^j \leftarrow (X \mod (2sk + k))$
    $X \leftarrow \lfloor X/(2sk + k) \rfloor$
  end for
  for $i = 2$ **to** s **do**
    for $j = 0$ **to** $k - 1$ **do**
      $n_i^j \leftarrow (X \mod (2sk + k))$
      $X \leftarrow \lfloor X/(2sk + k) \rfloor$
    end for
  end for

Given X in the enumeration of $Red(s, k)$, Algorithm 3 returns the numbers for the transitions that correspond to the machine M number X. To construct the initial transition, Algorithm 2 can be used. Non-initial transitions can be obtained with Algorithm 1.

2.4 From Reduced to Complete Enumeration

The experiments require to obtain the output strings of all halting machines in complete (s, k) spaces, or at least in large samples. But running only the machines in $Red(s, k)$ supposes a great reduction on the number of machines to run.

Suppose that it is run a set $N \subseteq Red(s, k)$ of machines with '0' as the blank symbol, and S is the function that assigns for every string x (built from the k symbols) the number $S(x) \in \mathbb{N}$ of apparitions of string x when running the set of machines N. Let n be the number of non halting machines in N.

This section explains how to modify S to get S' which represents not the output strings of N but of $N' \subseteq (s, k)$, when running all machines in N' with all possible blank symbols. If $N = Red(s, k)$ then $N' = (s, k)$, so S' will represent the output strings for the complete (s, k) space using all possible blank symbols. In case N is a random sample of $Red(s, k)$ the result N' will be a greater random sample of (s, k). The number n' of non-halting machines in N' will also be obtained.

Note that the main characteristic of $Red(s, k)$ is that the starting transition considers only $k(s-1)$ possibilities out of the $2sk + k$ possibilities in (s, k). So the symmetries are used to complete the output data S of the machines in N by with the machines using the other $sk + 2k$ starting transitions. The completion is done in three steps, constructing new versions of S and n as follows:

1. $k(s-1)$ transitions moving to left to a different state than the halting and starting ones were avoided. Call $rev(x)$ to the function that reverses a string x. Then S_1 and n_1 are constructed, such that for all possible x built from the k symbols

$$S_1(x) = S(x) + S(rev(x))$$
$$n_1 = 2n$$

   This is justified because for any machine producing x starting to the right there is a left-right symmetric machine producing $rev(x)$ (look to Fig. 2.4 and the associated explanation). This is also valid for non-halting machines.
2. k transitions (writing one of the k symbols) from the initial to the halting state were also avoided. Then define S_2 and n_2 as

$$S_2(x) = \begin{cases} \frac{|N|}{k(s-1)} \text{ iff } |x| = 1 \\ S_1(x) \text{ otherwise} \end{cases}$$
$$n_2 = n_1$$

   Note that these machines halting at the first transition produce strings "k'" with length $|k'| = 1$ for all possible symbols "k'". There are no machines in $Red(s, k)$ producing strings with length 1, as halting machines in $Red(s, k)$ visit at least two different cells (the initial cell and its right neighbor).
3. Finally, $2k$ transitions from the initial state to itself (2 movements and k symbols) were avoided. Machines starting this way do not halt because they remain always at

Table 2.4 Output strings in $Red(2, 2)$ and $(2, 2)$

x	$S(x)$	$S'(x)$	x	$S(x)$	$S'(x)$	x	$S(x)$	$S'(x)$
0	0	2000	100	1	6	0100	0	2
1	0	2000	110	2	6	0110	0	2
00	132	508	000	0	4	1001	1	2
01	132	508	010	1	4	1011	1	2
10	122	508	101	1	4	1101	1	2
11	122	508	111	2	4	1111	1	2
001	1	6	0000	0	2	$n = 1\,478$		
011	2	6	0010	0	2	$n' = 13\,912$		

the initial state. So the completion is done by including $\frac{2|N|}{s-1}$ non-halting machines. Then,

$$S_3(x) = S_2(x)$$
$$n_3 = n_2 + \frac{2\,N|}{s-1}$$

With these completions, $S_3(x)$ is the number of machines in (s, k) that when starting in a '0'-filled tape produce string x upon halting. In the same way, n_3 is the number of non-halting machines in (s, k). But k different possible blank symbols may be considered. Let $x_{k_1 \leftrightarrow k_2}$ be the result of interchanging in x the occurrences of k_1 and k_2 (for example: $0110_{0\leftrightarrow 1} = 1001$, $0011_{0\leftrightarrow 2} = 2211$, $1111_{0\leftrightarrow 2} = 1111$). Then,

$$S'(x) = S_3(x) + \sum_{i=1}^{k-1} S_3(x_{0\leftrightarrow i})$$
$$n' = k \cdot n_3$$

Look at Fig. 2.3 and the associated explanation to justify this last step.

As an example, Table 2.4 shows at the x columns the strings x found in $(2, 2)$. The value $S(x)$ is the number of occurrences of each string when running all machines in $Red(s, k)$ using '0' as the blank symbol. The value $S'(x)$ is the number of occurrences in $(2, 2)$ running the machines for both '0' and '1' as blank symbols. The values of n and n' are also provided.

2.5 Simulating Turing Machines

Appendix A shows the source code of the main version of the Turing machine simulator used for the experiments. The most relevant parts of the code are commented here, with some details of the variations used to study runtime distributions and to make samples.

2.6 Decoding the Enumeration

The simulator uses the reduced enumeration (Sect. 2.3) to avoid running many trivial or predictable machines. Given a machine number in the reduced enumeration, Algorithm 3 gets the number corresponding to each transition and then Algorithm 2 decodes the initial transition and Algorithm 1 the others.

The function `init_turing_machine(int s, int k, mpz_t numberTM)` at line 114 creates the `turing_machine` structure that is simulated. Its arguments are two integers with the number of states `s` and symbols `k` and a big integer with the machine number in the reduced enumeration.

At line 125 the creation of the transition table starts. All instructions are contained in the array `transition_table`. Each one of the transitions is contained in a structure `instruction` (line 12), with elements `head_state` (for the new state), `write_symbol` (new symbol to write) and `dir` (the direction to move the head).

The initial transition is decoded in lines 150–156, that implement Algorithm 2. When decoding other transitions from the initial state (lines 158–179) and other states (lines 182–205) the simulator implements Algorithm 1.

2.7 Detecting Non-halting Machines

It is useful to avoid running machines when it is easy to check that will not stop. These machines will consume the runtime without yielding any output. The simulator implements some filters that work after the machines are generated, in order to detect non-halting computations and skip running them. Some of these are detected when filling the transition table, others at runtime.

2.7.1 Machines Without Transitions to the Halting State

When the simulator starts to fill the transition table, it creates a flag `halts` (line 144) with initial value -2. Then, if a certain transition goes to the halting state, the value of `halts` changes to -1 (lines 167 and 192). Then, if after completing the transition table the flag value is still -2, then the machine will not stop. Thus, at line 206, the `running_state` of the Turing machine is set to `halts`. As the state code -2 is used for non-halting machines and -1 for running machines, it is possible to detect, just after the machine is generated, if it contains a transition to the halting state. Machines with `running_state` equal to -2 are not run, the simulator just counts them as non-halting machines.

As an example of the relevance of this filter, for s states and 2 symbols, the reduced enumeration contains $2(s-1)(4s)^{2s-1}$ machines without transition to the halting state. When computing $Red(5, 2)$, this is 4.096×10^{12} machines. It represents 42.41% of $|Red(5, 2)|$.

2.7.2 Detecting Escapees

Escapees are machines that run infinitely in the same direction over the tape. Some kinds are simple to check in the simulator. The detection of escapees is implemented in the function `run_step(turing_machine *m)` (line 39) that simulates a step of the Turing machine. The simulator uses a counter `escapees` in the machine state (line 24) that indicates the number of consecutive not-previously-visited tape positions that the machines visits. When the machine visits a new blank cell, the boolean flag `escap` (line 58), originally `false` is set to `true` (lines 67 and 75).

If the counter `escapees` exceeds the number of states, then there is a loop that will repeat infinitely. To justify this, suppose that at some stage the machine is visiting a certain tape-position for the first time, moving in a specific direction (the direction that points toward new cells). If the machine continues moving in the same direction for $s+1$ steps, and thus reading blank symbols, then it has repeated some state s_i in two transitions. As it is always reading (not previously visited) blank symbols, the machine has repeated the transition for $(s_i, blank)$ twice. But the behavior is deterministic, so if the machine has used the transition for $(s, blank)$ and after some steps in the same direction visiting blank cells, it has repeated the same transition, it will continue doing so forever, because it will always find the same symbols.

There is another possible direction in which this filter may apply: if the read symbol is a blank one not previously visited, the shift is in the direction of new cells and there is no modification of state. In fact this would be deemed an escapee, because the machine runs for $s+1$ new positions over the tape. But it is an escapee that is especially simple to detect, in just one step and not $s+1$. The machines detected by this simple filter are called *short escapees*, to distinguish them from other, more general escapees.

The simulator detects escapees at lines 80–93. The filter is only activated if `escap` is true. In that case, the value of `escapees` is incremented. The filter first looks for short escapees (line 83) and then regular escapees (line 87). The codes for short and regular escapees are respectively -3 and -4. When the escapee is detected, the simulation of the machine step finishes with return code 0 (lines 85 and 89). When `escap` was false, the value of `escapees` is set to 0 (line 92).

2.7.3 Detecting Cycles

The simulator detects cycles of period two. They are produced when in steps t and $t + 2$ the tape is identical and the machine is in the same state and the same position. When this is the case, the cycle will be repeated infinitely. In a cycle of period two, the machine cannot change any symbol on the tape, because if it did, the tape would be different after two steps. Then the filter would be activated when there is a transition that does not change the tape, for instance

$$(s, k) \rightarrow (s', k, d)$$

where $d \in \{-1, 1\}$ is some direction (left, right) and the head is at position i on tape T, which is to say, reading the symbol T_i. Then, there is a cycle of period two if and only if the transition that corresponds to (s', T_{i+d}) is

$$(s', T_{i+d}) \rightarrow (s, T_{i+d}, -d)$$

Cycles are also detected in the function `run_step(turing_machine *m)`. The filter works at lines 96–107. If the cycle is detected, the state code is set to -5 and the simulation of the machine step finishes with return code 0 (line 105). If no filter applies, it finishes with code 1 (line 108).

2.8 Running Machines and Storing the Output Strings

The `main` function (line 239) uses a `map<string, unsigned long long>` to store the output strings and their occurrences. The `main` function uses the arguments that are given to the simulator when executed from the command-line. These are, in order:

1. Number of states (integer)
2. Number of symbols (integer)
3. Runtime bound (integer)
4. Number of the first machine to simulate (big integer)
5. Number of the last machine to simulate (big integer)

Arguments 4 and 5 define a block of consecutive machines that are simulated. The code at lines 262–275 simulates all the machines in the block. Each machine runs until it halts or a filter detects that it will not halt (the value of `run_step` is only 0 when some filter detects a non-halting machine or the machine halts) or the runtime bound is reached (line 267).

The output of the machine is obtained at line 268 by calling the function `outputTM`. This function (lines 214–227) returns the output string of the Turing machine for halting computations (lines 217–221, when `running_state` is 0) or the state code of non-halting computations (line 223).

The returned string or non-halting code is stored in the map `results` (line 269). Machines without transitions to the halting state are not simulated, so their code -2 is stored at line 271.

When all machines are simulated, the content of `results` is returned (lines 281–285). In each line, a string (or non-halting code) and its number of occurrences are printed is CSV format.[1]

2.8.1 Producing Random Machines

Appendix A shows the main version of the Turing machine simulator. But in the experiments, some modifications have been used for different purposes. The key changes of such modifications are now presented.

When making samples, the simulator does not run blocks of consecutive machines as explained above, but a great number of random machines. Random machines are generated instruction by instruction, using the implementation of the Mersenne Twister in the Boost C++ library.

As explained in this chapter, the initial transition of a machine $Red(s, k)$ can be represented by an integer from 0 to $k(s-1)-1$. The other transitions go from 0 to $2sk+k-1$. Then, to generate a random machine, it is generated a random integer from 0 to $k(s-1)-1$ and its value is given to `rest` at line 151 for the initial transition. For the rest of transitions, random integers from 0 to $2sk+k-1$ are generated and assigned to `rest` (line 162 for transitions from the initial state and 186 from other states).

Finally, the `main` function (line 239) does not receive a block of consecutive machines, but the number of random machines to simulate. The mail loop (lines 262–275) simulates the indicated number of machines.

2.9 Halting and Runtime Distributions

Calude and Stay showed that "long-running" Turing machines can only halt at non-random times; the density of non-random times near n is about $1/n$. "Long-running" means that if we have a universal Turing machine U and machine M is implemented by a program m for U of length n, then $U(m)$ runs for more than $c \times 2^n$ steps, where c is some uncomputable constant depending on U.

[1] The resulting files are available at the download section of the Online Algorithmic Complexity Calculator, http://complexitycalculator.com/download.html.

t	k_t	$p(k_t)$
–	6544	0.65
1	2000	0.20
2	800	0.080
3	160	0.016
4	56	0.0056
5	362	0.036
6	78	0.0078

Fig. 2.6 Runtime distribution at which all machines halt (those that don't are indicated by "—"). Where t is the number of steps, k_t the number of machines that halted at t (out of a total of 3456 that halt), and $p(k_t)$ is the halting probability of a machine to halt (or not) in time t

2.9.1 Halting History of (2, 2) *and* (3, 2) *Turing Machines*

We know that a machine halts if it enters the halting state before reaching the known Busy Beaver value $S(n)$. If it does not, then it never halts. The halting problem and the halting probability problem are closely related to the Busy Beaver problem in that a solution to any one of them would yield a solution to each of the others.

Consider the halting space of all (2, 2) Turing machines (with an extra halting state) provided with an empty tape. Figure 2.6 shows the runtime distribution at which all machines in (2, 2) halt (or do not). There are 10 000 2-state, 2-symbol Turing machines (the 10 000 figure comes simply from the formula giving the number of Turing machines with $n = 2$ states $(4n + 2)^{2n}$). No other Turing machine halts after 6 steps in (2, 2). Machines that never halt are 6544 in number, representing around 0.65 of the total.

Figure 2.7 shows the runtime distribution in (3, 2) and Fig. 2.8 compares this distribution with the exponential function $100 \times 2^{14-t}$.

Figure 2.9 shows the accumulated number of machines that halt in at most each possible runtime in (2, 2) (top) and (3, 2) (bottom). Interesting output distribution facts:

- Out of 7 529 536 machines only 2 146 912 halt.
- There are 5 382 624 machines that do not halt.
- Those machines that halt only produce 126 different output strings, with the largest being 6 digits in length (the Busy Beavers).
- Exactly 0.2 of the Turing machines produce a 0 or a 1 as output.

2.9.2 Returning the Runtime

Some experiments require to get runtime distributions, specially when running spaces without known busy beavers. As the maximum halting runtime is not known in those

t	k_t	$100 \times 2^{14-t}$	$p(k_t)$
–	5382624		0.71
1	1075648	819200	0.14
2	614656	409600	0.082
3	263424	204800	0.035
4	97216	102400	0.013
5	53760	51200	0.0071
6	20800	25600	0.0028
7	12512	12800	0.0017
8	4264	6400	0.00057
9	2424	3200	0.00032
10	1064	1600	0.00014
11	536	800	0.000071
12	304	400	0.000040
13	176	200	0.000023
14	128	100	0.000017

Fig. 2.7 Runtime distribution in (3, 2), where t is the number of steps, k_t the number of machines that halted at t, and $p(k_t)$ is the halting probability calculated from t and k_t. $100 \times 2^{14-t}$ is a good fit to the limit behavior as a function relating runtimes and the number of Turing machines halting at a certain runtime for the 14 runtimes at which Turing machines halt

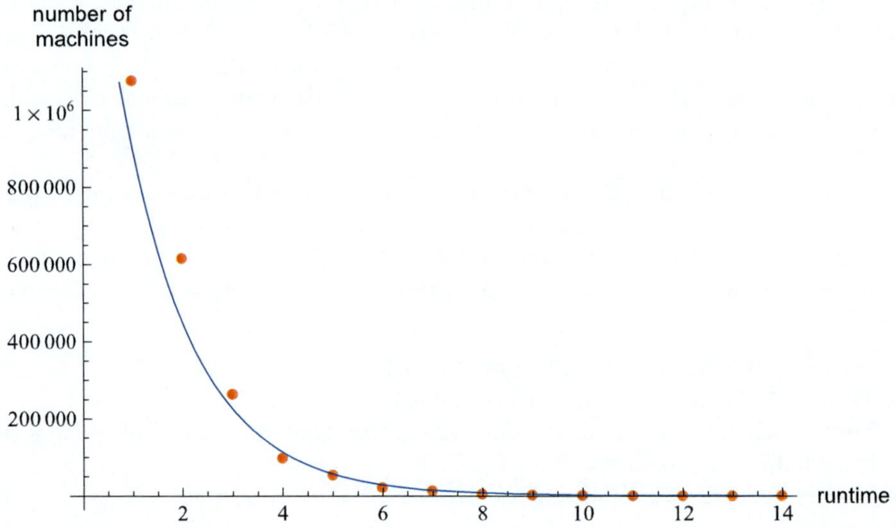

Fig. 2.8 Number of machines in (3, 2) that halt step by step versus $100 \times 2^{14-t}$ (dark line (blue in color version))

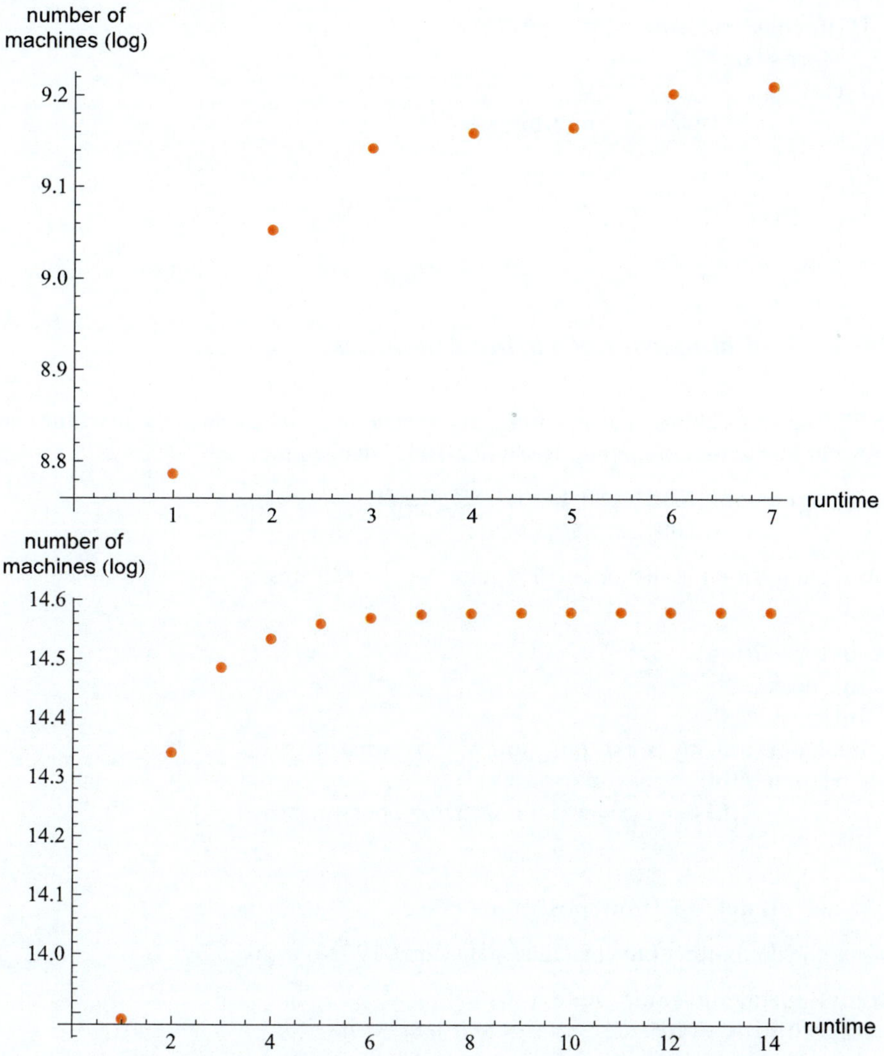

Fig. 2.9 Accumulated number of machines in (2, 2) (top) and (3, 2) (bottom) that halt step by step

cases (or when it is believed to be so high to allow running all machines until that runtime) runtime samples are done with randomly generated machines to choose a proper runtime that ensures that almost all output strings will be obtained.

To return the runtime the changes are minimal. At line 244, the map `results` is defined as `map<int, unsigned long long>`, to store runtime values and their frequency. Then, at the main loop, lines 265–271 are replaced with:

```
if(machine.running_state==-1){
  for(i=1; i<=runtime && run_step(&machine); i++){};
```

```
  if(machine.running_state ==0){
    ++results[i];
  } else {
    ++results[machine.running_state];
  }
} else
  ++results[-2];
```

2.9.3 Two-dimensional Turing Machines

In 2D Turing machines, the tape is a two dimensional grid in which the machine can move in four different directions. So line 10 is changed to:

```
enum direction {STOP, DIR_LEFT, DIR_RIGHT,
                DIR_UP, DIR_DOWN};
```

Also, the position in the tape will require two coordinates, so a specific structure is used:

```
struct position {
  int posx;
  int posy;
  bool operator<( const position & n ) const {
    return (this->posx < n.posx) ||
           (this->posx==n.posx && this->posy<n.posy);
  }
};
typedef struct position position;
```

Then, the definition of the machine state (lines 19–26) is given by:

```
struct turing_machine_state {
  int control_state;          // current state
  int max_head_x_position;    // most-right visited cell
  int min_head_x_position;    // most-left visited cell
  int max_head_y_position;    // most-up visited cell
  int min_head_y_position;    // most-down visited cell
  int escapees; // consecutive shifts to blank symbols
  position head_position;
  map<position, int> tape;
};
typedef struct turing_machine_state
               turing_machine_state;
```

When filling the transition table, the four possible non-halting directions are considered. The code to get the transitions for states different to the initial one (lines

182–205) is now shown (the code to get transitions from the initial state (lines 158–179) is modified accordingly):

```
for (i=1; i<states; i++) {
  for(j=0;j<colors;j++){
    mpz_fdiv_qr_ui(gc, gr, numberTM,
                   colors*((4*states)+1));
    mpz_set(numberTM, gc);
    rest = mpz_get_ui(gr);
    if(rest<colors){
      m.transition_table[i][j].control_state =
             m.halting_state;
      m.transition_table[i][j].write_symbol = rest;
      m.transition_table[i][j].dir = STOP;
       halts = -1;
    } else {
       rest = rest - colors;
      m.transition_table[i][j].write_symbol =
             (rest % colors);
       rest = rest/colors;
       switch(rest % 4){
         case 0:
           m.transition_table[i][j].dir = DIR_RIGHT;
           break;
         case 1:
           m.transition_table[i][j].dir = DIR_LEFT;
           break;
         case 2:
           m.transition_table[i][j].dir = DIR_UP;
           break;
         case 3:
           m.transition_table[i][j].dir = DIR_DOWN;
           break;
    }
       rest = rest / 4;
      m.transition_table[i][j].control_state =
             (rest % states);
    }
  }
}
```

Also, when executing a machine step, the four directions are considered. The function `run_step` (lines 40–109) now becomes:

```
int run_step(turing_machine *m){
  int state1 = m->state.control_state;
```

```
position cell = m->state.head_position;
symbol s1 = m->state.tape[cell];
transition_result tr =
    m->transition_table[m->state.control_state][s1];
m->state.tape[cell] = tr.write_symbol;
m->state.control_state = tr.control_state;

// Stops (without moving the head) if halting state
if(m->state.control_state == m->halting_state){
  m->running_state=0;
  return 0;
}
// escapees detector
bool escap = false;

switch(tr.dir){ // moving the head
case DIR_LEFT:  // LEFT
  m->state.head_position.posx--;
  if(m->state.min_head_x_position >
     m->state.head_position.posx){
    m->state.min_head_x_position--;
    escap = true;
  };
  break;
case DIR_RIGHT: // RIGHT
  m->state.head_position.posx++;
  if(m->state.max_head_x_position <
     m->state.head_position.posx){
    m->state.max_head_x_position++;
    escap = true;
  };
  break;
case DIR_UP:  // UP
  m->state.head_position.posy++;
  if(m->state.max_head_y_position <
     m->state.head_position.posy){
    m->state.max_head_y_position++;
    escap = true;
  };
  break;
case DIR_DOWN:  // DOWN
  m->state.head_position.posy--;
  if(m->state.min_head_y_position >
     m->state.head_position.posy){
    m->state.min_head_y_position--;
```

```
      escap = true;
    };
    break;
  }

  if(escap){  // detecting escapees
    m->state.escapees++;
    if(s1 == m->blank_symbol &&
       m->state.control_state == state1){
      m->running_state = -3; // short escapee
      return 0;
    }
    if(m->state.escapees > m->number_states){
      m->running_state = -4; // regular escapee
      return 0;
    }
  } else{
    m->state.escapees = 0;
  }

  // Detecting cycles of order two
  if(tr.write_symbol == s1){
    symbol s2 = m->state.tape[m->state.head_position];
    transition_result tr2 =
        m->transition_table[m->state.control_state][s2];
    if (tr2.control_state == state1 &&
        tr2.write_symbol == s2 &&
        reverseDir(tr2.dir,tr.dir)){
      m->running_state = -5;
      return 0;
    }
  }
  return 1;
}
```

Now the escapees are detected with the four parameters that store the maximum and minimum visited horizontal and vertical positions at the tape. To detect cycles of order two, the function `reverseDir(direction d1, direction d2)` is called. This function returns `true` only when `d1` and `d2` are opposite directions.

Finally, to return the output array of a halting computation, the simulator visits the minimal rectangle that contains all visited cells. The function `outputTM(turing_machine *m)` (lines 214–227) is now:

```
string outputTM(turing_machine *m){
  string sout = "";
  position pos;
```

```
  char symb;
  if(m->running_state == 0){
    int i;
    int j;
    for (i=m->state.max_head_y_position;
         i>=m->state.min_head_y_position; i--){
      pos.posy = i;
      for(j=m->state.min_head_x_position;
          j<=m->state.max_head_x_position; j++){
        pos.posx = j;
        symb = (m->state.tape[pos])+'0';
        sout = sout+symb;
      }
      sout = sout+'-';
    }
  } else {
    stringstream strm;
    strm << m->running_state;
    strm >> sout;
  }
  return sout;
}
```

So output arrays are represented as strings with one dash '–' after each row.

References

1. Levin, L.: Laws of information conservation (non-growth) and aspects of the foundation of probability theory. Problems in Form. Transmission **10**, 206–210 (1974)
2. W. Kirchherr, M. Li, and P. Vitnyi. "The miraculous universal distribution". In: The Mathematical Intelligencer 19.4 (1997), pp. 7–15
3. Solomonoff, R.J.: A formal theory of inductive inference: Parts 1 and 2. In: Information and Control, vol. 7.1–22, pp. 224–254 (1964)
4. A Turing: On Computable Numbers, with an Application to the Entscheidungsproblem. In: Proceedings of the London Mathematical Society, vol. 2.42, pp. 230–265 (1936)
5. Rad, T.: On non-computable functions. Bell Syst. Tech. J. **41.3**, 877–884 (1962)
6. Soler-Toscano, F. et al.: Small Turing Machines with Halting State: Enumeration and Running on a Blank Tape. In: Wolfram Demonstrations Project (2013). http://demonstrations.wolfram.com/SmallTuringMachinesWithHaltingStateEnumerationAndRunningOnAB/

Chapter 3
The Coding Theorem Method

The algorithmic complexity (hence also called *program-size* complexity) of a bit string is defined as the length of the shortest binary computer program that prints out the string (see (1.8) in Sect. 1.3). However, no general, finite and deterministic procedure exists to calculate algorithmic complexity. For a given string there are infinite many programs producing it. The most common approach to approximate the algorithmic complexity of a string is the use of compression algorithms exploiting the regularities of the string and producing shorter compressed versions. Compression algorithms have been used over the years, with the Lempel-Ziv algorithm [1] being one of the most prominent examples. The result of a compressed version of a string is an upper bound of the algorithmic complexity (denoted by $C(s)$) of the string s.

Attempts to estimate the uncomputable are always challenging, see for example [2–4] and more recently [5, 6]. This often requires combining theoretical and experimental results. In this chapter we describe a method to estimate the algorithmic complexity (hereafter denoted by $C(s)$) of (short) bit strings motivated by the concept of algorithmic probability by running a set of (relatively) large number of Turing machines for which the halting runtimes are known thanks to the Busy Beaver problem [2] or educated estimations of their halting time are calculated up to some confidence and uncertainty. The method describes a way to find the shortest program given a standard formalism of Turing machines (Chap. 2), executing all machines from the shortest (in number of states) to a certain (small) size one by one recording how many of them produce a string and then using a theoretical result linking this string frequency with the algorithmic complexity of a string that we use to approximate $C(s)$.

We use the theoretical notion of resource-bounded Kolmogorov complexity that for practical purposes is approximated from above not by the length of the compressed version of the data involved as traditionally done, but by applying the Coding Theorem to an approximation of a Universal Distribution under the assumption of machine

H. Zenil et al., *Methods and Applications of Algorithmic Complexity*,
Emergence, Complexity and Computation 44,
https://doi.org/10.1007/978-3-662-64985-5_3

optimality [7]. The approach we adopt here is different and independent of the direct machine size (small machines are used only because they allow us to calculate all of them up to a small size) and relies on the concept of algorithmic probability which averages over a large set of small computer programs. The result is a novel approach that we put forward for a numerical calculation of the complexity of short strings, as an alternative to the weaker and also indirect method that uses (and abuses [7]) popular statistical compression algorithms such as LZW. The procedure makes use of a combination of results from related areas of computation, such as the concept of halting probability [8], the Busy Beaver problem [2], algorithmic probability [9], Levin's semi-measure and Levin-Chaitin's coding theorem [10].

The approach consists in the thorough execution of all n-state, m-symbol Turing machines which, upon halting, generate a set of output strings from which a frequency distribution is calculated to obtain the algorithmic probability of a string to be the result of a halting machine. The algorithmic complexity of a string is then evaluated from the algorithmic probability using Levin-Chaitin's coding theorem.

The evaluation of the complexity of strings is key in many areas of science. For example, organization, structure, simplicity and randomness are all notions used to describe biological systems and functions, and are tools needed in bioinformatics for DNA sequence analysis, genotype and phenotype maps, and the study of regulatory networks.

In all the domains where formal mathematical definitions of complexity are needed, the field of algorithmic information theory (AIT) has been largely ignored as a potential source of applications. However, when researchers have chosen to apply the theory, which in principle was not supposed to be of any practical use [11], it has proven to be of great value, for example for genetic sequence analysis for DNA false positive repeat sequence detection [12], distance measures and classification methods [13], and many others applications [14]. But this effort has also been limited, both in terms of the number of people involved (a handful of senior researchers) and owing to the limitations of compressibility, for years the only method used to approximate the Kolmogorov complexity of strings. The method presented in this chapter aims to solve the problem of the evaluation of the complexity of short strings.

3.1 The Limits of Statistical Compression Algorithms

A fair compression algorithm is one that transforms a string into two components. The first of these is the compressed version while the other is the set of instructions for decompressing the string. Both together account for the final length of the compressed version. Thus the compressed string comes with its own decompression instructions. Paradoxically, lossless compression algorithms are more stable the longer the string. In fact the invariance theorem (see (1.9) in Sect. 1.3) guarantees that complexity values will only diverge by a constant c and will converge at the limit.

The use of lossless data compression algorithms as a method for estimating the Kolmogorov complexity of an object (e.g. a string) turns out to be accurate in direct

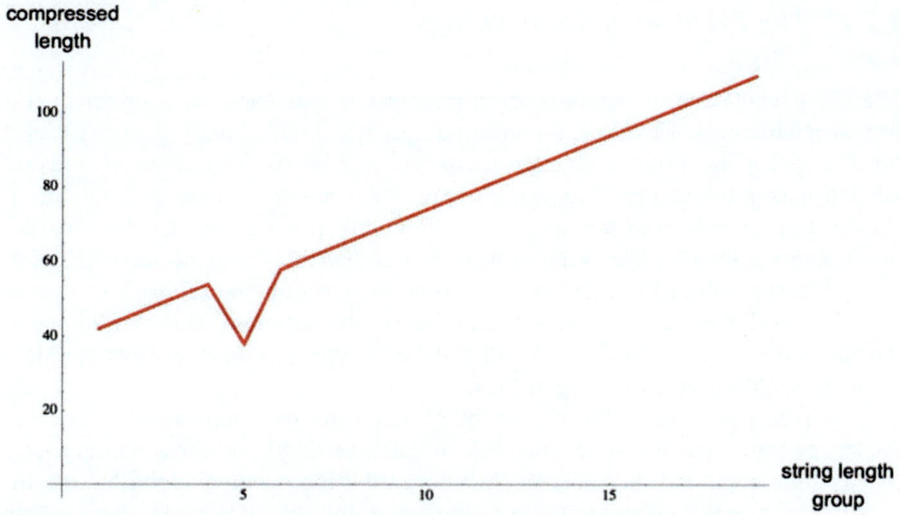

Fig. 3.1 The problem of short strings: On the x-axis are groups formed by strings of the form 1^n (that is a 1 followed by n 1s) that after compression have the same compressed length. The results are unsatisfactory for lengths $n < 53$ (all grouped in $x = 1$ for which $y = 42$). Strings of length $n < 42$ are compressed into strings longer than their original length. And it is also not uncommon to detect instabilities, such as for group $x = 5$, which for no apparent reason the compression algorithm was able to compress it better before returning to the average compression trend. This is not a malfunction of the particular compression algorithm (DEFLATE, used in most popular computer formats such as ZIP and PNG) or its implementation, but a common issue for lossless compression algorithms trying to compress short strings in general

proportion to the length of the string. In effect, the shorter the string, the greater the margin of error. Too great a margin of error means that one cannot really compare evaluations of the complexity of short strings. Populat statistical compression algorithms such as LZW are limited in their scope of application [7]. So if one wished to tell which of two strings is more or less randomly complex by approximating their algorithmic complexity using e.g. LZW or any of their cognates, it turns out that there is no way to do so if the strings are relatively short (as we can see from the example in Fig. 3.1, any string of length $n < 42$ bits is compressed at about the same length). The resulting compression lengths are greater than the length of the original strings, which, if taken as Kolmogorov complexity values, would mean that they are all random (even when the strings, such as those in the figure, are only simple repetitions of '1s').

3.1.1 The Problem of Short Strings

The chief advantage of compression algorithms is that they are a sufficient test of non-randomness. However, for short strings, which are usually the ones useful for practical applications, adding the decompression instructions to the compressed version makes the compressed string often, if not always, longer than the string itself, simply because the decompression instructions are at least equal in length to the original string (see Fig. 3.1). If the string is shorter than the size of the decompression algorithm, say, there will not be a way to compress the string into something shorter still. The result will be so dependent on the size of the decompression algorithm that the final value of the compressed length will be too unstable under different lossless compression/decompression algorithms.

Given the definition of algorithmic complexity based on compressibility, i.e., that the less compressible the more randomly complex a string is, it follows immediately that a single bit, 0 or 1, is surely random because it has maximal algorithmic complexity, since there is no way to further compress a single bit. It is hard to explain how 1 could seem more random than, say, any other possible string. If 1 is maximally random, how is it relatively more complex than 10 or 1011? (or indeed anything other than 1). So if a single bit is the most random among all finite strings, how can there be a phase transition from maximal randomness to extreme simplicity? (very low Kolmogorov complexity) for, say, strings of length 4, 6 or 10 bits? What if one asks how common 0 or 1 are as the output of a computer program? We will see that the method proposed herein addresses this issue in an alternative manner, providing some answers to these questions.

3.2 Approximating the Universal Distribution

One can attempt to approximate $m(s)$ (see (1.10) in Sect. 1.3) by running every Turing machine following a particular enumeration, for example, a quasi-lexicographical ordering, from shorter to longer (with number of states n and 2 fixed symbols). It is clear that in this fashion once a machine produces s for the first time, one can directly calculate an exact value of K under this model and under the assumption of optimality [7], because this is the length of the first Turing machine in the enumeration of programs of increasing size that produces s. Let us formalize this by using the function $D(n, m)$ as the function that assigns to every string s produced in (n, m) the quotient: (number of times that a machine in (n, m) produces s)/(number of machines that halt in (n, m)) as defined in [15, 16]. More formally,

$$D(n, m)(s) = \frac{|\{T \in (n, m) : T(p) = s\}|}{|\{T \in (n, m) : T \ halts \}|} \tag{3.1}$$

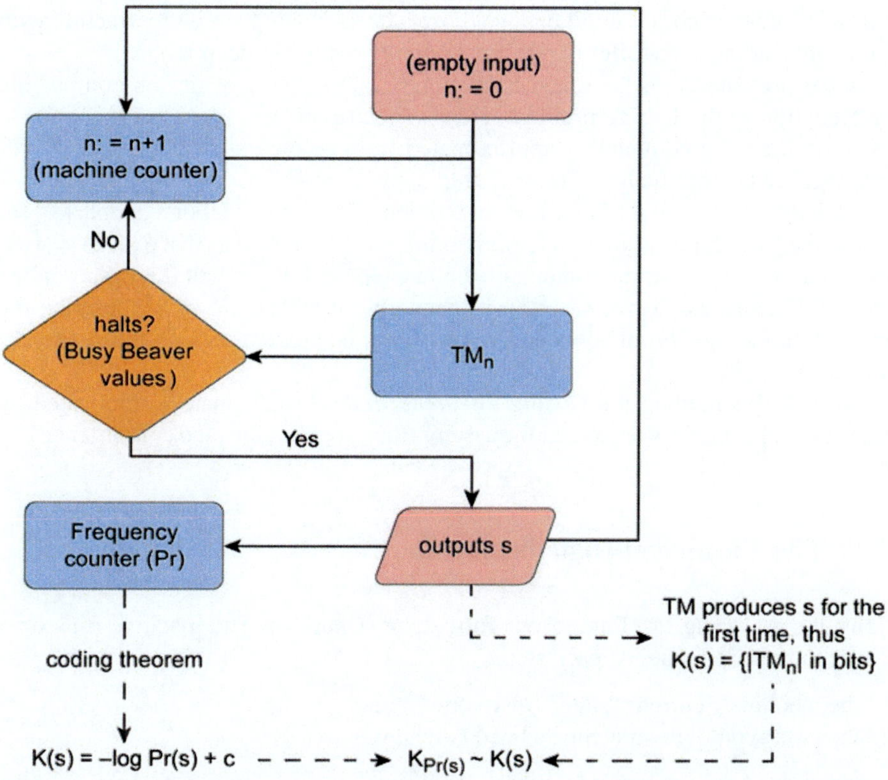

Fig. 3.2 A flow chart illustrating the *Coding Theorem Method*, a never-ending algorithm for evaluating the (Kolmogorov) complexity of a (short) string making use of several concepts and results from theoretical computer science, in particular the halting probability, the Busy Beaver problem, Levin's semi-measure and the Coding Theorem. The Busy Beaver values can be used up to 4 states for which they are known, for more than 4 states an informed maximum runtime is used (see Sect. 3.5.1). Notice that Pr are the probability values calculated dynamically by running an increasing number of Turing machines. Pr is intended to be an approximation to $m(s)$ out of which we build $D(n)$ after application of the Coding theorem under the assumption of optimality [7]

where $T(p)$ is the Turing machine with number p (and empty input) that produces s upon halting and $|A|$ is, in this case, the cardinality of the set A. A variation of this formula closer to the definition of m is given by:

$$D'(n, m)(s) = \frac{|\{T \in (n, m) : T(p) = s\}|}{|\{T \in (n, m)\}|} \tag{3.2}$$

Given that D' is strictly smaller than 1 (because of the Turing machines that never halt) just as m is but unlike D which for fixed n and m the sum will always be 1. We will use Eq. 3.1 for practical reasons, because it makes the frequency values more readable (most machines don't halt, so those halting would have a tiny fraction with too many leading zeros after the decimal point if written in decimal).

It was proven in [15, 16] that the function $(n, m) \to D(n, m)$ is non-computable by reduction to the halting problem. However, $D(n, m)$ is lower semi-computable meaning it can be computably approximated from below under the assumption of convergence and optimality [7], for example, by running small Turing machines for which known values of the Busy Beaver problem [2] are known. For example [3], for $n = 4$, the Busy Beaver function for maximum runtime S, tells us that $S(4, 2) = 107$, so we know that a machine running on a blank tape will never halt if it hasn't halted after 107 steps, and so we can stop it manually. In what follows we describe the exact methodology. From now on, $D(n)$ with a single parameter will mean $D(n, 2)$ (Fig. 3.2).

We call this method the *Coding Theorem Method* to estimate K motivated by algorithmic probability and under the assumptions of optimality and convergence [7].

3.3 The Empirical Distribution *D*

Consider[1] a Turing machine in our formalism (Chap. 2). The machine runs on a 2-way unbounded tape. At each step:

1. the machine's current "state" (instruction); and
2. the tape symbol the machine's head is scanning

define each of the following:

1. a unique symbol to write (the machine can overwrite a 1 on a 0, a 0 on a 1, a 1 on a 1, and a 0 on a 0);
2. a direction to move in: -1 (left), 1 (right) or 0 (none, when halting); and
3. a state to transition into (may be the same as the one it was in).

The machine halts if and when it reaches the special halt state 0. Since the domain of the program has size $2n$ and the target space has size $4n + 2$, we can easily count the number of Turing machines. There are $(4n + 2)^{2n}$ Turing machines with n states and 2 symbols according to the formalism described above.

No transition starting from the halting state exists, and the blank symbol is one of the 2 symbols (0 or 1) in the first run, while the other is used in the second run (in order to avoid any asymmetries due to the choice of a single blank symbol). In other words, we run each machine twice, one with 0 as the blank symbol (the symbol with which the tape starts out and is filled with), and an additional run with 1 as the

[1] The material in this section was adapted from **Numerical evaluation of algorithmic complexity for short strings: A glance into the innermost structure of randomness**, Authors: Jean-Paul Delahaye and Hector Zenil, Published in *Applied Mathematics and Computation*, Pages 63–77, Copyright 2012, with permission from Elsevier.

blank symbol.[2] The output string is taken from the number of contiguous cells on the tape the head of the halting n-state machine has gone through. A machine produces a string upon halting.

Some examples of $D(n)$ for $n = 1, n = 2$:

$$D(1) = \{0 \to 0.5,\ 1 \to 0.5\}$$
$$D(2) = \{0 \to 0.328,\ 1 \to 0.328,\ 00 \to .0834,\ \ldots\}$$

$D(n)$ is the probability distribution of the strings produced by all 2-symbol halting Turing machines with n states. Tables in Sect. 3.4.2 show the full results for $D(1)$, $D(2)$ and $D(3)$, and the top ranking of $D(4)$.

The distribution $D(n)$ is not computable. The proof can be done by contradiction given the uncomputability of the busy beaver problem. Suppose that $n \to D(n)$ is computable. Let n be a given integer. Let $D(n)$ be computed. By definition $D(n)$ is the function that assigns to every finite bit string s the quotient (number of times that a machine (0n, 2) produces s upon halting)/(number of halting machines in $(n, 2)$). To calculate D we use the values of the Busy Beaver function $S(n)$. Turing machines that run more than $S(n)$ steps will therefore never halt. Computing $D(n)$ for any n would therefore mean that the Busy Beaver functions $S(n)$ and $\Sigma(n)$ can be known for any n providing a method to calculate Rado's Busy Beaver functions. Which is a contradiction with the non-computability of Rado's Busy Beaver functions.

Exact values can be, however, calculated for small Turing machines up to $n = 4$ states thanks to the Busy Beaver values of $S(n)$ for $n < 5$ and also by taking educated guesses following a resource-bounded approach where runtime is cut short. For example, for $n = 4$, $S(4) = 107$, so 107 is the maximum runtime we have to run each machine in (4, 2) in order to get the all the outputs, but we also know that for Turing machines with $n = 4$, most machines will either halt soon or never halt, and for those still running, the closer to the upper runtime limit the more machines will halt and only a small quantifiable fraction will not [17, 18].

For each Busy Beaver candidate with $n > 4$ states, a sample of Turing machines running up to the candidate $S(n)$ is possible. As for Rado's Busy Beaver functions $\sum(n)$ and $S(n)$, D is also approachable *from above*. For larger n sampling methods asymptotically converging to $D(n)$ can be used to approximate $D(n)$. In Sect. 3.4.2 we provide exact values of $D(n)$ for $n < 5$ for which the Busy Beaver functions are known.

Another common property between $D(n)$ and Rado's Busy Beaver functions is that $D(4)$ is well-defined in the sense that the calculation of the digits of $D(n)$ are fully determined as is in the decimal expansion of the mathematical constant π, but

[2] Due to the symmetry of the computation, there is no real need to run each machine twice; one can *complete* the string frequencies assuming that each string produced its reversed and complemented version with the same frequency, and then group and divide by symmetric groups. See the discussion about the reduced enumeration in Chap. 2.

the calculation of $D(n)$ rapidly becomes impractical to determine for even a slightly larger number of states.

3.4 Methodology

The approach for evaluating the complexity $C(s)$ of a string s presented herein is limited by (1) the halting problem and (2) computing time constraints. Restriction (1) was overcome because the values of the Busy Beaver problem provided the halting times for all (4, 2) Turing machines that started with a blank tape. Restriction (2) represented a challenge in terms of computing time and programming skills. It is also the same restriction that has kept others from attempting to solve the Busy Beaver problem for a greater number of states. Constraint (2) and the knowledge of the values of the Busy Beaver function permitted us to systematically study machines up to 4 states. We were able to compute up to about 1.3775×10^9 machines per day or 15 943 per second, taking us about 9 days[3] to run all (4, 2) Turing machines each up to the number of steps bounded by the Busy Beaver values.

Our quest is similar in several respects to the Busy Beaver problem or the calculation of the digits of Chaitin's Ω number. The main underlying difficulty in analyzing thoroughly a given class of machines is their uncomputability. Just as it is done for solving small values of the Busy Beaver problem, we rely on the experimental approach to analyze and describe a computable fraction of the uncomputable. A similar quest for the calculation of the digits of a Chaitin's Ω number was undertaken by Calude et al. [5], but unlike Chaitin's Ω, the calculation of $D(n)$ does not depend on the enumeration of Turing machines. In $D(n)$, however, one obtains different probabilities for the same string for each n, but the relative order seems to be preserved. In fact, every $(2, n)$ machine contributing to $D(n)$ is included in $D(n+1)$ simply because every machine rule in $(2, n)$ is in $(2, n+1)$.

3.4.1 *Numerical Calculation of* $D(4)$

We consider the space (n, 2) of Turing machines with $0 < n < 5$. The halting "history" and output probability followed by their respective runtimes, presented in Tables 3.1, 3.2, 3.3 and 3.4 show the frequency of the strings produced.

We provide exact values for $n = \{2, 3, 4\}$. We derive $D(n)$ for $n < 5$ from counting the number of n-strings produced by all (n, 2) Turing machines upon halting. We

[3] Running on a single computer on a MacBook Intel Core Duo at 1.83 Ghz, 2 Gb. of memory and a solid state hard drive, using the TuringMachine[] function available in *Mathematica* 8 for $n < 4$ and a C++ program for $n = 4$. Since for $n = 4$ there were 2.56×10^8 machines involved, running on both 0 and 1 as blank, further optimizations were required. The use of a bignum library and an actual enumeration of the machines rather than producing the rules beforehand (which would have meant overloading the memory even before the actual calculation) was necessary.

consider D an *empirical universal distribution* in Levin's sense, and calculate the algorithmic complexity C of a string s in terms of D using the coding theorem (Sect. 1.3), from which we won't escape to an additive constant introduced by the application of the coding theorem, but the additive constant is common to all values and therefore should not impact the relative order. One has to bear in mind, however, that the tables in Sect. 3.4.2 should be read as dependent of this last-step additive constant because using the coding theorem as an approximation method fixes a prefix-free UTM via that constant, but according to the choices we make this seems to be the most natural way to do so as an alternative to other indirect numerical methods.

We calculated the 72, 20 000, 15 059 072 and 22 039 921 152 two-way tape Turing machines started with a tape filled with 0s and 1s for $D(2)$, $D(3)$ and $D(4)$.[4] The number of Turing machines to calculate grows exponentially with the number of states. For $D(5)$ there are 53 119 845 582 848 machines to calculate, which makes the task as difficult as finding the Busy Beaver values for $\sum(5)$ and $S(5)$, Busy Beaver values which are currently unknown but for which the best candidate may be $S(5) = 47\,176\,870$ which makes the exploration of (5,2) quite a challenge.

Although several ideas exploiting symmetries to reduce the total number of Turing machines have been proposed and used for finding Busy Beaver candidates [3, 19, 20] in large spaces such as $n \geq 5$, to preserve the structure of the data we couldn't apply all of them. This is because, unlike the Busy Beaver challenge, in which only the maximum values are important, the construction of a probability distribution requires every output to be equally considered. Some reduction techniques were, however, utilized, such as running only one-direction rules with a tape only filled with 0s and then completing the strings by reversion and complementation to avoid running every machine a second time with a tape filled with 1s (see the reduced enumeration in Chap. 2).

3.4.2 *Algorithmic Probability Tables*

$D(1)$ is trivial. (1,2) Turing machines produce only two strings, with the same number of machines producing each. The Busy Beaver values for $n = 1$ are $\sum(1) = 1$ and $S(1) = 1$. That is, all machines that halt do so after 1 step, and print at most one symbol.

The Busy Beaver values for $n = 2$ are $\sum(1) = 4$ and $S(1) = 6$. $D(2)$ is quite simple but starts to display some basic structure, such as a clear correlation between string length and occurrence, following what may be an exponential decrease:

[4] We ran the experiment on several computers and cores in parallel, which allowed us to shorten the time by about a fourth of that calculated in a single processor. The space occupied by the machine outputs was 77.06 GB (of which only 38.53 GB was actually necessary by taking advantage of machine rule symmetries that could be later compensated without having to run them).

Table 3.1 Complete $D(1)$ from 24 (1, 2)-Turing machines that halt out of a total of 64

$0 \to 0.5$
$1 \to 0.5$

Table 3.2 Complete $D(2)$ (22 bit-strings) from 6 088 (2, 2)-Turing machines that halt out of 20 000. Each string is followed by its probability (from the number of times produced), sorted from highest to lowest

$0 \to 0.328$	$010 \to 0.00065$
$1 \to 0.328$	$101 \to 0.00065$
$00 \to 0.0834$	$111 \to 0.00065$
$01 \to 0.0834$	$0000 \to 0.00032$
$10 \to 0.0834$	$0010 \to 0.00032$
$11 \to 0.0834$	$0100 \to 0.00032$
$001 \to 0.00098$	$0110 \to 0.00032$
$011 \to 0.00098$	$1001 \to 0.00032$
$100 \to 0.00098$	$1011 \to 0.00032$
$110 \to 0.00098$	$1101 \to 0.00032$
$000 \to 0.00065$	$1111 \to 0.00032$

$$P(|s| = 1) = 0.657$$
$$P(|s| = 2) = 0.333$$
$$P(|s| = 3) = 0.0065$$
$$P(|s| = 4) = 0.0026$$

Among the various facts one can draw from $D(2)$, there are:

- The relative string order in $D(1)$ is preserved in $D(2)$.
- A fraction of $1/3$ of the total machines halt while the remaining $2/3$ do not. That is, 24 among 72 (running each machine twice with tape filled with 1 and 0 as explained before).
- The longest string produced by $D(2)$ is of length 4.
- $D(2)$ does not produce all $\sum_1^4 2^n = 30$ strings shorter than 5, only 22. The missing strings are 0001, 0101 and 0011 never produced, hence neither were their complements and reversions: 0111, 1000, 1110, 1010 and 1100.

Given the number of machines to run, $D(3)$ constitutes the first non trivial probability distribution to calculate. The Busy Beaver values for $n = 3$ are $\sum(3) = 6$ and $S(3) = 14$. Among the various facts for $D(3)$:

- There are 4 294 368 machines that halt among the 15 059 072 in (3, 2). That is a fraction of 0.2851.
- The longest string produced in (3, 2) is of length 7.

Table 3.3 Complete $D(3)$ (128 bit-strings) produced by all the 15 059 072 (3, 2)-halting Turing machines

0 → 0.250	11110 → 0.0000470	100101 → 1.43×10^{-6}
1 → 0.250	00100 → 0.0000456	101001 → 1.43×10^{-6}
00 → 0.101	11011 → 0.0000456	000011 → 9.313×10^{-7}
01 → 0.101	01010 → 0.0000419	000110 → 9.313×10^{-7}
10 → 0.101	10101 → 0.0000419	001100 → 9.313×10^{-7}
11 → 0.101	01001 → 0.0000391	001101 → 9.313×10^{-7}
000 → 0.0112	01101 → 0.0000391	001111 → 9.313×10^{-7}
111 → 0.0112	10010 → 0.0000391	010001 → 9.313×10^{-7}
001 → 0.0108	10110 → 0.0000391	010010 → 9.313×10^{-7}
011 → 0.0108	01110 → 0.0000289	010011 → 9.313×10^{-7}
100 → 0.0108	10001 → 0.0000289	011000 → 9.313×10^{-7}
110 → 0.0108	00101 → 0.0000233	011101 → 9.313×10^{-7}
010 → 0.00997	01011 → 0.0000233	011110 → 9.313×10^{-7}
101 → 0.00997	10100 → 0.0000233	100001 → 9.313×10^{-7}
0000 → 0.000968	11010 → 0.0000233	100010 → 9.313×10^{-7}
1111 → 0.000968	00011 → 0.0000219	100111 → 9.313×10^{-7}
0010 → 0.000699	00111 → 0.0000219	101100 → 9.313×10^{-7}
0100 → 0.000699	11000 → 0.0000219	101101 → 9.313×10^{-7}
1011 → 0.000699	11100 → 0.0000219	101110 → 9.313×10^{-7}
1101 → 0.000699	000000 → 3.733×10^{-6}	110000 → 9.313×10^{-7}
0101 → 0.000651	111111 → 3.733×10^{-6}	110010 → 9.313×10^{-7}
1010 → 0.000651	000001 → 2.793×10^{-6}	110011 → 9.313×10^{-7}
0001 → 0.000527	011111 → 2.793×10^{-6}	111001 → 9.313×10^{-7}
0111 → 0.000527	100000 → 2.793×10^{-6}	111100 → 9.313×10^{-7}
1000 → 0.000527	111110 → 2.793×10^{-6}	0101010 → 9.313×10^{-7}
1110 → 0.000527	000100 → 2.333×10^{-6}	1010101 → 9.313×10^{-7}
0110 → 0.000510	001000 → 2.333×10^{-6}	001110 → 4.663×10^{-7}
1001 → 0.000510	110111 → 2.333×10^{-6}	011100 → 4.663×10^{-7}
0011 → 0.000321	111011 → 2.333×10^{-6}	100011 → 4.663×10^{-7}
1100 → 0.000321	000010 → 1.863×10^{-6}	110001 → 4.663×10^{-7}
00000 → 0.0000969	001001 → 1.863×10^{-6}	0000010 → 4.663×10^{-7}
11111 → 0.0000969	001010 → 1.863×10^{-6}	0000110 → 4.663×10^{-7}
00110 → 0.0000512	010000 → 1.863×10^{-6}	0100000 → 4.663×10^{-7}
01100 → 0.0000512	010100 → 1.863×10^{-6}	0101110 → 4.663×10^{-7}
10011 → 0.0000512	011011 → 1.863×10^{-6}	0110000 → 4.663×10^{-7}
11001 → 0.0000512	100100 → 1.863×10^{-6}	0111010 → 4.663×10^{-7}
00010 → 0.0000489	101011 → 1.863×10^{-6}	1000101 → 4.663×10^{-7}
01000 → 0.0000489	101111 → 1.863×10^{-6}	1001111 → 4.663×10^{-7}
10111 → 0.0000489	110101 → 1.863×10^{-6}	1010001 → 4.663×10^{-7}
11101 → 0.0000489	110110 → 1.863×10^{-6}	1011111 → 4.663×10^{-7}
00001 → 0.0000470	111101 → 1.863×10^{-6}	1111001 → 4.663×10^{-7}
01111 → 0.0000470	010110 → 1.43×10^{-6}	1111101 → 4.663×10^{-7}
10000 → 0.0000470	011010 → 1.43×10^{-6}	

- $D(3)$ has not all $\sum_1^7 2^n = 254$ strings shorter than 7 but 128 only, half of all the possible strings up to that length.
- $D(3)$ preserves the string order of $D(2)$.

$D(3)$ ratifies the tendency of classifying strings by length with exponentially decreasing values. The distribution comes sorted by length blocks from which one cannot easily say whether those at the bottom are more random-looking than those in the middle, but one can definitely say that the ones at the top, both for the entire distribution and by length block, are intuitively the simplest. Both 0^k and its reversed 1^k for $n \leq 8$ are always at the top of each block, with 0 and 1 at the top of them all. There is a single exception in which strings were not sorted by length, this is the string group 0101010 and 1010101 that are found four places further away from their length block, which we take as a second indication of a complexity classification becoming more visible since these 2 strings correspond to what one would intuitively consider less random-looking because they are easily described as the repetition of two bits.

$D(4)$ with 22 039 921 152 machines to run was the first true challenge, both in terms of programming specification and computational resources. The Busy Beaver values for $n = 4$ are $\sum(3) = 13$ and $S(n) = 107$. Evidently every machine in (n, 2) for $n \leq 4$ is in (4, 2) because a rule in (n, 2) with $n \leq 4$ is a rule in (4, 2) in which a part of it is never used and halts, which is guaranteed to exist because the computation is exhaustive over (4, 2).
Among the various facts from these results:

- There are 5 970 768 960 machines that halt among the 22 039 921 152 in (4, 2). That is a fraction of 0.27.
- A total number of 1824 strings were produced in (4,2).
- The longest string produced is of length 16 (only 8 among all the 2^{16} possible were generated).
- The Busy Beaver machines (writing more 1s than any other and halting) found in (4, 2) had very low probability among all the halting machines: pr(111111111 11101) $= 2.01 \times 10^{-9}$. Because of the reverted string (10111111111111), the total probability of finding a Busy Beaver in (4, 2) is therefore 4.02×10^{-9} only (or twice that number if the complemented string with the maximum number of 0s is taken).
- The longest strings in (4, 2) were in the string group formed by: 1101010 101010101, 1101010100010101, 1010101010101011 and 1010100010101011, each with 5.4447×10^{-10} probability, i.e. an even smaller probability than for the Busy Beavers, and therefore the most random in the classification.
- (4,2) produces all strings up to length 8, then the number of strings larger than 8 decreases. The following are the number of strings by length $|\{s : |s| = l\}|$ generated and represented in $D(4)$ from a total of 1 824 different strings. From $i = 1, \ldots, 15$ the values l of $|\{s : |s| = n\}|$ are 2, 4, 8, 16, 32, 64, 128, 256, 486, 410, 252, 112, 46, 8, and 0, which indicated all 2^l strings where generated for $n \leq 8$.

Table 3.4 The top 129 strings from $D(4)$ with highest probability from a total of 1832 different produced strings

0 → 0.205	01101 → 0.000145	110111 → 0.0000138
1 → 0.205	10010 → 0.000145	111011 → 0.0000138
00 → 0.102	10110 → 0.000145	001001 → 0.0000117
01 → 0.102	01010 → 0.000137	011011 → 0.0000117
10 → 0.102	10101 → 0.000137	100100 → 0.0000117
11 → 0.102	00110 → 0.000127	110110 → 0.0000117
000 → 0.0188	01100 → 0.000127	010001 → 0.0000109
111 → 0.0188	10011 → 0.000127	011101 → 0.0000109
001 → 0.0180	11001 → 0.000127	100010 → 0.0000109
011 → 0.0180	00101 → 0.000124	101110 → 0.0000109
100 → 0.0180	01011 → 0.000124	000011 → 0.0000108
110 → 0.0180	10100 → 0.000124	001111 → 0.0000108
010 → 0.0171	11010 → 0.000124	110000 → 0.0000108
101 → 0.0171	00011 → 0.000108	111100 → 0.0000108
0000 → 0.00250	00111 → 0.000108	000110 → 0.0000107
1111 → 0.00250	11000 → 0.000108	011000 → 0.0000107
0001 → 0.00193	11100 → 0.000108	100111 → 0.0000107
0111 → 0.00193	01110 → 0.0000928	111001 → 0.0000107
1000 → 0.00193	10001 → 0.0000928	001101 → 0.0000101
1110 → 0.00193	000000 → 0.0000351	010011 → 0.0000101
0101 → 0.00191	111111 → 0.0000351	101100 → 0.0000101
1010 → 0.00191	000001 → 0.0000195	110010 → 0.0000101
0010 → 0.00190	011111 → 0.0000195	001100 → 9.943×10^{-6}
0100 → 0.00190	100000 → 0.0000195	110011 → 9.943×10^{-6}
1011 → 0.00190	111110 → 0.0000195	011110 → 9.633×10^{-6}
1101 → 0.00190	000010 → 0.0000184	100001 → 9.633×10^{-6}
0110 → 0.00163	010000 → 0.0000184	011001 → 9.3×10^{-6}
1001 → 0.00163	101111 → 0.0000184	100110 → 9.3×10^{-6}
0011 → 0.00161	111101 → 0.0000184	000101 → 8.753×10^{-6}
1100 → 0.00161	010010 → 0.0000160	010111 → 8.753×10^{-6}
00000 → 0.000282	101101 → 0.0000160	101000 → 8.753×10^{-6}
11111 → 0.000282	010101 → 0.0000150	111010 → 8.753×10^{-6}
00001 → 0.000171	101010 → 0.0000150	001110 → 7.863×10^{-6}
01111 → 0.000171	010110 → 0.0000142	011100 → 7.863×10^{-6}
10000 → 0.000171	011010 → 0.0000142	100011 → 7.863×10^{-6}
11110 → 0.000171	100101 → 0.0000142	110001 → 7.863×10^{-6}
00010 → 0.000166	101001 → 0.0000142	001011 → 6.523×10^{-6}
01000 → 0.000166	001010 → 0.0000141	110100 → 6.523×10^{-6}
10111 → 0.000166	010100 → 0.0000141	000111 → 6.243×10^{-6}
11101 → 0.000166	101011 → 0.0000141	111000 → 6.243×10^{-6}
00100 → 0.000151	110101 → 0.0000141	0000000 → 3.723×10^{-6}
11011 → 0.000151	000100 → 0.0000138	1111111 → 3.723×10^{-6}
01001 → 0.000145	001000 → 0.0000138	0101010 → 2.393×10^{-6}

Table 3.5 Probabilities of finding n 1s (or 0s) in (4, 2)

Number	
n of 1s	pr(n)
1	0.472
2	0.167
3	0.0279
4	0.00352
5	0.000407
6	0.0000508
7	6.5×10^{-6}
8	1.31×10^{-6}
9	2.25×10^{-7}
10	3.62×10^{-8}
11	1.61×10^{-8}
12	1.00×10^{-8}
13	4.02×10^{-9}

- While the probability of producing a string with an odd number of 1s is the same than the probability of producing a string with an even number of 1s (and therefore the same for 0s), the probability of producing a string of odd length is 0.559 and 0.441 for even length.
- As in $D(3)$, where we report that one string group (0101010 and its reversion), in $D(4)$ 399 strings climbed to the top and were not sorted among their length groups.
- In $D(4)$ string length was no longer a determinant for string positions. For example, between positions 780 and 790, string lengths are: 11, 10, 10, 11, 9, 10, 9, 9, 9, 10 and 9 bits.
- $D(4)$ preserves the string order of $D(3)$ except in 17 places out of 128 strings in $D(3)$ ordered from highest to lowest string frequency. The maximum rank distance among the farthest two differing elements in $D(3)$ and $D(4)$ was 20, with an average of 11.23 among the 17 misplaced cases and a standard deviation of about 5 places. The Spearman's rank correlation coefficient between the two rankings had a critical value of 0.98, meaning that the order of the 128 elements in $D(3)$ compared to their order in $D(4)$ were in an interval confidence of high significance with almost null probability to have produced by chance (Tables 3.5 and 3.6).

Table 3.6 String groups formed by reversion and complementation followed by the total machines producing them

String group	Occurrences
0, 1	1224440064
01, 10	611436144
00, 11	611436144
001, 011, 100, 110	215534184
000, 111	112069020
010, 101	102247932
0001, 0111, 1000, 1110	23008080
0010, 0100, 1011, 1101	22675896
0000, 1111	14917104
0101, 1010	11425392
0110, 1001	9712752
0011, 1100	9628728
00001, 01111, 10000, 11110	2042268
00010, 01000, 10111, 11101	1984536
01001, 01101, 10010, 10110	1726704
00000, 11111	1683888
00110, 01100, 10011, 11001	1512888
00101, 01011, 10100, 11010	1478244
00011, 00111, 11000, 11100	1288908
00100, 11011	900768
01010, 10101	819924
01110, 10001	554304
000001, 011111, 100000, 111110	233064
000010, 010000, 101111, 111101	219552
000000, 111111	209436
010110, 011010, 100101, 101001	169896
001010, 010100, 101011, 110101	167964
000100, 001000, 110111, 111011	164520
001001, 011011, 100100, 110110	140280
010001, 011101, 100010, 101110	129972

3.4.2.1 Same Length String Distribution

The following are the top 10 string groups (i.e. with their reverted and complemented counterparts) in $D(4)$ appearing sooner than expected and getting away from their length blocks. That is, their lengths were greater than the next string in the classification order): 11111111, 11110111, 000000000, 111111111, 000010000, 111101111, 111111110, 010101010, 101010101, 000101010. This means these string groups

Table 3.7 The probability of producing a string of length l exponentially decreases as l linearly increases. The slowdown in the rate of decrease for string length $l > 8$ is due to the few longer strings produced in (4, 2)

Length n	pr(n)
1	0.410
2	0.410
3	0.144
4	0.0306
5	0.00469
6	0.000818
7	0.000110
8	0.0000226
9	4.69×10^{-6}
10	1.42×10^{-6}
11	4.9×10^{-7}
12	1.69×10^{-7}

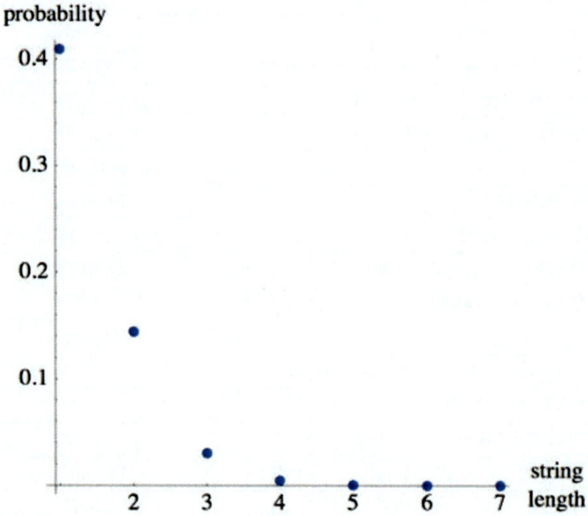

Fig. 3.3 (4, 2) frequency distribution by string length

had greater algorithmic probability and therefore less algorithmic complexity than shorter strings (Table 3.7).

Figure 3.3 displays some statistical information of the distribution. The distribution is skewed to the right, the mass of the distribution is therefore concentrated on the left with a long right tail, as shown in Fig. 3.4 (Table 3.8).

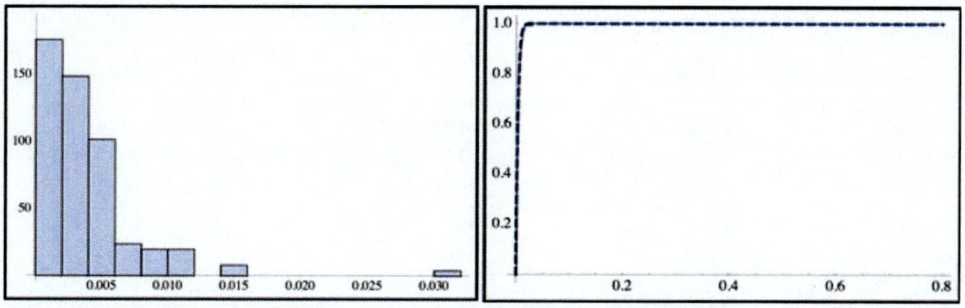

Fig. 3.4 Probability density function of bit strings of length $l = 8$ from (4, 2). The histogram (left) shows the probabilities to fall within a particular region. The cumulative version (right) shows how well the distribution fits a Pareto distribution (dashed) with location parameter $k = 10$. The reader may see but a single curve, that is because the lines overlap. $D(4)$ (and the sub-distributions it contains) is therefore log-normal

Table 3.8 Statistical values of the empirical distribution function $D(4)$ for strings of length $l = 8$

	Value
Mean	0.00391
Median	0.00280
Variance	0.0000136
Kurtosis	23
Skewness	3.6

3.4.3 Derivation and Calculation of Algorithmic Complexity

Algorithmic complexity values are calculated from the output probability distribution $D(4)$ through the application of the coding theorem (Table 3.9).

The largest complexity value max $\{C(s) : s \in D(4)\} = 29$ bits. The output strings (11111111111101 and 10111111111111) produced by the Busy Beavers in (4,2) were close to the maximal complexity in this space for the number of printed 1s among all the produced strings, with a program-size complexity of 28 bits (Fig. 3.5).

3.4.3.1 Same Length String Complexity

The classification Table 3.10 allows to make a comparison of the structure of the strings related to their calculated complexity among all the strings of the same length extracted from $D(4)$ (Fig. 3.6).

Table 3.9 Top 180 strings from $D(4)$ sorted from lowest to highest algorithmic complexity

0→2.29	10110→12.76	100100→16.38	0100000→19.10
1→2.29	01010→12.83	110110→16.38	1011111→19.10
00→3.29	10101→12.83	010001→16.49	1111101→19.10
01→3.29	00110→12.95	011101→16.49	0000100→19.38
10→3.29	01100→12.95	100010→16.49	0010000→19.38
11→3.29	10011→12.95	101110→16.49	1101111→19.38
000→5.74	11001→12.95	000011→16.49	1111011→19.38
111→5.74	00101→12.98	001111→16.49	0001000→19.45
001→5.79	01011→12.98	110000→16.49	1110111→19.45
011→5.79	10100→12.98	111100→16.49	0000110→19.64
100→5.79	11010→12.98	000110→16.52	0110000→19.64
110→5.79	00011→13.18	011000→16.52	1001111→19.64
010→5.87	00111→13.18	100111→16.52	1111001→19.64
101→5.87	11000→13.18	111001→16.52	0101110→19.68
0000→8.64	11100→13.18	001101→16.59	0111010→19.68
1111→8.64	01110→13.39	010011→16.59	1000101→19.68
0001→9.02	10001→13.39	101100→16.59	1010001→19.68
0111→9.02	000000→14.80	110010→16.59	0010001→20.04
1000→9.02	111111→14.80	001100→16.62	0111011→20.04
1110→9.02	000001→15.64	110011→16.62	1000100→20.04
0101→9.03	011111→15.64	011110→16.66	1101110→20.04
1010→9.03	100000→15.64	100001→16.66	0001001→20.09
0010→9.04	111110→15.64	011001→16.76	0110111→20.09
0100→9.04	000010→15.73	100110→16.76	1001000→20.09
1011→9.04	010000→15.73	000101→16.80	1110110→20.09
1101→9.04	101111→15.73	010111→16.80	0010010→20.11
0110→9.26	111101→15.73	101000→16.80	0100100→20.11
1001→9.26	010010→15.93	111010→16.80	1011011→20.11
0011→9.28	101101→15.93	001110→16.96	1101101→20.11
1100→9.28	010101→16.02	011100→16.96	0010101→20.15
00000→11.79	101010→16.02	100011→16.96	0101011→20.15
11111→11.79	010110→16.10	110001→16.96	1010100→20.15
00001→12.51	011010→16.10	001011→17.23	1101010→20.15
01111→12.51	100101→16.10	110100→17.23	0100101→20.16
10000→12.51	101001→16.10	000111→17.29	0101101→20.16
11110→12.51	001010→16.12	111000→17.29	1010010→20.16
00010→12.55	010100→16.12	0000000→18.03	1011010→20.16
01000→12.55	101011→16.12	1111111→18.03	0001010→20.22
10111→12.55	110101→16.12	0101010→18.68	0101000→20.22
11101→12.55	000100→16.15	1010101→18.68	1010111→20.22
00100→12.69	001000→16.15	0000001→18.92	1110101→20.22
11011→12.69	110111→16.15	0111111→18.92	0100001→20.26
01001→12.76	111011→16.15	1000000→18.92	0111101→20.26
01101→12.76	001001→16.38	1111110→18.92	1000010→20.26
10010→12.76	011011→16.38	0000010→19.10	1011110→20.26

3.4.3.2 Halting Summary

In summary, among the (running over a tape filled with 0 only): 12, 3 044, 2 147 184 and 2 985 384 480 Turing machines in (n,2), $n < 5$, there were 36, 10 000, 7 529 536 and 11 019 960 576 that halted, that is slightly decreasing fractions of 0.333..., 0.3044, 0.2851 and 0.2709 respectively. Full results can be found online at http://www.algorithmicnature.org/.

3.4.4 Runtimes Investigation

Runtimes much longer than the lengths of their respective halting programs are rare and the empirical distribution approaches the a priori computable probability distribution on all possible runtimes predicted in [21]. As reported in [21] "long" runtimes are effectively rare. The longer it takes to halt, the less likely it is to stop (Fig. 3.7 and Tables 3.11, 3.12).

Among the various miscellaneous facts from these results:

- All 1-bit strings were produced at $t = 1$.
- 2-bit strings were produced at all $2 < t < 14$ times.
- $t = 3$ was the time at which the first 2 bit strings of different lengths were produced ($n = 2$ and $n = 3$).
- Strings produced before 8 steps account for 49% of the strings produced by all (4, 2) halting machines.
- There were 496 string groups produced by (4, 2), that is strings that are not symmetric under reversion or complementation.
- There is a relation between t and n; no n-bit string is produced before $t < n$. This is obvious because a machine needs at least t steps to print t symbols.
- At every time t there was at least one string of length n for $1 < n < t$.

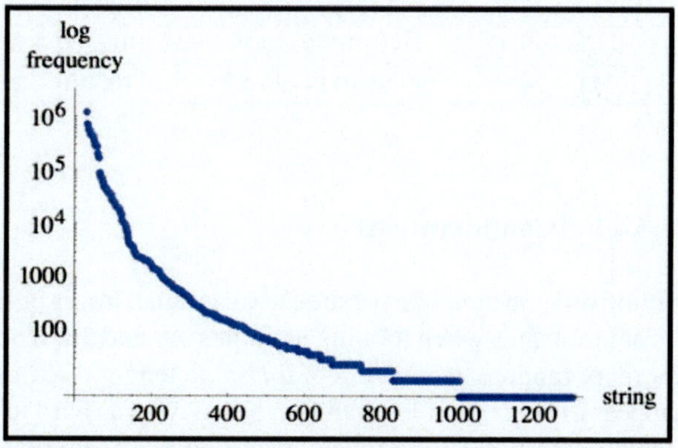

Fig. 3.5 (4, 2) output log-frequency plot, ordered from most to less frequent string, the slope is clearly exponential

Table 3.10 Classification–from less to more random–for 7-bit strings extracted from $D(4)$

0000000→18.03	1001000→20.09	0101001→20.42	0000111→20.99
1111111→18.03	1110110→20.09	0110101→20.42	0001111→20.99
0101010→18.68	0010010→20.11	1001010→20.42	1110000→20.99
1010101→18.68	0100100→20.11	1010110→20.42	1111000→20.99
0000001→18.92	1011011→20.11	0001100→20.48	0011110→21.00
0111111→18.92	1101101→20.11	0011000→20.48	0111100→21.00
1000000→18.92	0010101→20.15	1100111→20.48	1000011→21.00
1111110→18.92	0101011→20.15	1110011→20.48	1100001→21.00
0000010→19.10	1010100→20.15	0110110→20.55	0111110→21.03
0100000→19.10	1101010→20.15	1001001→20.55	1000001→21.03
1011111→19.10	0100101→20.16	0011010→20.63	0011001→21.06
1111101→19.10	0101101→20.16	0101100→20.63	0110011→21.06
0000100→19.38	1010010→20.16	1010011→20.63	1001100→21.06
0010000→19.38	1011010→20.16	1100101→20.63	1100110→21.06
1101111→19.38	0001010→20.22	0100010→20.68	0001110→21.08
1111011→19.38	0101000→20.22	1011101→20.68	0111000→21.08
0001000→19.45	1010111→20.22	0100110→20.77	1000111→21.08
1110111→19.45	1110101→20.22	0110010→20.77	1110001→21.08
0000110→19.64	0100001→20.26	1001101→20.77	0010011→21.10
0110000→19.64	0111101→20.26	1011001→20.77	0011011→21.10
1001111→19.64	1000010→20.26	0010110→20.81	1100100→21.10
1111001→19.64	1011110→20.26	0110100→20.81	1101100→21.10
0101110→19.68	0000101→20.29	1001011→20.81	0110001→21.13
0111010→19.68	0101111→20.29	1101001→20.81	0111001→21.13
1000101→19.68	1010000→20.29	0001101→20.87	1000110→21.13
1010001→19.68	1111010→20.29	0100111→20.87	1001110→21.13
0010001→20.04	0000011→20.38	1011000→20.87	0011100→21.19
0111011→20.04	0011111→20.38	1110010→20.87	1100011→21.19
1000100→20.04	1100000→20.38	0011101→20.93	0001011→21.57
1101110→20.04	1111100→20.38	0100011→20.93	0010111→21.57
0001001→20.09	0010100→20.39	1011100→20.93	1101000→21.57
0110111→20.09	1101011→20.39	1100010→20.93	1110100→21.57

3.4.4.1 Considerations

Intuitively, one may be persuaded to assign a lower or higher algorithmic complexity to some strings when looking at Tables 3.9 and 3.10, because they may seem simpler or more random than others of the same length. We think that very short strings may appear to be more or less random but may be as hard to produce as others of the same length, because Turing machines producing them may require the same quantity of

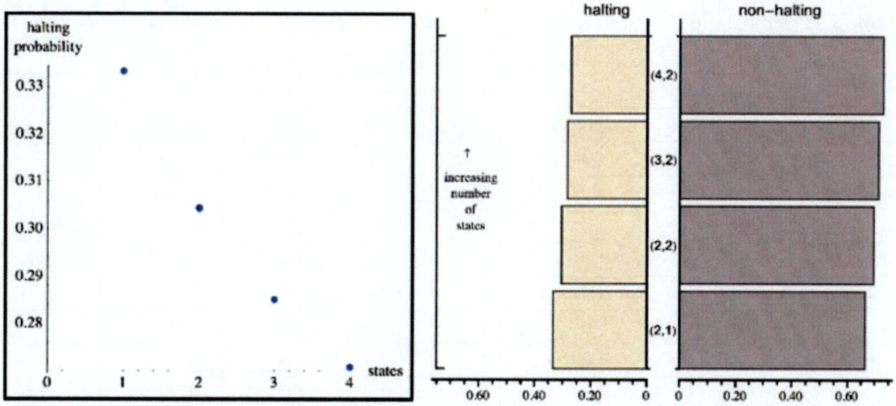

Fig. 3.6 Graphs showing the halting probabilities among (n,2), $n < 5$. The list plot on the left shows the decreasing probability of the number of halting Turing machines while the paired bar chart on the right allows a visual comparison between both halting and non-halting machines side by side

resources to print them out and halt as they would with others of the same (very short) length.

For example, is 0101 more or less complex than 0011? Is 001 more or less complex than 010? The string 010 may seem simpler than 001 to us because we may picture it as part of a larger sequence of alternating bits, forgetting that such is not the case and that 010 actually was the result of a machine that produced it when entering into the halting state, using this extra state to somehow delimit the length of the string. No satisfactory argument may exist to say whether 010 is really more or less random than 001, other than actually running the machines and looking at their objective ranking

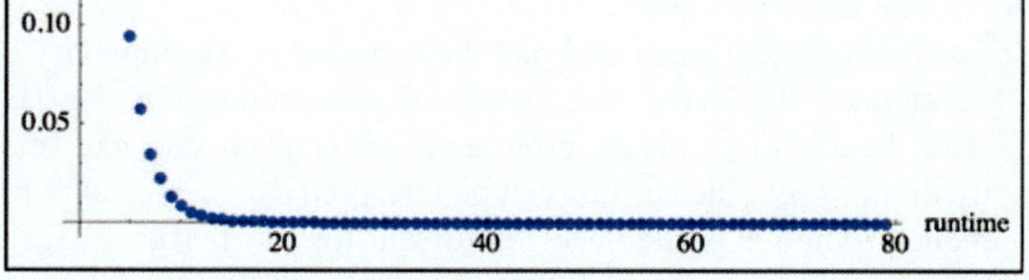

Fig. 3.7 Runtimes distribution in (4, 2)

Table 3.11 Probability that a n-bit string among all $n < 10$ bit strings is produced at times $t < 8$

	$t = 1$	$t = 2$	$t = 3$	$t = 4$	$t = 5$	$t = 6$	$t = 7$	
n = 1	1.0	0	0	0	0	0	0	0
n = 2	0	1.0	0.60	0.45	0.21	0.11	0.052	0.025
n = 3	0	0	0.40	0.46	0.64	0.57	0.50	0.36
n = 4	0	0	0	0.092	0.15	0.29	0.39	0.45
n = 5	0	0	0	0	0	0.034	0.055	0.16
n = 6	0	0	0	0	0	0	0	0.0098
n = 7	0	0	0	0	0	0	0	0
n = 8	0	0	0	0	0	0	0	0
n = 9	0	0	0	0	0	0	0	0
n = 10	0	0	0	0	0	0	0	0
Total	1	1	1	1	1	1	1	1

Table 3.12 Probability that a n-bit string with $n < 10$ is produced at time $t < 7$

	$t = 1$	$t = 2$	$t = 3$	$t = 4$	$t = 5$	$t = 6$	$t = 7$	Total
n = 1	0.20	0	0	0	0	0	0	0.20
n = 2	0	0.14	0.046	0.016	0.0045	0.0012	0.00029	0.20
n = 3	0	0	0.030	0.017	0.014	0.0063	0.0028	0.070
n = 4	0	0	0	0.0034	0.0032	0.0031	0.0022	0.012
n = 5	0	0	0	0	0	0.00037	0.00031	0.00069
n = 6	0	0	0	0	0	0	0	0
n = 7	0	0	0	0	0	0	0	0
n = 8	0	0	0	0	0	0	0	0
n = 9	0	0	0	0	0	0	0	0
n = 10	0	0	0	0	0	0	0	0
Total	0.21	0.14	0.076	0.037	0.021	0.011	0.0057	

according to the formalism and method described herein. The situation changes for larger strings, when an alternating string may in effect strongly suggest that it should be less random than other strings because a short description is possible in terms of the simple alternation of bits. Some strings will assume their correct rank when the calculation is taken further to $D(5)$.

On the other hand, it may seem odd that the program size complexity of a string of length l is systematically larger than l when l can be produced by a *print* function of length 1+{the length of the print program}, and indeed one can interpret the results exactly in this way. The surplus can be interpreted as a constant product of a *print* phenomenon which is particularly significant for short strings. But since it is a constant, one can subtract it from all the strings. For example, subtracting 1 from all values brings the complexity results for the shortest strings to exactly their size, which is what one would expect from the values for algorithmic complexity. On the other hand, subtracting the constant preserves the relative order, even if larger strings continue having algorithmic complexity values larger than their lengths. What we provide herein, besides the numerical values, is a hierarchical structure from which one can tell whether a string is of greater, lesser or equal algorithmic complexity.

The *print program* assumes the implicit programming of the halting configuration. In C language, for example, this is delimited by the semicolon. The fact then that a single bit string requires a 2 bit "program" may be interpreted as the additional information represented by the length of the string; the fact that a string is of length n is not the result of an arbitrary decision but it is encoded in the producing machine. In other words, the string not only carries the information of its n bits, but also of the delimitation of its length. This is different to, for example, approaching the algorithmic complexity by means of cellular automata–there being no encoded halting state, one has to manually stop the computation upon producing a string of a certain arbitrary length according to an arbitrary stopping time. This is a research program that we have explored before [16].

It is important to point out that after the application of the coding theorem one often gets a non-integer value when calculating $C(s)$ from $m(s)$. Even though when interpreted as the size in bits of the program produced by a Turing machine it should be an integer value because the size of a program can only be given in an integer number of bits. The non-integer values are, however, useful to provide a finer structure providing information on the exact places in which strings have been ranked.

A question addressed in the next sections is how much of the relative string order (hence the relative algorithmic probability and the relative algorithmic complexity) of $D(n)$ will be preserved when calculating $D(i)$ for larger Turing machine spaces such that $0 < n < i$. As reported here, $D(n)$ preserves most of the string orders of $D(n-1)$ for $1 < n < 5$. While each space $(n,2)$ contains all $(n-1,2)$ machines, the exponential increase in number of machines when adding states may easily produce strings such that the order of the previous distribution is changed. What the results presented here show, however, is that each new space of larger machines contributes in the same proportion to the number of strings produced in the smaller spaces, in such a way that they preserve much of the previous string order of the distributions of smaller spaces, as shown by calculating the Spearman coefficient indicating a very strong ranking correlation. In fact, some of the ranking variability between the distributions of spaces of machines with different numbers of states occurred later in the classification, likely due to the fact that the smaller spaces missed the production of some strings. For example, the first rank difference between $D(3)$ and $D(4)$ occurred in place 20, meaning that the string order in $D(3)$ was strictly preserved in $D(4)$ up to the top 20 strings sorted from higher to lower frequency. Moreover, one may ask whether the actual frequency values of the strings converge.

3.5 Calculating $D(5)$

According to the formalism in Chap. 2, there are $(4n + 2)^{2n}$ different machines with n states. There are therefore 26 559 922 791 424 Turing machines with 5 states. That is, for comparison, about the same number of red cells in the blood of an average adult. Because there is a large enough number of machines to run even for a small number of machine states, applying the coding theorem provides a finer and increasingly stable evaluation of $K(s)$ based on the frequency of production of a large number of Turing machines, but the number of Turing machines grows exponentially, and producing $D(5)$ requires considerable computational resources.

3.5.1 *Setting the Runtime*

The Busy Beaver for Turing machines with 4 states is known to be 107 steps [3], that is, any Turing machine with 2 symbols and 4 states running longer than 107 steps will never halt. However, the exact number is not known for Turing machines with 2 symbols and 5 states, although it is believed to be 47 176 870, as there is a candidate machine that runs for this long and halts and no machine greater runtime has yet been found.

So we decided to let the machines with 5 states run for 4.6 times the Busy Beaver value for 4-state Turing machines (for 107 steps), knowing that this would constitute a sample significant enough to capture the behavior of Turing machines with 5 states. The chosen runtime was rounded to 500 steps, which was used to build the output frequency distribution for $D(5)$. The theoretical justification for the pertinence and significance of the chosen runtime is provided in Sect. 3.7.

3.6 A Glance at $D(5)$

Running all the Turing machines with 5 states in the reduced enumeration (see Sect. 2.3) up to 500 steps for the calculation of $D(5)$ took 18 days using 25×86–64 CPUs running at 2128 MHz with 4 GB of memory each.[5] In order to save space in the output of D(5), our C++ simulator produced partial results every 10^9 consecutive machines according to the reduced enumeration. Every 10^9 machines, the counters for each string produced were updated. The final classification is only 4.1 Megabytes but we can estimate the size of the output had we not produced partial results on the order of 1.28 Terabytes for the reduced space and 6.23 Terabytes for the full one. If we were to include in the output an indication for non-halting machines, the files would grow an extra 1.69 Terabytes for the reduced enumeration and 8.94 Terabytes for the full one.

[5] A supercomputer located at the Centro Informático Científico de Andalucía (CICA), Spain.

Table 3.13 The 147 most frequent strings from $D(5)$ (by row). The first column is a counter to help locate the rank of each string

1	1	0	11	10	01	00	111
8	000	110	100	011	001	101	010
15	1111	0000	1110	1000	0111	0001	1101
22	1011	0100	0010	1010	0101	1100	0011
29	1001	0110	11111	00000	11110	10000	01111
36	00001	11101	10111	01000	00010	11011	00100
43	10110	10010	01101	01001	10101	01010	11010
50	10100	01011	00101	11100	11000	00111	00011
57	11001	10011	01100	00110	10001	01110	111111
64	000000	111110	100000	011111	000001	111101	101111
71	010000	000010	101010	010101	101101	010010	111011
78	110111	001000	000100	110101	101011	010100	001010
85	101001	100101	011010	010110	110110	100100	011011
92	001001	111100	110000	001111	000011	101110	100010
99	011101	010001	110010	101100	010011	001101	111001
106	100111	011000	000110	111010	101000	010111	000101
113	100110	011001	110011	001100	100001	011110	110100
120	001011	111000	000111	110001	100011	011100	001110
127	1111111	0000000	1111110	1000000	0111111	0000001	1010101
134	0101010	1111101	1011111	0100000	0000010	1111011	1101111
141	0010000	0000100	1110111	0001000	1111100	1100000	0011111

Table 3.13 provides a glance at $D(5)$ showing the 147 most frequent (and therefore simplest) calculated strings out of 99 608. The top strings of $D(5)$ conform to an intuition of simplicity. Table 3.14 shows all the 2^n strings for $n = 7$, hence displaying what $D(5)$ suggests are the strings sorted from lowest to highest complexity, which seems to agree well with the intuition of simple (top left) to random-looking (bottom right).

3.7 Reliability of the Approximation of $D(5)$

Not all 5-state Turing machines have been used to build $D(5)$, since only the output of machines that halted at or before 500 steps were taken into consideration. As an experiment to see how many machines we were leaving out, we ran 1.23×10^{10} random Turing machines for up to 5000 steps (see Fig. 3.8d). Among these, only 50 machines halted after 500 steps and before 5000 (that is less than 1.75164×10^{-8} because in the reduced enumeration we don't include those machines that halt in one

Table 3.14 All the 2^n strings for $n = 7$ from $D(5)$ sorted from highest frequency (hence lowest complexity) to lowest frequency (hence highest (random) complexity). Strings in each row have the same frequency (hence the same Kolmogorov complexity). There are 31 different groups representing the different complexities of the $2^7 = 128$ strings

1	1111111	0000000						
2	1111110	1000000	0111111	0000001				
3	1010101	0101010						
4	1111101	1011111	0100000	0000010				
5	1111011	1101111	0010000	0000100				
6	1110111	0001000						
7	1111100	1100000	0011111	0000011				
8	1011010	1010010	0101101	0100101				
9	1101101	1011011	0100100	0010010	1111001	1001111	0110000	0000110
10	1110101	1010111	0101000	0001010				
11	1101110	1000100	0111011	0010001				
12	1101010	1010100	0101011	0010101				
13	1010110	1001010	0110101	0101001				
14	1111010	1010000	0101111	0000101				
15	1110110	1001000	0110111	0001001				
16	1010001	1000101	0111010	0101110				
17	1011110	1000010	0111101	0100001				
18	1011101	0100010						
19	1101011	0010100	1001001	0110110				
20	1110011	1100111	0011000	0001100				
21	1100101	1010011	0101100	0011010				
22	1011001	1001101	0110010	0100110	1000001	0111110		
23	1111000	1110000	0001111	0000111	1101001	1001011	0110100	0010110
24	1110010	1011000	0100111	0001101	1101100	1100100	0011011	0010011
25	1100010	1011100	0100011	0011101				
26	1100110	1001100	0110011	0011001				
27	1001110	1000110	0111001	0110001				
28	1100001	1000011	0111100	0011110				
29	1110001	1000111	0111000	0001110				
30	1100011	0011100						
31	1110100	1101000	0010111	0001011				

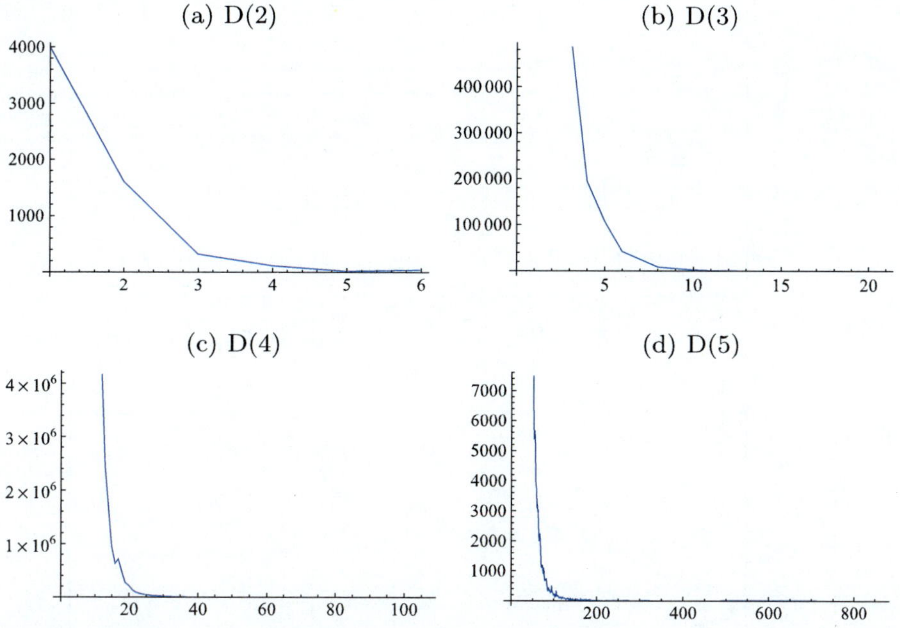

Fig. 3.8 Distribution of runtimes from $D(2)$ to $D(5)$. On the y-axes are the number of Turing machines and on the x-axes the number of steps upon halting (notice that given that for 5-states not known Busy Beaver values are known, $D(5)$ (Fig. d) was produced by all the Turing machines with 5 states that run for at most $t = 500$ steps. These plots show that the tendencies suggest that the runtime cutoff $t = 500$ for the production of $D(5)$ covers most of the halting Turing machines, hence the missed machines are neglectible.)

step or that we know won't halt before generating them, so it's a smaller fraction), with the remaining 1 496 491 379 machines not halting at 5000 steps. As far as these are concerned–and given the unknown values for the Busy Beavers for 5 states–we do not know after how many steps they would eventually halt, if they ever do. According to the following analysis, our election of a runtime of 500 steps therefore provides a good estimation of $D(5)$.

3.7.1 Exponential Regression

The frequency of runtimes of (halting) Turing machines has theoretically been proven to drop exponentially [17], and our experiments are closer to the theoretical behavior (see Fig. 3.8). To estimate the fraction of halting machines that were missed because Turing machines with 5 states were stopped after 500 steps, we hypothesize that the number of steps S a random halting machine needs before halting is an exponential RV (random variable), defined by $\forall k \geq 1,\ P(S = k) \propto e^{-\lambda k}$. We do not have direct

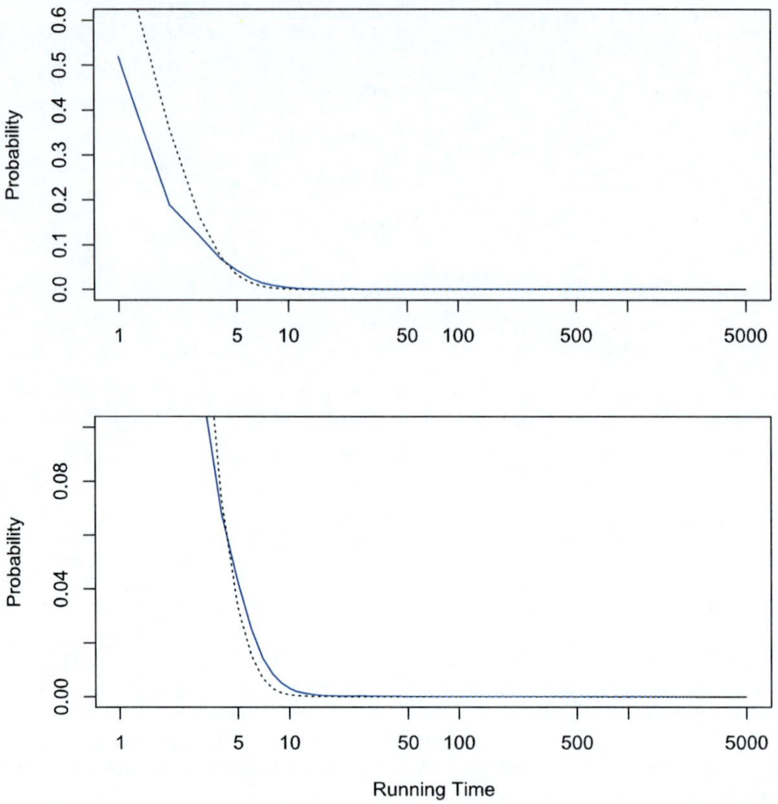

Fig. 3.9 Observed (solid) and theoretical (dotted) $P(S = k|S \leq 5000)$ against k. The x-axis is logarithmic. Two different scales are used on the y-axis to allow for a more precise visualization

access to an evaluation of $P(S = k)$, since we only have data for those machines for which $S \leq 5000$. But we may compute an approximation of $P(S = k|S \leq 5000)$, $1 \leq k \leq 5000$, which is proportional to the desired distribution.

A non-linear regression using ordinary least-squares gives the approximation $P(S = k|S \leq 5000) = \alpha e^{-\lambda k}$ with $\alpha = 1.12$ and $\lambda = 0.793$. The residual sum-of-squares is 3.392×10^{-3}, the number of iterations 9 with starting values $\alpha = 0.4$ and $\lambda = 0.25$. Figure 3.9 helps to visualize how the model fits the data.

The model's λ is the same λ appearing in the general law $P(S = k)$, and may be used to estimate the number of machines we lose by using a 500 step cut-off point for running time: $P(k > 500) \approx e^{-500\lambda} \approx 6 \times 10^{-173}$. This estimate is far below the point where it could seriously impair our results: the less probable (non-impossible) string according to $D(5)$ has an observed probability of 1.13×10^{-9}.

Although this is only an estimate, it suggests that missed machines are few enough to be considered negligible.

3.7.2 *The Exponential Fit*

The exponential model is not totally satisfactory because the real runtimes distribution seems to decrease too slowly for an exponential. For $k \geq 6$, the real frequency of runtime k seems to remain above the theoretical one [17]. This could mean that our estimate for the missed machines in our sample is optimistic.

To overcome this problem, it may be advisable to compare the observed runtime distribution with another theoretical distribution that will not drop so fast. Let us use the distribution μ defined by $\forall k \geq 1, \mu(k) = b \times e^{-\lambda\sqrt{x-1}}$ (same λ as before). We must have $b = \lambda^2/2$. With this distribution, the conditional probability is such that

$$\forall k \in \{1, ..., 5000\}, \mu(k|k \leq 5000) = \frac{be^{-\lambda\sqrt{k-1}}}{1 - \lambda e^{-\lambda\sqrt{4999}}(\sqrt{4999} + \lambda^{-1})}$$

Figure 3.10 shows that the actual distribution decreases faster than μ. Then, μ would give us a more conservative estimate of the proportion of missed machines. This estimate is given by

$$\lambda e^{-\lambda\sqrt{4999}}(\sqrt{4999} + \lambda^{-1}) \approx 2.55 \times 10^{-23}.$$

Although this is much higher than our previous estimate, it remains negligible for practical purposes.

3.8 Features of $D(5)$

3.8.1 *Lengths*

5-state Turing machines produced 99,608 different binary strings (to be compared to the 1832 strings for $D(4)$). While the largest string produced for $D(4)$ was of length 16 bits and only all 2^n strings for $n = 8$ were produced, the strings in $D(5)$ have lengths from 1 to 49 bits (excluding lengths 42 and 46 that never occur) and include every possible string of length $l < 12$. Among the 12 bit strings, only two were not produced (000110100111 and 111001011000). Of $n = 13, \ldots, 15$ about half the 2^n strings were produced (and therefore have frequency and complexity values). Figure 3.11 shows the proportion of n-long strings appearing in $D(5)$ outputs, for $1 \leq n \leq 49$.

The cumulative probability of every n-long string gives a probability law on $\mathbb{N}^*$. Figure 3.12 shows such a law obtained with $D(5)$, with $D(4)$, and with the theoretical 2^{-n} appearing in Levin's semi-measure. The most important difference may be the fact that this law does not decrease for $D(5)$, since length 2 is more likely than length 1.

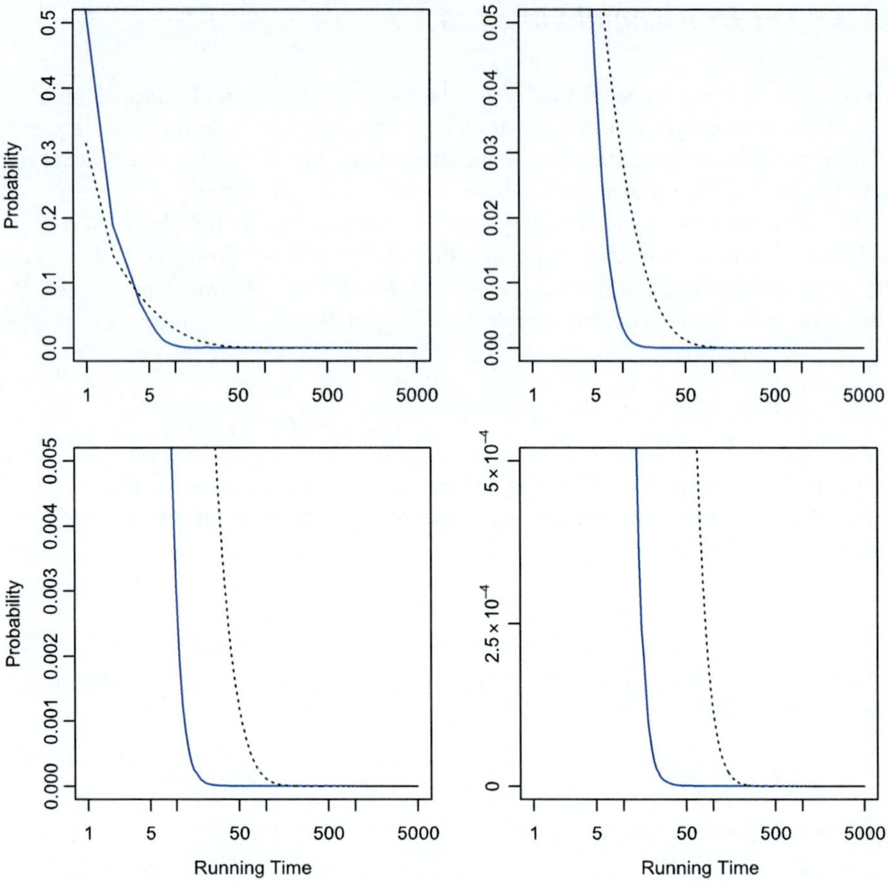

Fig. 3.10 Observed (solid) and theoretical (according to μ, dotted) $P(S = k|S \leq 5000)$ against k. The x-axis is logarithmic. 4 different scales are shown on the y-axis to allow for a more precise visualization

3.8.2 Global Simplicity

Some binary sequences may seem simple from a global point of view because they show symmetry (1011 1101) or repetition (1011 1011). Let us consider the string $s = 1011$ as an example. We have $P_{D(5)}(s) = 3.267414 \times 10^{-3}$. The repetition $ss = 10111011$ has a much lower probability $P_{D(5)}(ss) = 4.645999 \times 10^{-7}$. This is not surprising considering the fact that ss is much longer than s, but we may then wish to consider other strings based on s. In what follows, we will consider three methods (repetition, symmetrization, 0-complementation). The repetition of s is $ss = 10111011$, the "symmetrized" $s\bar{s} = 10111101$, and the 0-complementation 10110000. These three strings of identical length have different probabilities (4.645999×10^{-7}, 5.335785×10^{-7} and 3.649934×10^{-7} respectively).

Let us now consider all strings of length 3–6, and their symmetrization, 0-complementation and repetition. Figure 3.13 is a visual presentation of the results. In each case, even the minimum mean between the mean of symmetrized, complemented and repeated patterns (dotted horizontal line) lies in the upper tail of the $D(5)$ distribution for $2n$-length strings. And this is even more obvious with longer strings. Symmetry, complementation and repetition are, on average, recognized by $D(5)$.

Another method for finding "simple" sequences is based on the fact that the length of a string is negatively linked to its probability of appearing in $D(5)$. When ordered by decreasing probability, strings show increasing lengths. Let's call those sequences for which length is greater than that of the next string "climbers". The first 50 climbers appearing in $D(5)$ are given in Table 3.15 and show subjectively simple patterns, as expected.

Strings are not sorted by length but follow an interesting distribution of length differences that agrees with our intuition of simplicity and randomness and is in keeping with the expectation from an approximation to $m(s)$ and therefore $K(s)$.

3.8.3 Binomial Behavior

In a random binary string of length n, the number of '0s' conforms to a binomial law given by $P(k) = 2^{-n}\binom{n}{k}$. On the other hand, if a random Turing machine is drawn, simpler patterns are more likely to appear. Therefore, the distribution arising from Turing machine should be more scattered, since most simple patterns are often

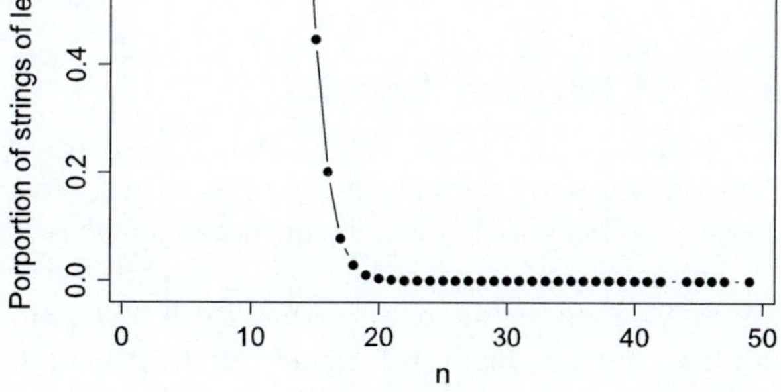

Fig. 3.11 Proportion of all n-long strings appearing in $D(5)$ against n

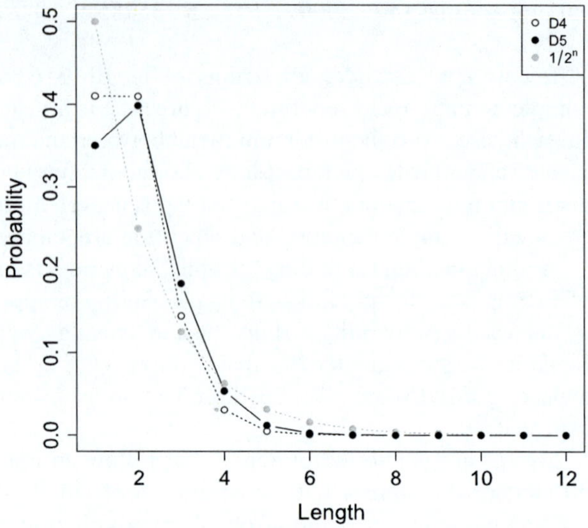

Fig. 3.12 Cumulative probability of all n-long strings against n

Table 3.15 Minimal examples of emergence: the first 50 climbers

00000000	000000000	000000001	000010000	010101010
000000010	000000100	0000000000	0101010101	0000001010
0010101010	00000000000	0000000010	0000011010	0100010001
0000001000	0000101010	01010101010	0000000011	0101010110
0000000100	0000010101	000000000000	0000110000	0000110101
0000000110	0110110110	00000010000	0000001001	00000000001
0010101101	0101001001	0000011000	00010101010	01010010101
0010000001	00000100000	00101010101	00000000010	00000110000
00000000100	01000101010	01010101001	01001001001	010101010101
01001010010	000000000001	00000011000	00000000101	0000000000000

unbalanced (such as 0000000). This is indeed what Fig. 3.14 shows: compared to truly random sequences of length n, $D(5)$ yields a larger standard deviation.

3.8.4 A Bayesian Approach

$D(5)$ allows us to determine, using a Bayesian approach, the probability that a given sequence is random: Let s be a sequence of length l. This sequence may be produced by a machine, let's say a 5-state Turing machine (event M), or by a random process (event R). Let's set the prior probability at $P(R) = P(M) = 1/2$. Because s does not have a fixed length, we cannot use the usual probability $P(s) = 1/2^l$, but we

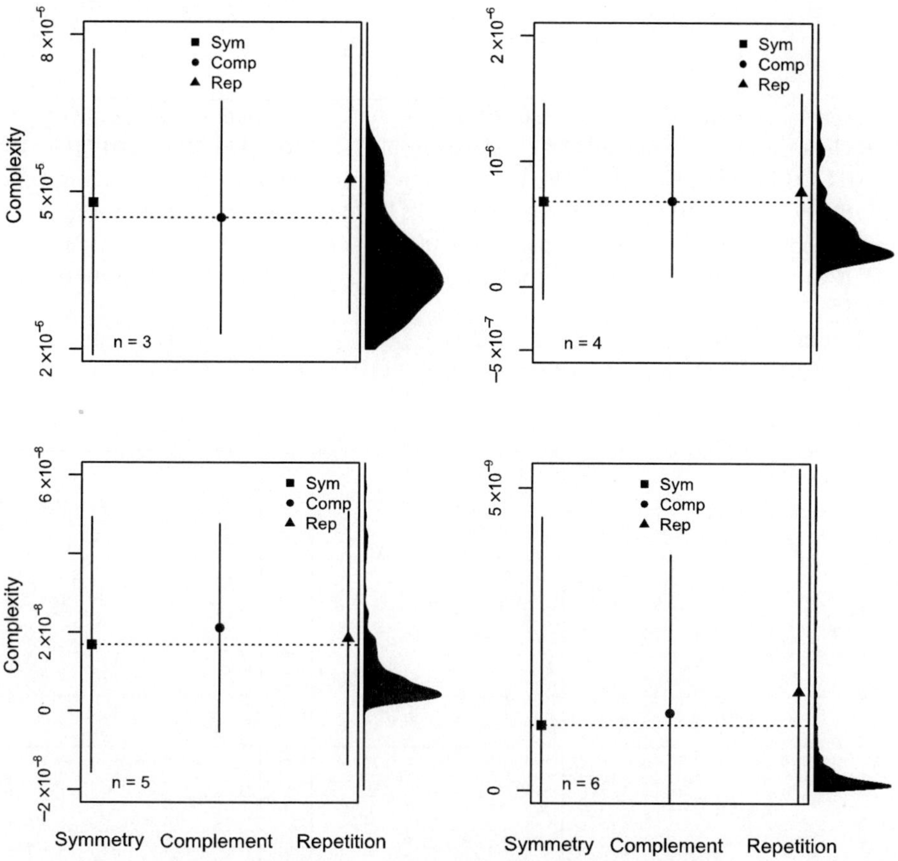

Fig. 3.13 Mean ± standard deviation of $D(5)$ of $2n$-long strings given by process of symmetrization (Sym), 0-complementation (Comp) and repetition (Rep) of all n-long strings. The dotted horizontal line shows the minimum mean among Sym, Comp and Rep. The density of $D(5)$ (smoothed with Gaussian kernel) for all $2n$-long strings is given in the right-margin

may, following Levin's idea, use $P(s|R) = 1/2^{2l}$. Given s, we can compute

$$P(R|s) = \frac{P(s|R)P(R)}{P(s)},$$

with

$$P(s) = P(s|M)P(M) + P(s|R)P(R) = \frac{P_{D(5)}(s)}{2} + \frac{1}{2^{2l+1}}.$$

Since $P(R) = 1/2$ and $P(s|R) = 1/2^{2l}$, the formula becomes

$$P(R|s) = \frac{1}{2^{2l} P_{D(5)}(s) + 1}.$$

There are 16 strings s such that $P(R|s) < 10^{-16}$ (the "least random strings"). Their lengths lie in the range [47–49]. An example is: 111011101110 111011101110 1110111011101111010101. The fact that the "least random" strings are long can intuitively be deemed correct: a sequence must be long before we can be certain it is not random. A simple sequence such as 00000000000000000000 (twenty '0s') gives a $P(R|s) = 0.006$. The size does matter, and there is no sequence of 40 or more '0s' in $D(5)$.

A total of 192 strings achieve a $P(R|s) > 1 - 1.7 \times 10^{-4}$. They all are of length 12 or 13. Examples are the strings 1110100001110, 1101110000110 or

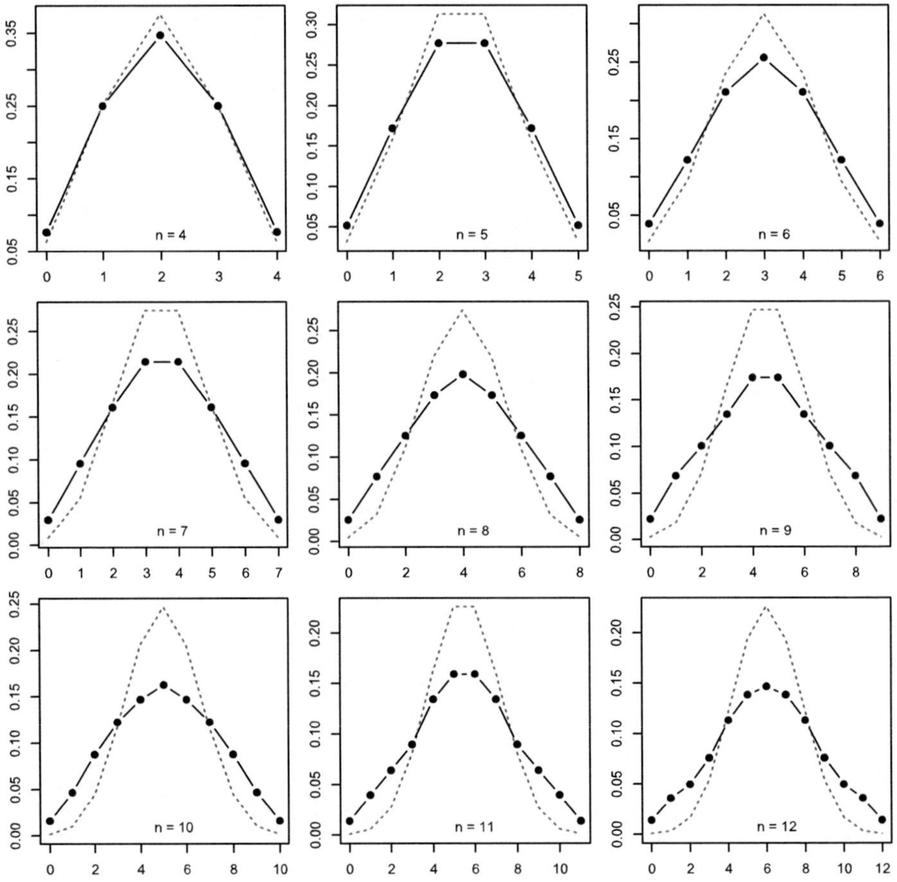

Fig. 3.14 Distributions of the number of zeros in n-long binary sequences according to a truly random drawing (red, dotted), or a $D(5)$ drawing (black, solid) for length 4 to 12

1100101101000. This is consistent with our idea of a random sequence. However, the fact that only lengths 12 and 13 appear here may be due to the specificity of $D(5)$.

3.9 Comparing $D(4)$ and $D(5)$

Every 4-state Turing machine may be modeled by a 5-state Turing machine whose fifth state is never attained. Therefore, the 1832 strings produced by $D(4)$ calculated in [16] also appear in $D(5)$. We thus have 1832 ranked elements in $D(4)$ to compare with. The basic idea at the root of this work is that $D(5)$ is a refinement (and major extension) of $D(4)$, previously calculated in an attempt to understand and evaluate algorithmic complexity. This would be hopeless if $D(4)$ and $D(5)$ led to totally different measures and rankings of simplicity versus complexity (randomness).

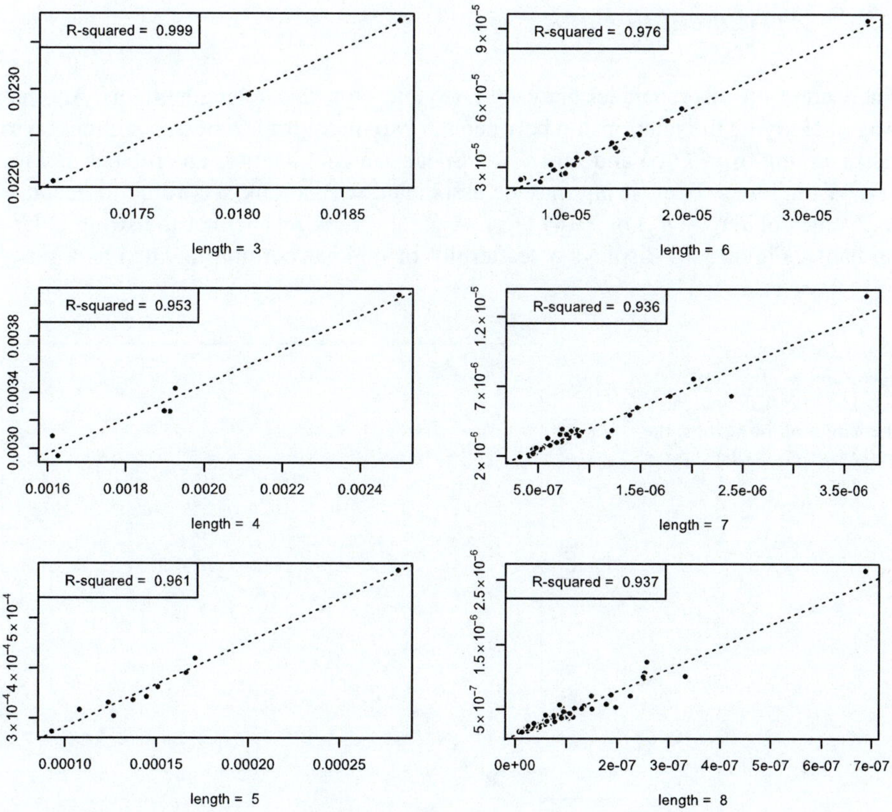

Fig. 3.15 $D(5)$ against $D(4)$, for n-long strings

3.9.1 Agreement in Probability

The link between $D(4)$ and $D(5)$ seen as measures of simplicity may be measured by the determination coefficient r^2, r being the Pearson correlation coefficient. This coefficient is $r^2 = 99.23\%$, which may be understood as "$D(4)$ explains 99.23% of the variations of $D(5)$". The scatterplot in Fig. 3.15 displays $D(5)(s)$ against $D(4)(s)$ for all strings s of length $n = 3, \ldots, 8$ (8 being the largest integer l such that $D(4)$ comprises every l-long sequence).

The agreement between $D(5)$ and $D(4)$ is almost perfect, but there are still some differences. Possible outliers may be found using a studentized residual in the linear regression of $D(5)$ against $D(4)$. The only strings giving absolute studentized residuals above 20 are 0 and 1. The only strings giving absolute studentized residuals lying between 5 and 20 are all the 3-long strings. All 4-long strings fall between 2 and 5. This shows that the differences between $D(5)$ and $D(4)$ may be explained by the relative importance given to the diverse lengths, as shown above (Fig. 3.12).

3.9.2 Agreement in Rank

There are some discrepancies between $D(5)$ and $D(4)$ due to length effects. Another way of studying the relationship between the two measures is to turn our attention to ranks arising from $D(5)$ and $D(4)$. The Spearman coefficient is an efficient tool for comparing ranks. Each string may be associated with a rank according to decreasing values of $D(5)$ (R_5) or $D(4)$ (R_4). A greater rank means that the string is less probable. Figure 3.16 displays a scatterplot of ranks according to $D(5)$ as a func-

Fig. 3.16 R_5 (rank according to $D(5)$) against R_4. The grayscale indicates the length of the strings: the darker the point, the shorter the string

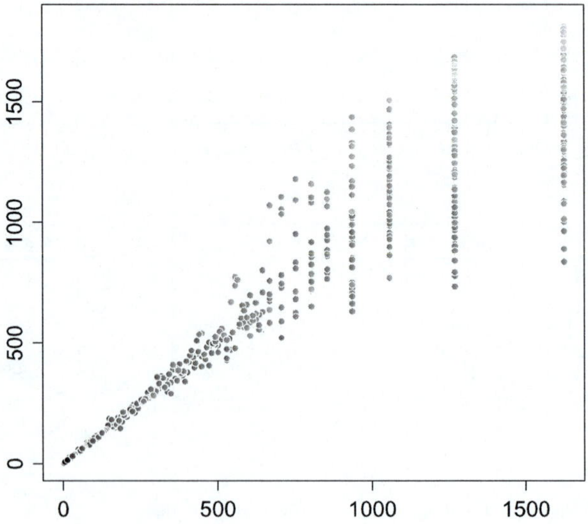

Table 3.16 The 20 strings for which $|R_4 - R_5| \geq 600$

Sequence	R_4	R_5
010111110	1625.5	837.5
011111010	1625.5	837.5
100000101	1625.5	837.5
101000001	1625.5	837.5
000011001	1625.5	889.5
011001111	1625.5	889.5
100110000	1625.5	889.5
111100110	1625.5	889.5
001111101	1625.5	963.5
010000011	1625.5	963.5
101111100	1625.5	963.5
110000010	1625.5	963.5
0101010110	1625.5	1001.5
0110101010	1625.5	1001.5
1001010101	1625.5	1001.5
1010101001	1625.5	1001.5
0000000100	1625.5	1013.5
0010000000	1625.5	1013.5
1101111111	1625.5	1013.5
1111111011	1625.5	1013.5

tion of $D(4)$-rank. Visual inspection shows that the ranks are similar, especially for shorter sequences. The Spearman correlation coefficient amounts to 0.9305, indicating strong agreement.

Not all strings are equally ranked and it may be interesting to take a closer look at outliers. Table 3.16 shows the 20 strings for which $|R_4 - R_5| \geq 600$. All these sequences are ties in $D(4)$, whereas $D(5)$ distinguishes 5 groups. Each group is made up of 4 equivalent strings formed from simple transformations (reversing and complementation). This confirms that $D(5)$ is fine-grained compared to $D(4)$.

The shortest sequences such that $|R_4 - R_5| \geq 5$ are of length 6. Some of them show an intriguing pattern, with an inversion in ranks, such as 000100 ($R_5 = 85$, $R_4 = 77$) and 101001 with reversed ranks.

On the whole, $D(5)$ and $D(4)$ are similar measures of simplicity, both from a measurement point of view and a ranking point of view. Some differences may arise from the fact that $D(5)$ is more fine-grained than $D(4)$. Other unexpected discrepancies still remain: we must be aware that $D(5)$ and $D(4)$ are both approximations of a more general limit measure of simplicity versus randomness. Differences are inevitable, but the discrepancies are rare enough to allow us to hope that $D(5)$ is for the most part a good approximation of this properties.

Table 3.17 Top 20 strings in $D(5)$ with highest frequency and therefore lowest Kolmogorov (program-size) complexity. From frequency (middle column) to complexity (extreme right column) applying the coding theorem in order to get $K_{D(5)}$ through the Coding theorem

Sequence	Frequency ($m(s)$)	Complexity ($K(s)$)
1	0.175036	2.51428
0	0.175036	2.51428
11	0.0996187	3.32744
10	0.0996187	3.32744
01	0.0996187	3.32744
00	0.0996187	3.32744
111	0.0237456	5.3962
000	0.0237456	5.3962
110	0.0229434	5.44578
100	0.0229434	5.44578
011	0.0229434	5.44578
001	0.0229434	5.44578
101	0.0220148	5.50538
010	0.0220148	5.50538
1111	0.0040981	7.93083
0000	0.0040981	7.93083
1110	0.00343136	8.187
1000	0.00343136	8.187
0111	0.00343136	8.187
0001	0.00343136	8.187

3.10 Kolmogorov Complexity Calculation

It is now straightforward to apply the coding theorem (see (1.11) in Sect. 1.3) to convert string frequency (as a numerical approximation of algorithmic probability $m(s)$) to an approximation of Kolmogorov complexity $K_{D(n)}(s)$ (see Tables 3.17 and 3.18). Formally,

$$K_{D(n)}(s) = -\log_2 D(n)(s)$$

First it is worth noting that the calculated complexity values in Tables 3.17 and 3.18 are real numbers, when they are supposed to be the lengths of programs (in bits) that produce the strings, hence integers. The obvious thing to do is to round the values to the next closest integer, but this would have a negative impact as the decimal expansion provides a finer classification. Hence the finer structure of the classification is favored over the exact interpretation of the values as lengths of computer programs. It is also worth mentioning that the lengths of the strings (as shown in Table 3.18) are almost always smaller than their Kolmogorov (program-size) values, which is somehow to be expected from this approach. Consider the single bit. It not only encodes itself, but the length of the string (1 bit) as well, because it is produced by a Turing machine that has reached the halting state and produced this output upon halting.

Consider the Shannon entropy of a single bit calculated from $D(5)$. Given that $D(5)$ is not uniformly distributed (just as $m(s)$ is not, by definition), receiving a single bit does not carry only the information of the bit in question, because the chances of not getting any other bit are exponentially larger than getting one. Just as the occurrence of a particular letter in English determines with some probability the next letter (e.g. 3 "e"s in English never normally occur, thus "e" brings with it more information than just "e"). In the same way we think that strings produced by halting machines also encode information about their length.

Also worth noting is the fact that the strings 00, 01, 10 and 11 all have the same complexity, according to our calculations (this is the case from $D(2)$ to $D(5)$). It might just be the case that the strings are too short to really have different complexities, and that a Turing machine that can produce one or the other is of exactly the same length. To us the string 00 may look more simple than 01, but we do not have many arguments to validate this intuition for such short strings, and it may be an indication that such intuition is misguided (think in natural language, if spelled out in words, 00 doesn't seem to have a much shorter description than the shortest description of 01).

Compare this phenomenon of program-sizes being greater than the length of these short strings to the extent of the problem posed by compression algorithms (see Fig. 3.1), which collapse all strings of up to length 40 at least, producing the same complexity approximations for all of them. One way to overcome this minor inconvenience involved in using the alternative approach developed here is to subtract a constant (no greater than the smallest complexity value) from all the complexity values, which gives these strings lower absolute random complexity values (preserving the relative order). But even if left "random", this alternative technique can be used to distinguish and compare them, unlike the lossless compression approach that is unable to further compress short strings.

The phenomenon of complexity values greater than the lengths of the strings is transitional. Out of the 99,608 strings in $D(5)$, 212 have greater string lengths than program-size values. The first string to have a smaller program-size value than string length is the string 10101010101010101010 101010101010101010101 (and its complementation), of length 41 but program-size of 33.11 (34 if rounded). The mean of the strings with greater program-size than length is 38.3494, The string with the greatest difference between length and program-size in $D(5)$ are strings of low Kolmogorov complexity such as 0101010001000100010001000100010001000100010001010, of length 49 but with an approximated Kolmogorov complexity (program-size) value of 39.0642. Hence far from random, both in terms of the measure and in terms of the string's appearance.

Table 3.18 Twenty random strings (sorted from lowest to highest complexity values) from the first half of $D(5)$ to which the coding theorem has been applied (extreme right column) to approximate $K(s)$

String length	String	Complexity ($K(s)$)
11	11011011010	28.1839
12	101101110011	32.1101
12	110101001000	32.1816
13	0101010000010	32.8155
14	11111111100010	34.1572
12	011100100011	34.6045
15	001000010101010	35.2569
16	0101100000000000	35.6047
13	0110011101101	35.8943
15	101011000100010	35.8943
16	1111101010111111	25.1313
18	000000000101000000	36.2568
15	001010010000000	36.7423
15	101011000001100	36.7423
17	10010011010010011	37.0641
21	100110000000110111011	37.0641
14	11000010000101	37.0641
17	01010000101101101	37.4792
29	01011101111100011101111010101	37.4792
14	11111110011110	37.4792

3.10.1 Randomness in $D(5)$

Paradoxically, the strings at the bottom of $D(5)$ as sorted from highest to lowest frequency and therefore lowest to highest Kolmogorov (random) complexity are not very random looking, but this is to be expected, as the actual most random strings of these lengths would have had very low frequencies and would not therefore have been produced. In fact what we are looking at are some of the strings with the greatest structure (lowest Kolmogorov complexity) that made it into $D(5)$ (most of them produced by a single Turing machine). Table 3.19, however, shows the bottom of the length $n = 12$ classification extracted from $D(5)$, for which all 2^n binary strings were produced, hence displaying more apparent randomness.

Table 3.19 Bottom 21 strings of length $n = 12$ with smallest frequency in $D(5)$

100111000110	100101110001	100011101001
100011100001	100001110001	011110001110
011100011110	011100010110	011010001110
011000111001	000100110111	111000110100
110100111000	001011000111	000111001011
110100011100	110001110100	001110001011
001011100011	110000111100	001111000011

3.10.2 Robustness of $K_{D(n)}$

An important question is how robust is $K_{D(n)}$, that is how sensitive it is to n. We know that the invariance theorem (Sect. 1.3) guarantees that the values converge in the long term, but the invariance theorem tells nothing about the rate of convergence. In Sect. 3.9.2 we have shown that $D(n+1)$ respects the order of $D(n)$ except for very few and minor value discrepancies concerning the least frequent strings (and therefore the most unstable given the few machines generating them). This is not obvious despite the fact that all Turing machines with n states in (n, m) are included in the space of $(n+1, m)$ machines (that is, the machines that never reach one of the $n+1$ states), because the number of machines in $(n+1, m)$ overcomes by far the number of machines in (n, m), and a completely different result could have been then produced. However, the agreement between $D(n)$ and $D(n+1)$ seems to be similarly high among, and despite, the few cases $n \leq 6$ in hand to compare with. The only way for this behaviour to radically change for $n > 5$ is if for some n', $D(n')$ starts diverging in ranks from $D(n'-1)$ on before starting to converge again (by the invariance theorem). If one does not have any reason to believe in such a change of behavior, the rate of rank convergence of $D(n)$ is close to optimal very soon, even for the relatively "small" sets of Turing machines for small n.

One may ask how robust the complexity values and classifications may be in the face of changes in computational formalism (e.g. Turing machines with several tapes, and all possible variations). We have shown [22] that radical changes to the computing model produce reasonable (and correlated with different degrees of confidence) ranking distributions of complexity values (using even completely different computing models, such as unidimensional deterministic cellular automata and Post tag systems).

We have also calculated the maximum differences between the Kolmogorov complexity evaluations of the strings occurring in every 2 distributions $D(n)$ and $D(n+1)$ for $n = 2, \ldots, 4$. This provides estimations for the constant c in the invariance theorem determining the maximum difference in bits among all the strings evaluated with one or another distribution, hence shedding light on the robustness of the evaluations under this procedure. The smaller the values of c the more stable our method. The values of these bounding constants (in bits) among the different exact evaluations of K using $D(n)$ for $n = 2, \ldots, 5$ after application of the coding theorem are:

$$|K_{D(2)}(s) - K_{D(3)}(s)| \leq c = 4.090; 4.090; 3.448; 0.39$$
$$|K_{D(3)}(s) - K_{D(4)}(s)| \leq c = 4.10; 3.234; 2.327; 2.327$$
$$|K_{D(4)}(s) - K_{D(5)}(s)| \leq c = 5.022; 4.274; 3.40; 2.797$$

where $K_{D(n)}(s)$ means $K(s)$ evaluated using the output frequency distribution $D(n)$ after application of the coding theorem for $n = 2, \ldots, 5$ ($n = 1$ is a trivial non interesting case) and where every value of c is calculated by quartiles (separated by semicolons), that is, the calculation of c among all the strings in the 2 compared distributions, then among the top 3/4, then the top half and finally the top quarter by rank. Notice that the estimation of c between $D(2)$ and $D(3)$, and $D(3)$ and $D(4)$ remained almost the same among all strings occurring in both, at about 4 bits. This means one could write a "compiler" (or translator) among the two distributions for all their occurring strings of size only 4 bits providing one or the other complexity value for K based on one or the other distribution. The differences are considerably smaller for more stable strings (towards the top of the distributions). One may think that given that the strings with their occurrences in $D(n+1)$ necessarily contain those in $D(n)$ for all n (because the space of all Turing machines with an additional state always contain the computations of the Turing machines with less states), the agreement should be expected. However, $D(n)$ contributes to to $D(n+1)$ with about the logarithm of the number of strings in $D(n+1)$. For example, $D(4)$ contributes only 1832 strings to the 99 608 produced in $D(5)$ (that is less than 2%). All in all, the largest difference found between $D(4)$ and $D(5)$ is only of 5 bits of among all the strings occurring both in $D(4)$ and $D(5)$ (1832 strings), where the values of K in $D(4)$ are between 2.285 and 29.9.

3.11 Validation by the Number of Instructions Used

We are now interested in the relation of $K_{D(5)}(s)$ to the minimal number of instructions that a Turing machine producing a string s uses. Machines in $D(5, 2)$ have a transition table with 10 entries, corresponding to the different pairs with one of the five states and a symbol either "0" or "1". These are the 10 instructions that the machine can use. But for a fixed input not all instructions are necessarily used. Then, for a blank tape, not all machines that halt use the same number of instructions. The simplest cases are machines halting in just one step, that is, machines whose transition for $(init\_state, blank\_symbol)$ goes to the halting state, producing a string "0" or "1". So the simplest strings produced in $D(5, 2)$ are computed by machines using just one instruction. We expected a correlation between the $K_{D(5)}$-complexity of the strings and the number of instructions used. As we show, the following experiment confirmed this.

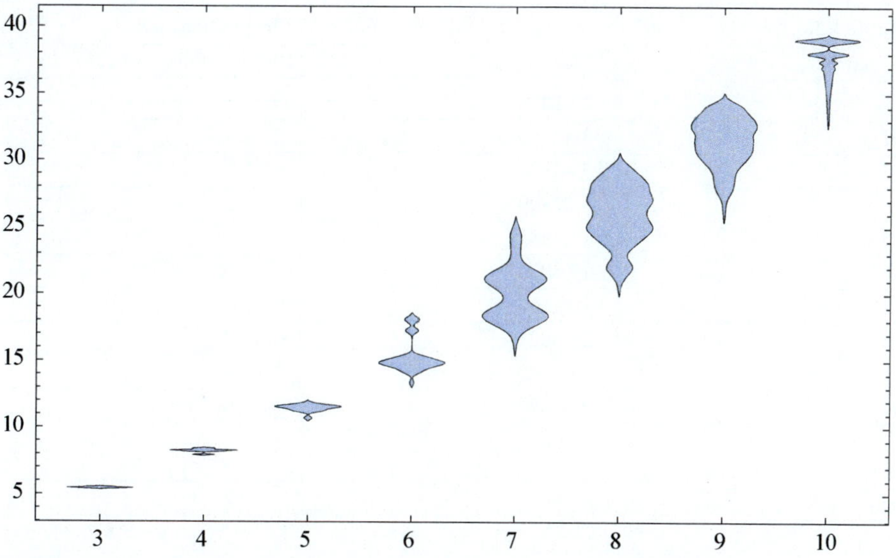

Fig. 3.17 Distribution of $K_{D(5)}$ values according to the number of instructions used

We used a sample of 2836×10^9 random machines in the reduced enumeration for $D(5, 2)$, that is, 29% the total number of machines. The output of the sample returns the strings produced by halting machines together with the number of instructions used, the runtime and the instructions for the Turing machine. To save space, we only saved the smallest number of instructions found for each string produced, and the smallest runtime corresponding to that particular number of instructions.

After doing the appropriate symmetry completions we have 99 584 different strings, which is to say almost all the 99 608 strings found in $D(5, 2)$. The number of instructions used goes from 1 to 10. When 1 instruction is used only "0" and "1" are generated, with a $K_{D(5)}$ value of 2.51428. With 2 instructions, all 2-bit strings are generated, with a $K_{D(5)}$ value of 3.32744. For 3 or more instructions, Fig. 3.17 shows the distribution of values of $K_{D(5)}$. Table 3.20 shows the mean $K_{D(5)}$ values for the different numbers of instructions used.

This accords with our expectations. Machines using a low number of instructions can be repeated many times by permuting the order of states. So the probability of producing their strings is greater, which means low $K_{D(5)}$ values.

We can also look at the relation between the number of instructions used and the length of the strings produced. For $1 \leq i \leq 5$, all strings of length i are produced by machines using i instructions. For a greater number of instructions used, Fig. 3.18 shows the distribution of string lengths. Table 3.20 shows the mean length for each number of instructions used.

The correlation $r_{K_{D(5)},N} = 0.83$ is a good indicator for quantifying the apparent relation between $K_{D(5)}$ and the number N of instructions used, proving a strong positive link. However, since the length L of outputs is linked with both variables,

Table 3.20 Mean $K_{D(5)}$ and string length for different numbers of instructions used

Used inst.	Mean $K_{D(5)}$	Mean length
1	2.51428	1
2	3.32744	2
3	5.44828	3
4	8.22809	4
5	11.4584	5
6	15.3018	6.17949
7	20.1167	7.76515
8	26.0095	9.99738
9	31.4463	12.6341
10	37.5827	17.3038

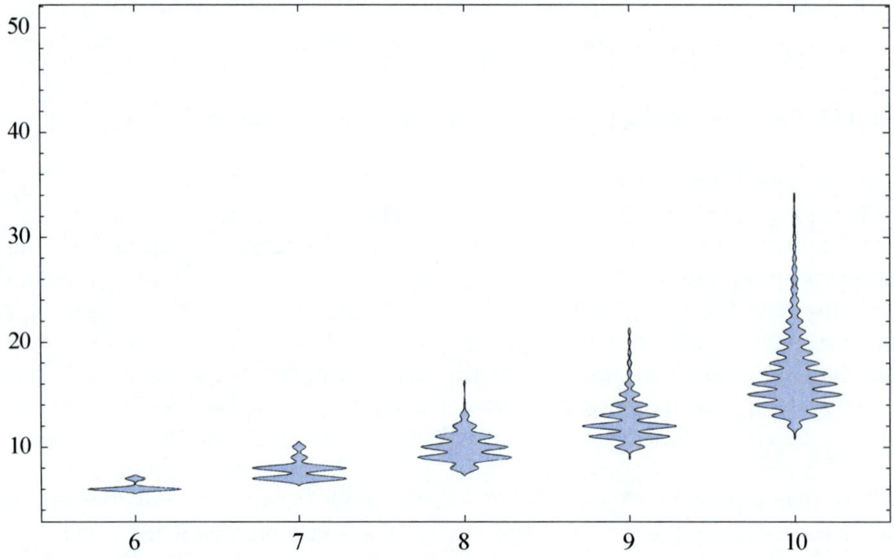

Fig. 3.18 Instructions used and string lengths

the partial correlation $r_{K_{D(5)},N.L} = 0.81$ is a better index. This value indicates a strong relation between $K_{D(5)}$ and N, even while controlling for L.

3.11.1 Logical Depth and $K_{D(5)}$

As explained above, we have also found that the machines which generate each string using the minimum number of instructions also have the minimum runtime. These

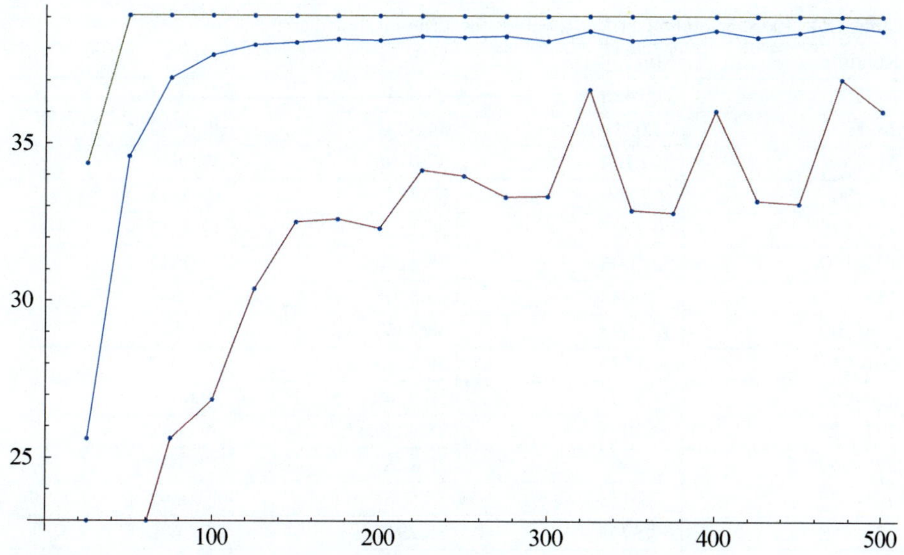

Fig. 3.19 LD and $K_{D(5)}$ (min, mean and max values)

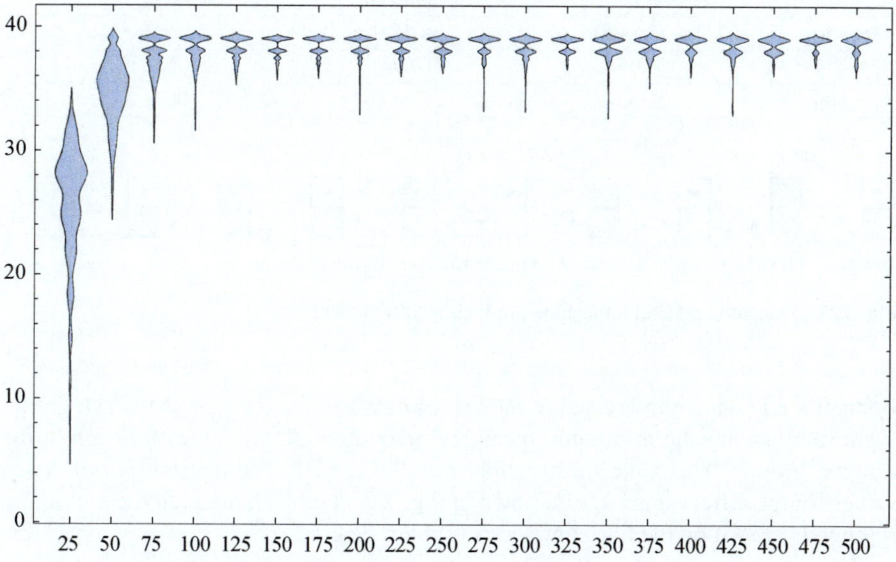

Fig. 3.20 LD and $K_{D(5)}$ (distribution)

runtimes are related to Bennett's logical depth (LD), as they are the shortest runtimes of the smallest Turing machines producing each string in $D(5, 2)$.

We have partitioned the runtime space from 1 to 500 (our runtime bound) into 20 groups of equal length (25 steps). In order to explore the relation of $K_{D(5)}$ to

Table 3.21 Extreme and mean $K_{D(5)}$ values for different runtime intervals

Runtime	Min $K_{D(5)}$	Mean $K_{D(5)}$	Max $K_{D(5)}$
1–25	2.51428	25.6049	34.3638
26–50	21.0749	34.5849	39.0642
51–75	25.6104	37.0796	39.0642
76–100	26.8569	37.8125	39.0642
101–125	30.3777	38.1337	39.0642
126–150	32.5096	38.2150	39.0642
151–175	32.6048	38.3208	39.0642
176–200	32.3093	38.2850	39.0642
201–225	34.1573	38.4213	39.0642
226–250	33.9767	38.3846	39.0642
251–275	33.3093	38.4249	39.0642
276–300	33.3363	38.2785	39.0642
301–325	36.7423	38.5963	39.0642
326–350	32.8943	38.2962	39.0642
351–375	32.8163	38.3742	39.0642
376–400	36.0642	38.6081	39.0642
401–425	33.2062	38.4035	39.0642
426–450	33.1100	38.5543	39.0642
451–475	37.0642	38.7741	39.0642
476–500	36.0642	38.6147	39.0642

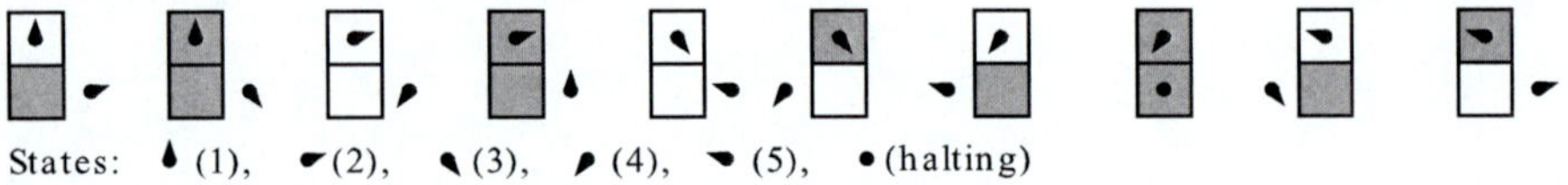

Fig. 3.21 Transition table of a machine producing “0011110001011”

Bennett’s LD we are interested in the values of $K_{D(5)}$ for the strings in each group. Figure 3.19 shows the minimum, mean and maximum $K_{D(5)}$ values for each of the runtime groups. The same information is in Table 3.20. The distribution of $K_{D(5)}$ values for the different groups is shown in Fig. 3.20. For each interval, the maximum runtime is shown on the horizontal axis (Table 3.21).

We now provide some examples of the discordance between $K_{D(5)}$ and LD. “0011110001011” is a string with high $K_{D(5)}$ and low LD. Figure 3.21 shows the transition table of the smallest machine found producing this string. The runtime is low–just 29 steps (of the 99 584 different strings found in our sample, only 3 360 are produced in fewer steps), but it uses 10 instructions and produces a string with complexity 39.0642. It is the greatest complexity we have calculated for $K_{D(5)}$. Figure 3.22 shows the execution of the machine.

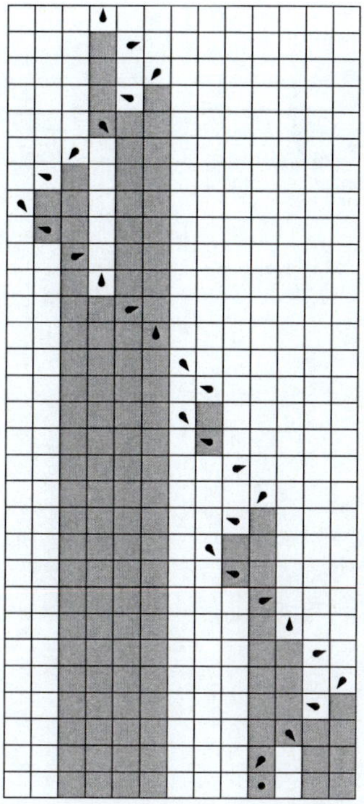

Fig. 3.22 Execution of the machine producing "0011110001011"

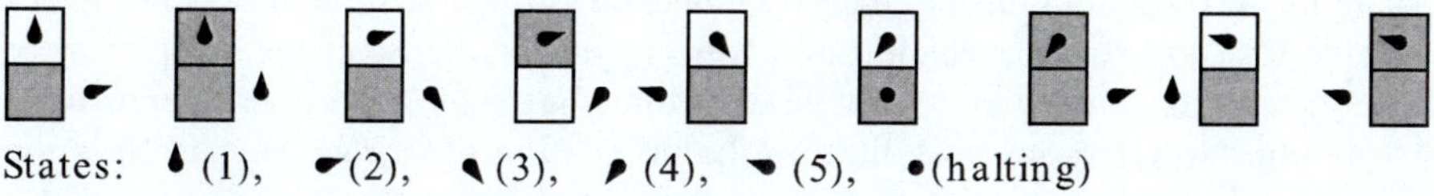

Fig. 3.23 Transition table of a machine producing "$(10)^{20}1$"

On the other hand, "$(10)^{20}1$" is a string with high LD but low $K_{D(5)}$ value. Figure 3.23 shows the transition table of the machine found producing this string, and Fig. 3.24 depicts the execution. The machine uses 9 instructions and runs for 441 steps (only 710 strings out of the 99 584 strings in our sample require more time) but its $K_{D(5)}$ value is 33.11. This is a low complexity if we consider that in $K_{D(5)}$ there are 99 608 strings and that 90 842 are more complex than this one.

We may rate the overall strength of the relation between $K_{D(5)}$ and LD by the correlation $r_{K_{D(5)},LD} = 0.41$, corresponding to a medium positive link. As we previously mentioned however, the fact that the length L of the strings is linked with both variables may bias our interpretation. A more relevant measure is thus $r_{K_{D(5)},LD.L} = -0.06$, a negative value indicating no significant relation between $K_{D(5)}$ and LD once L is controlled.

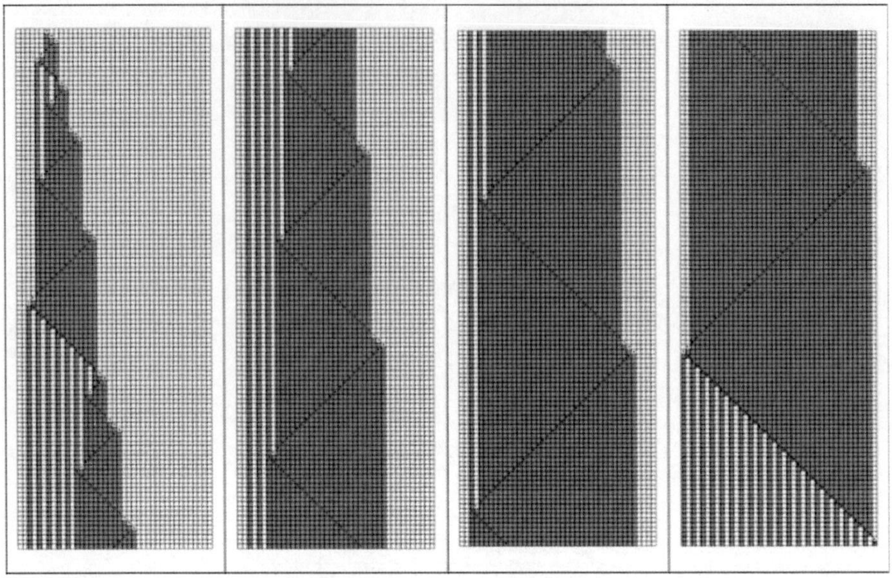

Fig. 3.24 Execution of the machine producing "$(10)^{20}1$"

3.12 Summary

Changing the programming language (or universal Turing machine) makes $K(s)$ very unstable for short strings, which prevents from getting a sense of the complexity of these strings. However, the concept of algorithmic probability and Levin's universal distribution ($m(s)$) varies very little in the production of strings, because it is the result of an operation that makes incremental changes to the frequencies from a very large number of calculations from a multitude of Turing machines. The chief advantage of evaluating $K_{m(s)}$ rather than $K(s)$ in general (for example, using an arbitrary universal Turing machine) is that $m(s)$ may be as sensitive to the additive constant involved in the invariance theorem for different choices, but it seems more natural to calculate $m(s)$ with an enumeration of Turing machines of increasing size (the traditional quasi-lexicographical order) than choosing or constructing an arbitrary universal Turing machine or Turing-complete programming language. In the case of the enumeration, for example, fixed a size of Turing machines, one can go through every machine in the enumeration, hence making the particular order of the enumeration irrelevant.

We've also shown that the procedure seems robust (small variation between different sets of Turing machines of different size) and that the results are in accordance with our intuitions of complexity and randomness. $D(5)$ is essentially as we expected; it provides a precise and objective notion of what a "simple string" is as compared to a "random (Kolmogorov complex) string", and it is not a trivial accomplishment to provide such a deep insight into the world of structure versus randomness while at the same time attaching a numerical and formal meaning to such concepts.

The method described here is computationally expensive, but it doesn't need to be executed more than once (and we have already done so for up to 5 state Turing machines), obviating the need to recalculate the complexity values of the strings that have already been calculated. The classification provides a prior distribution that should prove to be of general use and application. As a result we now have two complementary techniques for approximating K, the traditional lossless compression algorithm technique and now the one described in this chapter.[6] What remains to be done is to put these together and show that they actually work in harmony when they overlap over string lengths where both may provide reasonable approximations (Sect. 5.5). Also needed are extensions of the current model that provide insight into and formal approximations of n-dimensional objects other than unidimensional strings (e.g. images, for which we have also provided some important steps, see Sect. 5.1).

As we expected, the Kolmogorov-Chaitin complexity evaluated by means of Levin's Coding Theorem from the output distribution of small Turing machines correlates with the number of instructions used but not with logical depth. Logical depth yields a reasonable measure of complexity that is different from the measure obtained by considering algorithmic complexity (K) alone, and this investigation proves that all these three measures (Kolmogorov-Chaitin Complexity, Solomonoff-Levin Algorithmic Probability and Bennett's Logic Depth) are numerically approachable, sound and consistent with theoretical expectations, and may be used in real-world applications. K as a measure of program size is supposed to be an integer (the length of a program in bits). $K_{D(5)}$, however, yields non-integer values. Because $K_{D(5)}$ is shown to be a finer measure than the length of Turing machines, these results also justify the utility of non-integer values in the approximation of the algorithmic complexity of short strings, which also means being able to avoid the longer calculations that must be undertaken if only integer values were allowed.

In summary, this procedure seems to be a possible alternative to having to choose an universal Turing machine U in order to evaluate K_U, and is complementary to the compression approach that is ineffective for short strings. The technique in this chapter provides a reasonable and consistent approximation to $K(s)$ that is in agreement with the theory and represents evidence in confirmation of Levin's distribution and Solomonoff's universal induction.

---

[6] An Online Algorithmic Complexity Calculator implementing the technique presented herein and making the data available to the research community is accessible at http://www.complexitycalculator.com.

References

1. Lempel, A., Ziv. J.: On the complexity of finite sequences. IEEE Trans. Inform. Theory **22.1**, 75–81 (1976)
2. Rad, T.: On non-computable functions. Bell Syst. Tech. J. **41.3**, 877–884 (1962)
3. Brady, A.H.: The determination of the value of Rado's noncomputable function Sigma(k) for four-state Turning machines. Math. Comput. **40**, 162, 647–665 (1983)
4. Marxen, H., Buntrock, J.: Attacking the Busy Beaver 5. Bull. EATCS **40**, 247–251 (1990)
5. Calude, C.S., Dinneen, M.J., Shu, C.-K.: Computing a glimpse of randomness. Exp. Math. **11.2**, 369–378 (2002)
6. [Her09] Hertel, J.: Computing the uncomputable rado sigma function: an automated, symbolic induction prover for nonhalting turning machines. Mathematica J. **11.2** (2009)
7. Zenil, H.: A review of methods for estimating algorithmic complexity: options, challenges, and new directions. Entropy **22**, 612 (2020)
8. Chaitin, G.J.: On the length of programs for computing finite binary sequences: statistical considerations. J. ACM **16.1**, 145–159 (1969)
9. Solomonoff, R.J.: A formal theory of inductive inference: Parts 1 and 2. Inf. Control **7.1-22**, 224–254 (1964)
10. Levin, L.: Laws of information conservation (non-growth) and aspects of the foundation of probability theory. Probl Trans. **10**, 206–210 (1974)
11. Chaitin, G.J.: From Philosophy to Program Size (2003)
12. Rivals, E. et al.: Compression and genetic sequence analysis. Biochimie **78.5**, 315–322 (1996)
13. Cilibrasi, R., Vitanyi, P.: Clustering by compression. IEEE Trans. Inf. Theory **51.4**, 1523–1545 (2005)
14. Li, M., Vitnyi, P.: An Introduction to Kolmogorov Complexity and Its Applications. Springer (2008)
15. [Zen11] H. Zenil. "Une approche exprimentale la thorie algorithmique de la complexit, dissertation in fulfilment of the degree of Doctor in Computer Science (committee: J.-P". Highest honours. PhD thesis. Universit de Lille 1, June 2011
16. Delahaye, J.-P., Zenil, H.: Numerical evaluation of algorithmic complexity for short strings: a glance into the in-nermost structure of randomness. Appl. Math. Comput. **219.1**, 63–77 (2012)
17. Calude, C.S., Stay, M.A.: Most programs stop quickly or never halt. Adv. Appl. Math. **40**, 295–308 (2008)
18. Calude, C.S., Dumitrescu, M.: A statistical anytime algorithm for the Halting Problem. Computability **9**, 155–166 (2020)
19. Machlin, R., Stout, Q.F.: The complex behavior of simple machines. Physica **4**, 85–98 (1990)
20. Holkner, A.: Acceleration techniques for busy beaver candidates. In: Proceedings of the Second Australian Undergraduate Students'. Computing Conference (2004)
21. Calude, C.S.: Information and Randomness. Springer (2002)
22. Zenil, H., Delahaye, J.-P.: On the algorithmic nature of the world. Inf. Comput. **10** (2010). 9789814295482 0017

Chapter 4
Theoretical Aspects of Finite Approximations to Levin's Semi-measure

In this chapter we study the formal properties of a Levin-inspired measure m calculated from the output distribution of small Turing machines running on a blank tape. We introduce and justify finite approximations m_k that work in a similar way to our $D(k)$ distributions. We provide proofs of the relevant properties of both m and m_k and compare them to Levin's Universal Distribution. We calculate error estimations of m_k with respect to m.

4.1 Algorithmic Information Measures

Levin's semi-measure m_T defines the so-called Universal Distribution [1, 2], the value $m_T(s)$ being the probability that a random program halts and produces s running on a universal Turing machine T (see (1.10) in Sect. 1.3).

In this chapter we study the formal properties of m, a computable approximation to Levin's distribution. To avoid ambiguity, we will use $m_T(s)$ for Levin's semi-measure and $m(s)$ for our approximation, where there is no need to specify a T as we are considering for m the output distribution of all Turing machines (in the formalism of Sect. 2) running on a blank tape. Computing $m(s)$ requires to run a numerable infinite number of Turing machines, so we will focus on finite approximations $m_k(s)$ (for a natural number k) that require only machines up to k states. A key property of m_T and K_T is their universality: the choice of the Turing machine T used to compute the distribution is only relevant up to a (multiplicative or additive, respectively) constant. The computability of our measure $m(s)$ implies its lack of universality. The same happens when using common lossless compression algorithms to approximate K, but on top of their non-universality in the algorithmic sense, they are block entropy estimators as they traverse files in search of repeated patterns in a fixed-length window to build a replacement dictionary. Nevertheless, this does not prevent

H. Zenil et al., *Methods and Applications of Algorithmic Complexity*,
Emergence, Complexity and Computation 44,
https://doi.org/10.1007/978-3-662-64985-5_4

lossless compression algorithms to find useful applications in the same way as more algorithmic-based motivated measures can contribute even if also limited.

We start by presenting how our Turing machine formalism can be used to encode a prefix-free set of programs (Sect. 4.2). Then, in Sect. 4.3 we define a computable algorithmic probability measure m based on our Turing machine formalism and prove its main properties, both for m and for finite approximations m_k. In Sect. 4.4 we compute m_5, compare it with $D(5)$ and estimate the error in m_5 as an approximation to m.

4.2 Turing Machines as a Prefix-Free Set of Programs

Recall that $(n, 2)$ denotes the space of all n-state 2-symbol Turing machines (with the halting state not included among the n states) following the Busy Beaver Turing machine formalism as defined by Rado [3] (Chap. 2). We don not need here the specific details of our enumeration of Turing machines but only the following trivial proposition:

Proposition 4.1 *Turing Machines in the set* $(n, 2)$ *can be enumerated from* 0 *to* $(4n + 2)^{2n} - 1$

We use τ_t^n to denote the machine number t in $(n, 2)$ following the enumeration.

We now show that this set of Turing machines can be encoded as a prefix-free set of programs capable of generating any finite non-empty binary string.

Definition 4.2 (*Execution of a Turing machine*) Let $\tau \in (n, 2)$ be a Turing machine. We denote by $\tau(i)$ the execution of τ over an infinite tape filled with i (a blank symbol), where $i \in \{0, 1\}$. We write $\tau(i) \downarrow$ if $\tau(i)$ halts, and $\tau(i) \uparrow$ otherwise. We write $\tau(i) = s$ iff

- $\tau(i) \downarrow$, and
- s is the output string of $\tau(i)$, defined as the concatenation of the symbols in the tape of τ that were visited at some instant of the execution $\tau(i)$.

As Definition 4.2 establishes, we are only considering machines running over a blank tape with no input. Observe that the output of $\tau(i)$ considers the symbols in all cells of the tape written on by τ during the computation, so the output contains the entire fragment of the tape that was used. To produce a symmetrical set of strings, we consider both symbols 0 and 1 as possible blank symbols.

Definition 4.3 (*Program*) A *program* p is a triplet $\langle n, i, t\rangle$, where

- $n \geq 1$ is a natural number
- $i \in \{0, 1\}$
- $0 \leq t < (4n + 2)^{2n}$

We say that the *output* of p is s if, and only if, $\tau_t^n(i) = s$.

Programs can be executed by a universal Turing machine that reads a binary encoding of $\langle n, i, t\rangle$ (Definition 4.4) and simulates $\tau_t^n(i)$. Trivially, for each finite binary string s with length $|s| > 0$, there is a program p which outputs s.

Now that we have a formal definition of programs, we show that the set of valid programs can be represented as a prefix-free set of binary strings.

Definition 4.4 (*Binary encoding of a program*) Let $p = \langle n, i, t\rangle$ be a program (Definition 4.3). The *binary encoding* of p is a binary string with the following sequence of bits:

- First, $1^{n-1}0$, that is, $n-1$ repetitions of 1 followed by 0. This way we encode n.
- Second, a bit with value i encodes the blank symbol.
- Finally, t is encoded using $\lceil \log_2\left((4n+2)^{2n}\right)\rceil$ bits.

The use of $\lceil \log_2\left((4n+2)^{2n}\right)\rceil$ bits to represent t ensures that all programs with the same n are represented by strings of equal size. As there are $(4n+2)^{2n}$ machines in $(n, 2)$, with these bits we can represent any value of t. The process of reading the binary encoding of a program $p = \langle n, i, t\rangle$ and simulating $\tau_t^n(i)$ is computable, given the enumeration of Turing machines.

As an example, this is the binary representation of the program $\langle 2, 0, 185\rangle$:

1	0	0	0	0	0	0	0	0	1	0	1	1	1	0	0	1

The proposed encoding is prefix-free, that is, there is no pair of programs p, p' such that the binary encoding of p is a prefix of the binary encoding of p'. This is because the n initial bits of the binary encoding of $p = \langle n, i, t\rangle$ determine the length of the encoding. So p' cannot be encoded by a binary string having a different length but the same n initial bits.

Proposition 4.5 (Programming by coin flips) *Every source producing an arbitrary number of random bits generates a unique program (provided it generates at least one* 0*).*

Proof The bits in the sequence are used to produce a unique program following Definition 4.4. We start by producing the first n part, by selecting all bits until the first 0 appears. Then the next bit gives i. Finally, as we know the value of n, we take the following $\lceil \log_2\left((4n+2)^{2n}\right)\rceil$ bits to set the value of t. It is possible that constructing the program in this way, the value of t is greater than the maximum $(4n+2)^{2n} - 1$ in the enumeration. In which case we associate the program with some trivial non-halting Turing machine. For example a machine with the initial transition staying at the initial state.

The idea of programming by coin flips is very common in Algorithmic Information Theory. It produces a prefix-free coding system, that is, there is no string w encoding a program p which is a prefix of a string wz encoding a program $p' \neq p$. These coding systems make longer programs (for us, Turing machines with more states) exponentially less probable than short programs. In our case, this is because of the initial sequence of $n-1$ repetitions of 1, which are produced with probability $1/2^{n-1}$. This observation is important because when we later use machines in $\bigcup_{n=1}^{k}(n, 2)$ to reach a finite approximation of our measure, the greater k is, the exponentially smaller the error we will be allowing: the probability of producing by coin flips a random Turing machine with more than k states decreases exponentially with k [4].

4.3 A Levin-Style Algorithmic Measure

Definition 4.6 Given a Turing machine $\mathcal{A}$ accepting a prefix-free set of programs, the *probability distribution* of $\mathcal{A}$ is defined as

$$P_{\mathcal{A}}(s) = \sum_{p:\mathcal{A}(p)=s} \frac{1}{2^{|p|}} \tag{4.1}$$

where $\mathcal{A}(p)$ is equal to s if and only if $\mathcal{A}$ halts with input p and produces s. The length in bits of program p is represented by $|p|$.

$P_{\mathcal{A}}(s)$ measures how frequently the output s is generated when running random programs at $\mathcal{A}$. Given that the sum of $P_{\mathcal{A}}(s)$ for all strings is not 1 (non-halting programs not producing any strings are counted in $2^{|p|}$) it is said to be a semi-measure. If $\mathcal{A}$ is a universal Turing machine, then $P_{\mathcal{A}}(s)$ is Levin's distribution [1] for $\mathcal{A}$, also noted by $m_{\mathcal{A}}(s)$ (see (1.10) in Sect. 1.3). For universal Turing machines, the distribution is universal in the sense that the choice of $\mathcal{A}$ (among all the infinite possible universal reference universal Turing machines) is only relevant up to a multiplicative constant, and that the distribution is based on the universal model of Turing computability.

Definition 4.7 (*Distribution* $m(s)$) Let $\mathcal{M}$ be a Turing machine executing the programs introduced in Definition 4.3. Then, $m(s)$ is defined by

$$m(s) = P_{\mathcal{M}}(s).$$

Theorem 4.8 *For any binary string s,*

$$m(s) = \sum_{n=1}^{\infty} \frac{|\{\tau \in (n,2) \mid \tau(0) = s\}| + |\{\tau \in (n,2) \mid \tau(1) = s\}|}{2^{n+1+\lceil \log_2((4n+2)^{2n}) \rceil}} \tag{4.2}$$

Proof By Definition 4.4, the length of the encoding of program $p = \langle n, i, t \rangle$ is $n + 1 + \lceil \log_2\left((4n+2)^{2n}\right) \rceil$. It justifies the denominator of (4.2), as (4.1) requires it to be $2^{|p|}$. For the numerator, observe that the set of programs producing s with the same n value corresponds to all machines in $(n, 2)$ producing s with either 0 or 1 as blank symbol. Note that if a machine produces s both with 0 and 1, it is counted twice, as each execution is represented by a different program (that differ only as to the i digit).

4.3.1 Finite Approximations to m

The value of $m(s)$ for any string s depends on the output of an infinite set of Turing machines, so we have to manage ways to approximate it. The method proposed in Definition 4.9 approximates $m(s)$ by considering only a finite number of Turing machines up to a certain number of states.

Definition 4.9 (*Finite approximation $m_k(s)$*) The *finite approximation to $m(s)$ bound to k states, $m_k(s)$*, is defined as

$$m_k(s) = \sum_{n=1}^{k} \frac{|\{\tau \in (n,2) \mid \tau(0) = s\}| + |\{\tau \in (n,2) \mid \tau(1) = s\}|}{2^{n+1+\lceil \log_2((4n+2)^{2n}) \rceil}} \tag{4.3}$$

Proposition 4.10 (Convergence of $m_k(s)$ to $m(s)$)

$$\sum_{s\in(0+1)^*} \mid m(s) - m_k(s) \mid \ \leq \frac{1}{2^k}$$

Proof By (4.2) and (4.3),

$$\begin{aligned}
\sum_{s\in(0+1)^*} \mid m(s) - m_k(s) \mid &= \sum_{s\in(0+1)^*} m(s) \ - \sum_{s\in(0+1)^*} m_k(s) \\
&\leq \sum_{n=k+1}^{\infty} \frac{2(4n+2)^{2n}}{2^{n+1+\lceil \log_2((4n+2)^{2n}) \rceil}} \\
&\leq \sum_{n=k+1}^{\infty} \frac{2(4n+2)^{2n}}{2^n \cdot 2 \cdot 2^{\log_2((4n+2)^{2n})}} \\
&= \sum_{n=k+1}^{\infty} \frac{1}{2^n} = \frac{1}{2^k}
\end{aligned}$$

Proposition 4.10 ensures that the sum of the error in $m_k(s)$ as an approximation to $m(s)$, for all strings s, decreases exponentially with k. The bound of $1/2^k$ has only theoretical value; in practice we can find lower bounds. In fact, the proof counts all $2(4n+2)^{2n}$ programs of size n to bound the error (and many of them do not halt). In Sect. 4.4 we provide a finer error calculation for m_5 by removing from the count some very trivial machines that do not halt.

4.3.2 Properties of m and m_k

Levin's distribution is characterized by some important properties. First, it is lower semi-computable, that is, it is possible to compute lower bounds for it. Also, it is a semi-measure, because the sum of probabilities for all strings is smaller than 1. The key property of Levin's distribution is its *universality*: a semi-measure P is universal if and only if for every other semi-measure P' there exists a constant $c > 0$ (that may depend only on P and P') such that for every string s, $c \cdot P(s) \geq P'(s)$. That is, a distribution is universal if and only if it dominates (modulo a multiplicative constant) every other semi-measure. In this section we present some results pertaining to the computational properties of m and m_k.

Proposition 4.11 (Runtime bound) *Given any binary string s, a machine with k states producing s runs a maximum of $2^{|s|} \cdot |s| \cdot k$ steps upon halting or never halts.*

Proof Suppose that a machine τ produces s. We can trace back the computation of τ upon halting by looking at the portion of $|s|$ cells in the tape that will constitute the output. Before each step, the machine may be in one of k possible states, reading one of the $|s|$ cells. Also, the $|s|$ cells can be filled in $2^{|s|}$ ways (with a 0 or 1 in each cell). This makes for $2^{|s|} \cdot |s| \cdot k$ different possible instantaneous descriptions of the computation. So any machine may run, at most, that number of steps in order to produce s. Otherwise, it would produce a string with a greater length (visiting more than $|s|$ cells) or enter a loop.

Observe that a key property of our output convention is that we use all visited cells in the machine tape. This is what gives us the runtime bound which serves to prove the most important property of m_k, its computability (Theorem 4.12).

Theorem 4.12 (Computability of m_k) *Given k and s, the value of $m_k(s)$ is computable.*

Proof According to (4.3) and Proposition 4.1, there is a finite number of machines involved in the computation of $m_k(s)$. Also, Proposition 4.11 sets the maximum runtime for any of these machines in order to produce s. So an algorithm to compute $m_k(s)$ enumerates all machines in $(n, 2)$, $1 \leq n \leq k$ and runs each machine to the corresponding bound.

Corollary 4.13 *Given a binary string s, the minimum k with $m_k(s) > 0$ is computable.*

Proof Trivially, s can be produced by a Turing machine with $|s|$ states in just s steps. At each step i, this machine writes the ith symbol of s, moves to the right and changes to a new state. When all symbols of s have been written, the machine halts. So, to get the minimum k with $m_k(s) > 0$, we can enumerate all machines in $(n, 2)$, $1 \leq n \leq |s|$ and run all of them up to the runtime bound given by Proposition 4.11. The first machine producing s (if the machines are enumerated from smaller to larger size) gives the value of k.

Now, some uncomputability results of m_k

Proposition 4.14 *Given k, the length of the longest s with $m_k(s) > 0$ is non-computable.*

Proof We proceed by contradiction. Suppose that such a computable function as $l(k)$ gives the length of the longest s with $m_k(s) > 0$. Then $l(k)$, together with the runtime bound in Proposition 4.11, provides a computable function that gives the maximum runtime that a machine in $(k, 2)$ may run prior to halting. But it contradicts the uncomputability of the Busy Beaver [3]: the highest runtime of halting machines in $(k, 2)$ grows faster than any computable function.

Corollary 4.15 *Given k, the number of different strings s with $m_k(s) > 0$ is non-computable.*

Proof Also by contradiction: If the number of different strings with $m_k(s) > 0$ is computable, we can run in parallel all machines in $(k, 2)$ until the corresponding number of different strings has been found. This gives us the longest string, which is in contradiction to Proposition 4.14.

Now to the key property of m, its computability,

Theorem 4.16 (Computability of m) *Given any non-empty binary string, $m(s)$ is computable.*

Proof As we argued in the proof of Corollary 4.13, a non-empty binary string s can be produced by a machine with $|s|$ states. Trivially, it is then also produced by machines with more than $|s|$ states. So for every non-empty string s, the value of $m(s)$, according to (4.2), is the sum of enumerable infinite many rationals which produce a real number. A real number is computable if, and only if, there is some algorithm that, given n, returns the first n digits of the number. And this is what $m_k(s)$ does. Proposition 4.10 enables us to calculate the value of k such that $m_k(s)$ provides the required digits of $m(s)$, as $m(s) - m_k(s)$ is bounded by $1/2^k$.

The subunitarity of m and m_k implies that the sum of $m(s)$ (or $m_k(s)$) for all strings s is smaller than one. This is because of the non-halting machines:

Proposition 4.17 (Subunitarity) *The sum of $m(s)$ for all strings s is smaller than* 1, *that is,*

$$\sum_{s \in (0+1)^\star} m(s) < 1$$

Proof By using (4.2),

$$\sum_{s \in (0+1)^\star} m(s) = \sum_{n=1}^{\infty} \frac{|\{\tau \in (n, 2) \mid \tau(0) \downarrow\}| + |\{\tau \in (n, 2) \mid \tau(1) \downarrow\}|}{2^{n+1+\lceil \log_2((4n+2)^{2n}) \rceil}} \tag{4.4}$$

but $|\{\tau \in (n, 2) \mid \tau(0) \downarrow\}| + |\{\tau \in (n, 2) \mid \tau(1) \downarrow\}|$ is the number of machines in $(n, 2)$ that halt when starting with a blank tape filled with 0 plus the number of machines in $(n, 2)$ that halt when starting on a blank tape filled with 1. This number is at most twice the cardinality of $(n, 2)$, but we know that it is smaller, as there are very trivial machines that do not halt, such as those without transitions to the halting state, so

$$\sum_{s\in(0+1)^\star} m(s) < \sum_{n=1}^{\infty} \frac{2(4n+2)^{2n}}{2^{n+1+\lceil\log_2((4n+2)^{2n})\rceil}} = \sum_{n=1}^{\infty} \frac{(4n+2)^{2n}}{2^n \cdot 2^{\lceil\log_2((4n+2)^{2n})\rceil}}$$

$$\leq \sum_{n=1}^{\infty} \frac{(4n+2)^{2n}}{2^n(4n+2)^{2n}} = \sum_{n=1}^{\infty} \frac{1}{2^n} = 1$$

Corollary 4.18 *The sum of* $m_k(s)$ *for all strings* s *is smaller than* 1

Proof By Proposition 4.17, (4.2) and (4.3).

The key property of $m_k(s)$ and $m(s)$ is their computability, given by Propositions 4.12 and 4.16, respectively. So these distributions cannot be universal, as Levin's Universal Distribution is non-computable. In spite of this, the computability of our distributions (and the possibility of approximating them with a reasonable computational effort), as we have shown, provides us with a tool to approximate the algorithmic probability of short binary strings. In some sense this is similar to what happens with other (computable) approximations to (uncomputable) Kolmogorov complexity, such as common lossless compression algorithms, which in turn are estimators of the classical Shannon entropy rate (e.g. all those based in LZW, and unlike $m_k(s)$ and $m(s)$, are not able to find algorithmic content beyond statistical patterns, not even in principle, unless a compression algorithm is designed to seek a specific one. For example, the digital expansion of the mathematical constant π is believed to be normal and therefore will contain no statistical patterns of the kind that compression algorithms can detect, yet there will be a (short) computer program that can generate it, or at least finite (and small) initial segments of π.

4.4 Computing m_5

We have explored the sets of Turing machines in $(n, 2)$ for $n \leq 5$ (Sect. 3.5). For $n \leq 4$, the maximum time that a machine in $(n, 2)$ may run upon hating is known [5]. It allows us to calculate the exact values of m_4. For $n = 5$, we we use the same runtime bound of 500 steps used to calculate $D(5)$. We have the database of machines producing each string s for each value of $n \leq 5$. So we have applied (4.3) to estimate m_5.

The measure m_k is similar to $D(k)$, but the denominator of (4.3) is the number of (detected) halting machines in $(k, 2)$. Using $D(5)$ as an approximation to Levin's

Table 4.1 Top 10 strings in m_5 and $D(5)$ with their estimated complexity

s	$K_{m_5}(s)$	$K_{D(5)}(s)$	s	$K_{m_5}(s)$	$K_{D(5)}(s)$
0	3.7671	2.5143	11	6.8255	3.3274
1	3.7671	2.5143	000	10.4042	5.3962
00	6.8255	3.3274	111	10.4042	5.3962
01	6.8255	3.3274	001	10.4264	5.4458
10	6.8255	3.3274	011	10.4264	5.4458

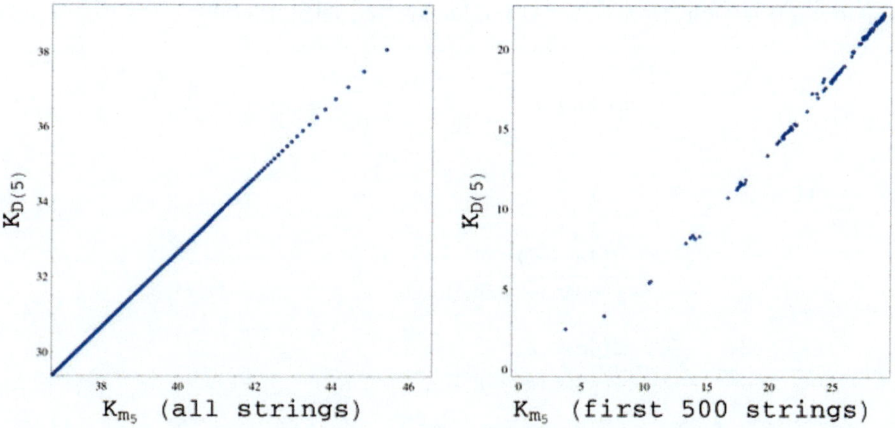

Fig. 4.1 Correlation of rank comparison between K_{m_5} and $K_{D(5)}$

distribution, algorithmic complexity is estimated by means of the algorithmic Coding Theorem as $K_{D(5)}(s) = -\log_2 D(5)(s)$. Now, m_5 provides us with another estimation: $K_{m_5}(s) = -\log_2 m_5(s)$. Table 4.1 shows the 10 most frequent strings in both distributions, together with their estimated complexity.

Figure 4.1 shows a rank comparison of both estimations of algorithmic complexity after application of the algorithmic Coding Theorem. With minor differences, there is an almost perfect agreement. So in classifying strings according to their relative algorithmic complexity, the two distributions are equivalent.

The main difference between m_k and $D(k)$ is that $D(k)$ is not computable, because computing it would require us to know the exact number of halting machines in $(k, 2)$, which is impossible given the halting problem. We work with approximations to $D(k)$ by considering the number of halting machines detected. In any case, though m_k is computable, it is computationally intractable, so in practice (approximations to) the two measures can be used interchangeably.

We can make some estimations about the error in m_5 with respect to m. "0" and "1" are two very special strings, both with the maximum m_5 value. These strings are the most frequent outputs in $(n, 2)$ for $n \leq 5$, and we may conjecture that they are the most frequent outputs for all values of n. These strings then have the greatest absolute error, because the terms in the sum of $m(\text{"0"})$ (the argument for $m(\text{"1"})$ is identical) not included in $m_5(\text{"0"})$ are always the greatest independent of n.

We can calculate the exact value of the terms for $m(\text{"0"})$ in (4.2). To produce "0", starting with a tape filled with $i \in \{0, 1\}$, a machine in $(n, 2)$ must have the transition corresponding to the initial state and read symbol i with the following instruction: write 0 and change to the halting state (thus not moving the head). The other $2n - 1$ transitions may have any of the $4n + 2$ possible instructions. So there are $(4n + 2)^{2n-1}$ machines in $(n, 2)$ producing "0" when running on a tape filled with i. Considering both values of i, we have $2(4n + 2)^{2n-1}$ programs of the same length $n + 1 + \lceil \log_2 \left((4n + 2)^{2n}\right)\rceil$ producing "0". Then, for "0",

$$m(\text{"0"}) = \sum_{n=1}^{\infty} \frac{2(4n+2)^{2n-1}}{2^{n+1+\lceil \log_2\left((4n+2)^{2n}\right)\rceil}} \tag{4.5}$$

This can be approximated by

$$\begin{aligned}
m(\text{"0"}) &= \sum_{n=1}^{\infty} \frac{2(4n+2)^{2n-1}}{2^{n+1+\lceil \log_2\left((4n+2)^{2n}\right)\rceil}} \\
&= \sum_{n=1}^{\infty} \frac{2(4n+2)^{2n-1}}{2^{n+1}2^{\lceil \log_2\left((4n+2)^{2n}\right)\rceil}} \\
&= \sum_{n=1}^{\infty} \frac{(4n+2)^{2n-1}}{2^{n}2^{\lceil \log_2\left((4n+2)^{2n}\right)\rceil}} \\
&= \sum_{n=1}^{2000} \frac{(4n+2)^{2n-1}}{2^{n}2^{\lceil \log_2\left((4n+2)^{2n}\right)\rceil}} + \sum_{n=2001}^{\infty} \frac{(4n+2)^{2n-1}}{2^{n}2^{\lceil \log_2\left((4n+2)^{2n}\right)\rceil}} \\
&< \sum_{n=1}^{2000} \frac{(4n+2)^{2n-1}}{2^{n}2^{\lceil \log_2\left((4n+2)^{2n}\right)\rceil}} + \sum_{n=2001}^{\infty} \frac{(4n+2)^{2n-1}}{2^{n}2^{\log_2\left((4n+2)^{2n}\right)}} \\
&= \sum_{n=1}^{2000} \frac{(4n+2)^{2n-1}}{2^{n}2^{\lceil \log_2\left((4n+2)^{2n}\right)\rceil}} + \sum_{n=2001}^{\infty} \frac{(4n+2)^{2n-1}}{2^{n}(4n+2)^{2n}} \\
&= \sum_{n=1}^{2000} \frac{(4n+2)^{2n-1}}{2^{n}2^{\lceil \log_2\left((4n+2)^{2n}\right)\rceil}} + \sum_{n=2001}^{\infty} \frac{1}{2^{n}(4n+2)} \simeq 0.0742024
\end{aligned}$$

we have divided the infinite sum into two intervals cutting at 2000 because the approximation of $2^{\lceil \log_2\left((4n+2)^{2n}\right)\rceil}$ to $(4n + 2)^{2n}$ is not good for low values of n, but has almost no impact for large n. In fact, cutting at 1000 or 4000 gives the same result with a precision of 17 decimal places. We have used *Mathematica* to calculate both the sum from 1 to 2000 and the convergence from 2001 to infinity. So the value $m(\text{"0"}) = 0.0742024$ is exact for practical purposes. The value of $m_5(\text{"0"})$ is 0.0734475, so the error in the calculation of $m(\text{"0"})$ is 0.0007549. If "0" and "1" are the strings with the highest m value, as we (informedly) conjecture, then this is the maximum error in m_5 as an approximation to m.

As a reference, K_{m_5}("0") is 3.76714. With the real m("0") value, the approximated complexity is 3.75239. The difference is not relevant for most practical purposes.

We can also provide an upper bound for the sum of the error in m_5 for strings different from "0" and "1". Our way of proceeding is similar to the proof of Proposition 4.10, but we count in a finer fashion. The sum of the error for strings different from "0" and "1" is

$$\sum_{n=6}^{\infty} \frac{|\{\tau \in (n,2) \mid \tau(0)\downarrow,\ \tau(0) \notin \{\text{"0"},\text{"1"}\}\}| + |\{\tau \in (n,2) \mid \tau(1)\downarrow,\ \tau(1) \notin \{\text{"0"},\text{"1"}\}\}|}{2^{n+1+\lceil \log_2\left((4n+2)^{2n}\right)\rceil}} \tag{4.6}$$

The numerators of the above sum contain the number of computations (with blank symbol "0" or "1") of Turing machines in $(n, 2)$, $n \geq 6$, that halt and produce an output different from "0" and "1". We can obtain an upper bound of this value by removing, from the set of computations in $(n, 2)$, those that produce "0" or "1" and some trivial cases of machines that do not halt.

First, the number of computations in $(n, 2)$ is $2(4n + 2)^{2n}$, as all machines in $(n, 2)$ are run twice for both blank symbols ("0" and "1"). Also, the computations producing "0" or "1" are $4(4n + 2)^{2n-1}$. Now, we focus on two sets of trivial non-halting machines:

- Machines with the initial transition staying at the initial state. For blank symbol i, there are $4(4n + 2)^{2n-1}$ machines that when reading i at the initial state do not change the state (for the initial transition there are 4 possibilities, depending on the writing symbol and direction, and for the other $2n - 1$ transitions there are $4n + 2$ possibilities). These machines will keep moving in the same direction without halting. Considering both blank symbols, we have $8(4n + 2)^{2n-1}$ computations of this kind.
- Machines without transition to the halting state. To keep the intersection of this and the above set empty, we also consider that the initial transition moves to a state different from the initial state. So for blank symbol i, we have $4(n - 1)$ different initial transitions (2 directions, 2 writing symbols and $n - 1$ states) and $4n$ different possibilities for the other $2n - 1$ transitions. This makes a total of $4(n - 1)(4n)^{2n-1}$ different machines for blank symbol i and $8(n - 1)(4n)^{2n-1}$ computations for both blank symbols.

Now, an upper bound for (4.6) is:

$$\sum_{n=6}^{\infty} \frac{2(4n+2)^{2n} - 4(4n+2)^{2n-1} - 8(4n+2)^{2n-1} - 8(n-1)(4n)^{2n-1}}{2^{n+1+\lceil \log_2\left((4n+2)^{2n}\right)\rceil}}$$

The result of the above sum is 0.0104282 (smaller than 1/32, as guaranteed by Proposition 4.10). This is an upper bound of the sum of the error $m(s) - m_5(s)$ for all infinite strings s different from "0" and "1". Smaller upper bounds can be found by removing from the above sum other kinds of predictable non-halting machines.

References

1. Levin, L.: Laws of information conservation (non-growth) and aspects of the foundation of probability theory. Prob. Form. Trans. **10**, 206–210 (1974)
2. Kircher, W., Li, M., Vitnyi, P.: The miraculous universal distribution. Math. Intell. **19**(4), 7–15 (1997)
3. Rad, T.: On non-computable functions. Bell Syst. Tech. J. **41**(3), 877–884 (1962)
4. Cover, T.M., Thomas, J.A.: Information Theory. Wiley (2006)
5. Brady, A.H.: The determination of the value of Rado's noncomputable function Sigma(k) for four-state Turing machines. Math. Comput. **40**(162), 647–665 (1983)

Chapter 5
Validation and Generalization of CTM

5.1 Multidimensional Complexity

This chapter introduces a generalization of CTM to higher dimension and thus to tensors. It requires an extension of Turing machines capable of operating on a 2-dimensional tape (by simply moving up and down in addition to the traditional left and right movements of the head). A reference to this kind of investigation and definition of 2D Turing machines can be found in [1, 2] and they are also known as Turmites, a whimsical combination of the words Turing and Termites, as their evolution over time may resemble the insect.

We define K_{2D}, an approximation to Kolmogorov complexity for binary arrays and present some results from the analysis of 2D Turing machines with 4 states and two symbols.

5.2 Deterministic Two-Dimensional Turing Machines

2-dimensional (2D) Turing machines run not on a 1-dimensional tape but in a 2-dimensional infinite array. At each step they can move in four different directions (*up*, *down*, *left*, *right*) or *stop*. In transitions $(s_1, k_1) \rightarrow (s_2, k_2, d)$ (the machine is in state s_1 and reads the symbol k_1, then it writes k_2, changes to state s_2 and moves to a contiguous cell following direction d) the only change affects to d. If s_2 is the halting state then d is $stop$. In other cases, d can be any of the other four directions.

Let $(n, m)_{2D}$ be the set of Turing machines with n states and m symbols. These machines have nm entries in the transition table, and for each entry (s_1, k_1) there are $4nm + m$ possible instructions, that is, m different halting instructions (writing one of the different symbols) and $4nm$ non-halting instructions (4 directions, n states and m different symbols). So the number of machines in $(n, m)_{2D}$ is $(4nm + m)^{nm}$. It is possible to enumerate all these machines in a similar way to what we did for one-dimensional Turing machines. We can assign one number to each entry in the

H. Zenil et al., *Methods and Applications of Algorithmic Complexity*,
Emergence, Complexity and Computation 44,
https://doi.org/10.1007/978-3-662-64985-5_5

$$
\begin{aligned}
\{1,1\} &\rightarrow \{0,0,stop\}\\
\{1,0\} &\rightarrow \{3,1,right\}\\
\{2,1\} &\rightarrow \{3,1,up\}\\
\{2,0\} &\rightarrow \{0,1,stop\}\\
\{3,1\} &\rightarrow \{0,0,down\}\\
\{3,0\} &\rightarrow \{2,0,left\}
\end{aligned}
$$

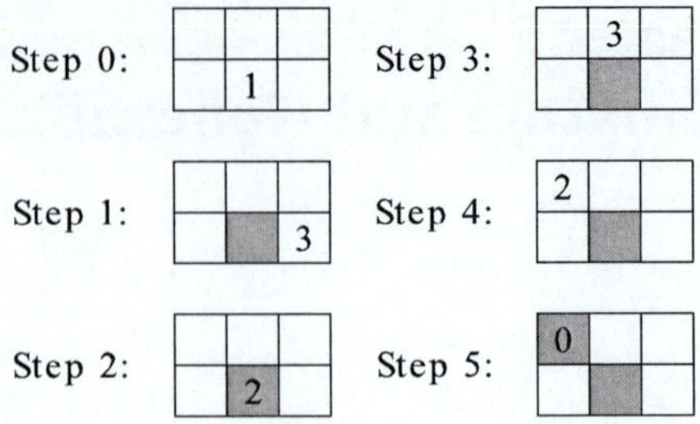

Fig. 5.1 Example of a deterministic 2-dimensional Turing machine

transition table. These numbers go from 0 to $4nm + m - 1$ (given that there are $4nm + m$ different instructions). The numbers corresponding to all entries in the transition table (irrespective of the convention followed in sorting them) form a number with nm digits in base $4nm + m$. Then, the translation of a transition table to a natural number and vice versa can be done through elementary arithmetical operations (Chap. 2).

We take as output for a 2D Turing machine the minimal rectangular array that includes all cells visited by the machine. Note that this probably includes cells that have not been visited, but it is the most natural way of producing output with some regular format and at the same time reducing the set of different outputs.

Figure 5.1 shows an example of the transition table of a Turing machine in $(3, 2)_{2D}$ and its execution over a '0'-filled grid. We show the portion of the grid that is returned as the output array. Two of the six cells have not been visited by the machine.

5.3 An Approximation to the Universal Distribution

We have run all machines in $(4, 2)_{2D}$ just as we did in Chap. 3 for deterministic one-dimensional Turing machines. Also, we considered the output of all different machines starting both in a '0'-filled grid and in a '1'-filled grid.

We also used a reduced enumeration to avoid running certain trivial machines whose behavior can be predicted from the transition table, as well as filters to detect non-halting machines before exhausting the entire runtime (see Sect. 2.3). In the reduced enumeration we considered only machines with an initial transition moving to the right and changing to a state different to the initial and halting states. Machines moving to the initial state at the starting transition run forever, and machines moving to the halting state produce single-character output. So we reduce the number of initial transitions in $(n, m)_{2D}$ to $m(n - 1)$ (the machine can write any of the m symbols and change to any state in $\{2, \cdots, n\}$). The set of different machines is reduced accordingly to $k(n - 1)(4nm + m)^{nm-1}$. To enumerate these machines we construct a mixed-radix number, given that the digit corresponding to the initial transition now goes from 0 to $m(n - 1) - 1$. To the output obtained when running this reduced

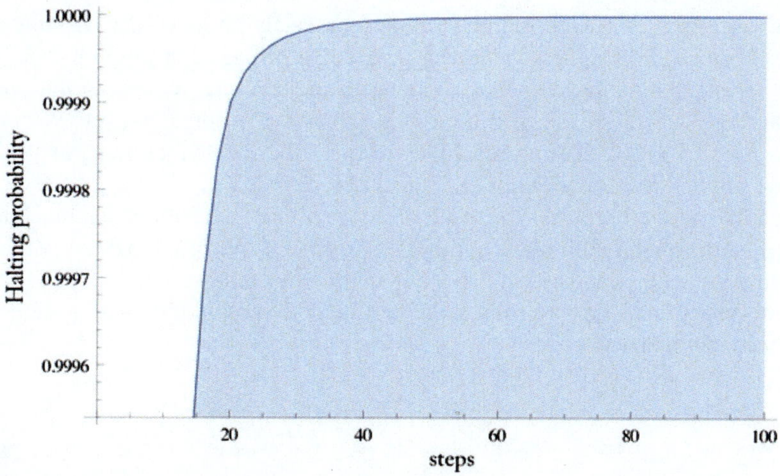

Fig. 5.2 Accumulated runtime distribution for $(4, 2)_{2D}$

enumeration we add the single-character arrays that correspond to machines moving to the initial state at the starting transition. These machines and their output can be easily quantified. Also, to take into account machines with the initial transition moving in a different direction than the right one, we consider the 90, 180 and 270 degree rotations of the strings produced, given that for any machine moving up (left/down) at the initial transition, there is another one moving right that produces the identical output but rotates (90, 180 or 270°).

To set the runtime, we took into account that the Busy Beaver runtime value for (4, 2) is 107 steps before halting. But no Busy Beavers are known for 2-dimensional Turing machines. So to set the runtime in our experiment we generated a sample of 334×10^8 random machines in the reduced enumeration. We used a runtime of 2000 steps for the runtime sample, this is 10.6% of the machines in the reduced enumeration for $(4, 2)_{2D}$, but 1500 steps for running all $(4, 2)_{2D}$. These machines were randomly generated instruction by instruction. As we have explained above, it is possible to assign a natural number to every instruction. So, generating a random machine in the reduced enumeration for $(n, m)_{2D}$ requires to produce a random number from 0 to $m(n-1)-1$ for the initial transition and from 0 to $4nm+m-1$ for the other $nm-1$ transitions. We used the implementation of the Mersenne Twister in the Boost C++ library. The output of this sample was the distribution of the runtime of the halting machines.

Figure 5.2 shows the resulting probability that a random halting machine (halting in at most 2000 steps) will halt in at most the number of steps indicated on the horizontal axis. For 100 steps this probability is 0.9999995273. Note that the machines in the sample are in the reduced enumeration, a large number of very trivial machines halting in just one step have been removed. So in the complete enumeration the probability of halting in at most 100 steps is even greater.

But we found some high runtime values—precisely 23 machines required more than 1000 steps. The highest value was a machine progressing through 1483 steps before halting. So we have enough evidence to believe that by setting the runtime at 2000 steps we have obtained almost all (if not all) output arrays. We ran all $6 \times 34^7 \approx 3.15 \times 10^{11}$ Turing machines in the reduced enumeration for $(4, 2)_{2D}$. Then we applied the completions explained before.

The final output represents the result of $2(4nm + m)^2$ executions (all machines in $(4, 2)_{2D}$ starting with both blank symbols '0' and '1'). We found 3 079 179 980 224 non-halting machines and 492 407 829 568 halting machines. A number of 1 068 618 different binary arrays were produced after 12 days of calculation with a supercomputer of medium size.[1]

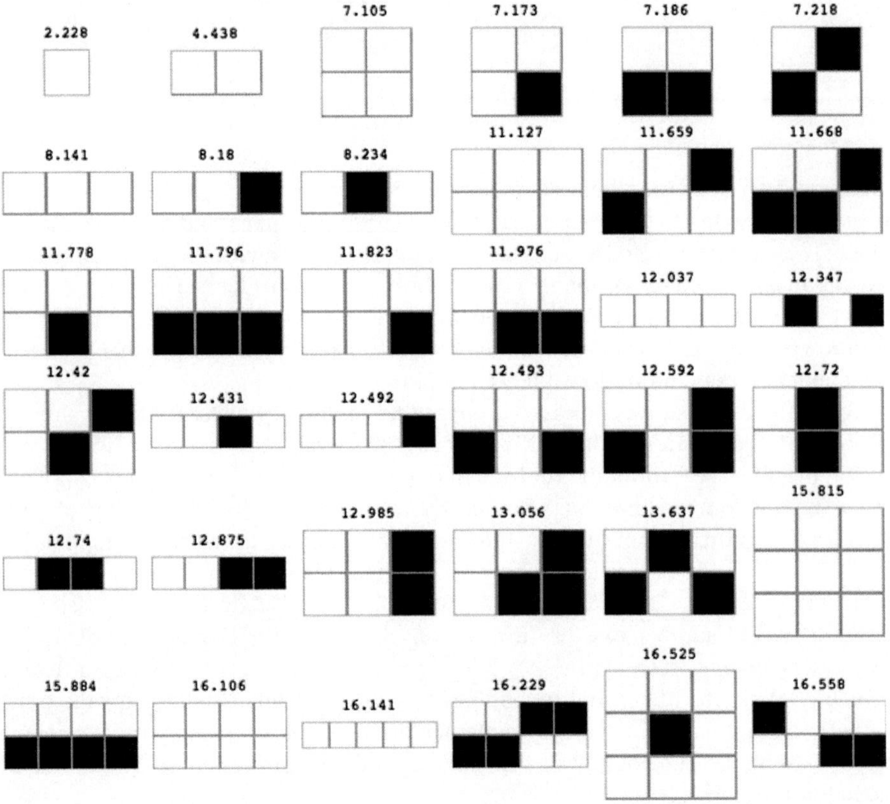

Fig. 5.3 The top 36 objects in $D(4, 2)_{2D}$ preceded by their K_{2D} values, sorted by higher to lower frequency and therefore from smaller to larger Kolmogorov complexity after application of the coding theorem). Only non-symmetrical cases are displayed

[1] Twenty five x86-64 CPUs running at 2128 MHz each with 4 GB of memory each, located at the Centro Informático Científico de Andalucía (CICA), Spain.

Let $D(4, 2)_{2D}$ be the distribution that assigns to every binary array s the result of dividing the occurrences of s produced by halting machines in $(4, 2)_{2D}$ by the total number of halting machines in $D(4, 2)_{2D}$. This is an extension of (3.1) (Sect. 3.2). By using the coding theorem (1.11) (Sect. 1.3) we can estimate the Kolmogorov complexity K_{2D}. For every binary array s,

$$K_{2D}(s) = -\log_2(D(4, 2)_{2D}(s)). \tag{5.1}$$

Figure 5.3 shows the top 36 objects in $D(4, 2)_{2D}$, that is, the objects with lowest Kolmogorov complexity values.

5.4 Evaluating 2-Dimensional Kolmogorov Complexity

K_{2D} distributes 1 068 618 arrays into 1272 different complexity values, with a minimum complexity value of 2.22882 bits, a maximum value of 36.2561 bits and a mean of 35.1201. Considering the number of possible square binary arrays given by the formula $2^{d \times d}$ (without considering any symmetries), $D(4, 2)_{2D}$ can be said to produce all square binary arrays of length up to 3×3, that is $\sum_{d=1}^{3} 2^{d \times d} = 530$ square arrays, and 60016 of the $2^{(4 \times 4)} = 65536$ square arrays with side of length (or dimension) $d = 4$. It only produces 84104 of the 33 554 432 possible square binary arrays of length $d = 5$ and only 11328 of the possible 68 719 476 736 of dimension $d = 6$. The largest square array produced in $D(4, 2)_{2D}$ is of side length $d = 13$ (see Fig. 5.4 left) out of a possible 748×10^{48}. It has a K_{2D} value equal to 34.2561 (Figs. 5.5, 5.6 and 5.7).

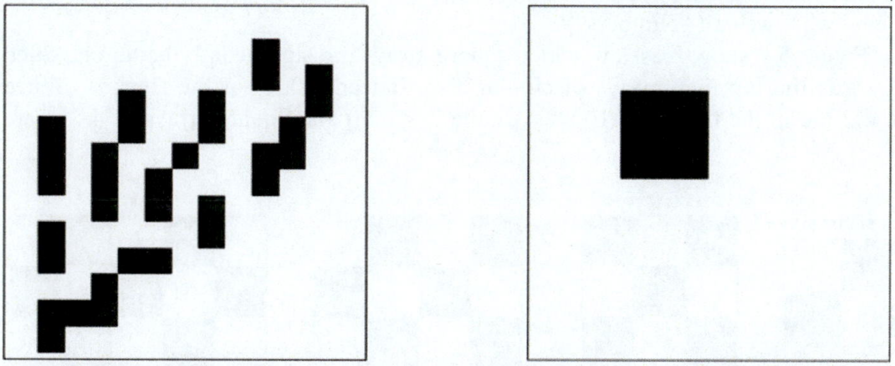

Fig. 5.4 Left: The only and most complex square array (with 15 other symmetrical cases) in $(4, 2)_{2D}$ with $K_{2D} = 34.2561$. Another way to see this array is as one among those of length 13 with low complexity given that it occurred once in the sampled distribution in the classification unlike all other square arrays of the same size that are missing in $D(4, 2)_{2D}$. Right: With a value of $K_{2D} = 6.7$ this is the simplest 4×4 square array after the preceding all-blank 4×4 array (with $K_{2D} = 6.4$) and before the 4×4 square array with a black cell in one of the array corners (with complexity $K_{2D} = 6.9$)

What one would expect from a distribution where simple patterns are more frequent (and therefore have lower Kolmogorov complexity after application of the coding theorem) would be to see patterns of the "checkerboard" type with high frequency and low random complexity (K), and this is exactly what we found (see Fig. 5.5), while random looking patterns were found at the bottom among the least frequent ones (Fig. 5.8).

We have coined the informal notion of a "climber" as an object in the frequency classification (from greatest to lowest frequency) that appears better classified among objects of smaller size rather than with the arrays of their size, this is in order to highlight possible candidates for low complexity, hence illustrating how the process make low complexity patterns to emerge. For example, "checkerboard" patterns (see Fig. 5.5) seem to be natural "climbers" because they come significantly early (more frequent) in the classification than most of the square arrays of the same size. In fact, the larger the checkerboard array, the more of a climber it seems to be. This is in agreement with what we have found in the case of strings [3–5] where patterned objects *emerge* (e.g. $(01)^n$, that is, the string 01 repeated n times), appearing relatively increasingly higher in the frequency classifications the larger n is, in agreement with the expectation that patterned objects should also have low Kolmogorov complexity.

For example, Fig. 5.9 shows arrays that come together among groups of much shorter arrays, thereby demonstrating, as expected from a measure of randomness, that array (or string) size is not what determines complexity (as we have shown in Chap. 3 for binary strings). The fact that square arrays may have low Kolmogorov complexity can be understood in several ways, some of which strengthen the intuition that square arrays should be less Kolmogorov random, such as for example, the fact that for square arrays one only needs the information of one of its dimensions to determine the other, either height or width.

In Fig. 5.7 more low complexity arrays are depicted, coming together in a block in the frequency distribution before the larger group of arrays of the same size (hence being what we call "climbers").

Figure 5.9 shows cases in which square arrays are significantly better classified towards the top than arrays of similar size. Indeed, 100% of the squares of size 2×2 are in the first fifth (F1), as are the 3×3 arrays. Square arrays of 4×4 are

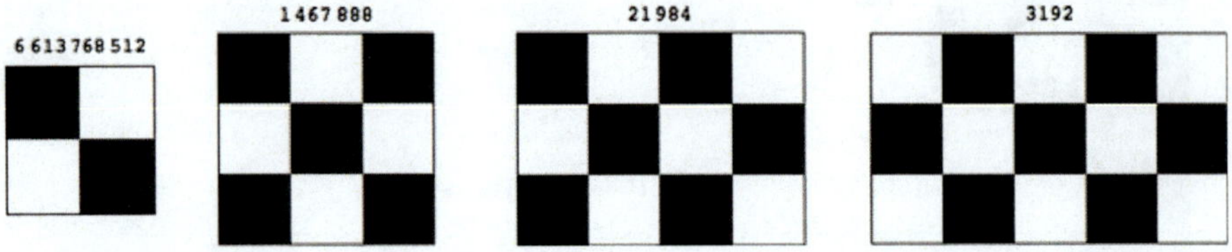

Fig. 5.5 Frequency of appearance of "checkerboard" patterns sorted from more to less frequent according to $D(4, 2)_{2D}$ (displayed only non-symmetrical cases under rotation and complementation). The checkerboard of size 4×4 does not occur. The number above each arrays indicates the occurrences in $(4, 2)_{2D}$

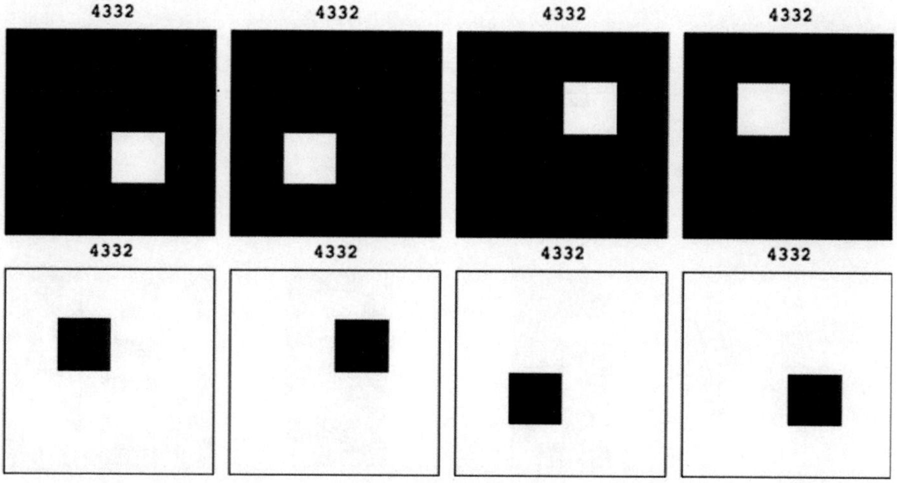

Fig. 5.6 With a value of $K_{2D} = 6.7$ this is the greatest 4×4 square array after the preceding all-blank 4×4 array (with $K_{2D} = 6.4$) and before the 4×4 square array with a black cell in one of the corners (with complexity $K_{2D} = 6.9$)

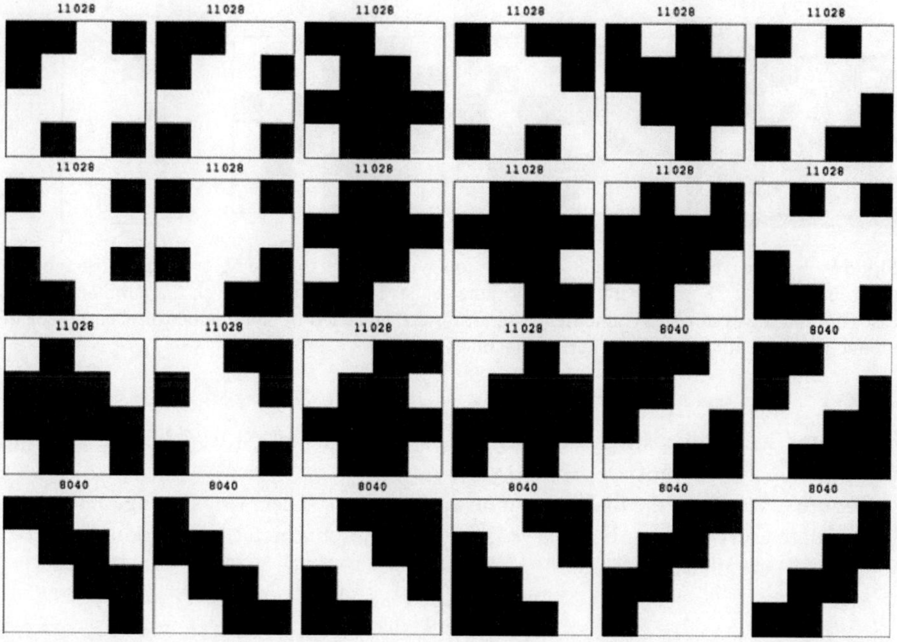

Fig. 5.7 Other "climber" arrays occurring in $(4, 2)_{2D}$. That is, arrays that appear significantly before (less random) in the classification than the rest of the arrays of about the same size

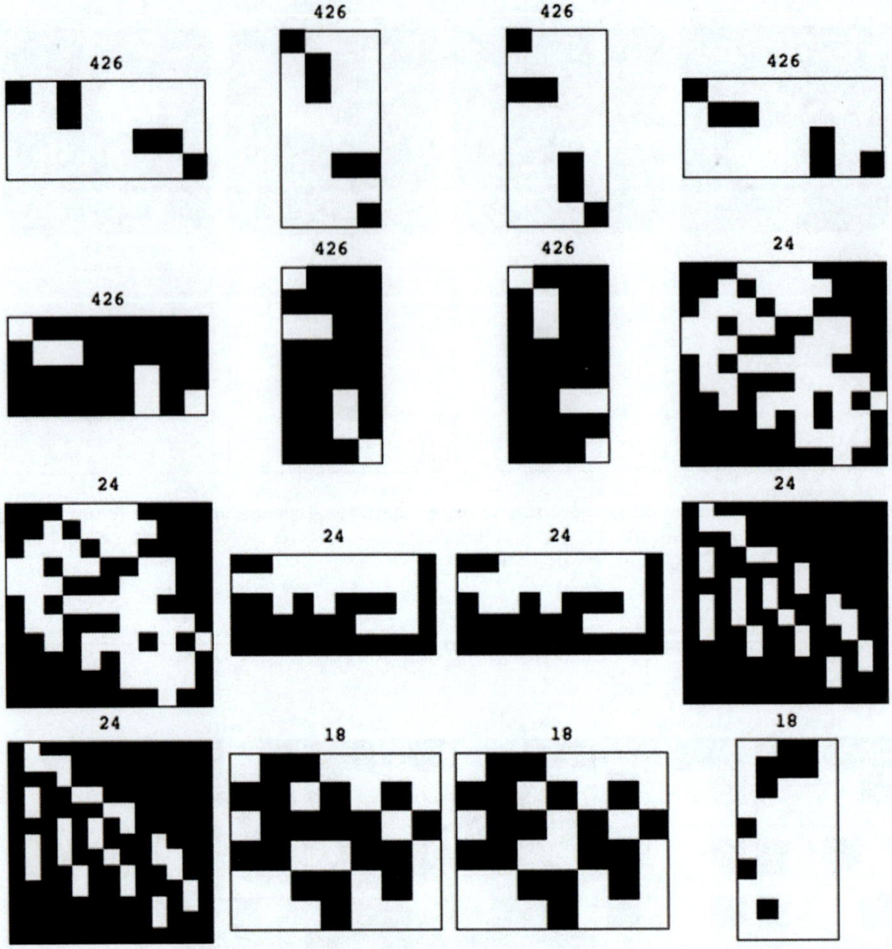

Fig. 5.8 Bottom 16 objects in the classification with lowest frequency, or being most random according to $D(4, 2)_{2D}$. It is interesting to note the strong similarities given that similar-looking cases are not always exact symmetries. The arrays are preceded by the number of occurrences of production from all the $(4, 2)_{2D}$ Turing machines

distributed as follows when dividing $(4, 2)_{2D}$ in 5 equal parts: 72.66%, 15.07%, 6.17359%, 2.52%, 3.56%.

Figure 5.10 shows the distribution of square arrays. The 3 rings in Fig. 5.10 show how square arrays of length $1 < d < 5$ in $(4, 2)_{2D}$, for which $(4, 2)_{2D}$ contains most $2^{d \times d}$ cases, are distributed.

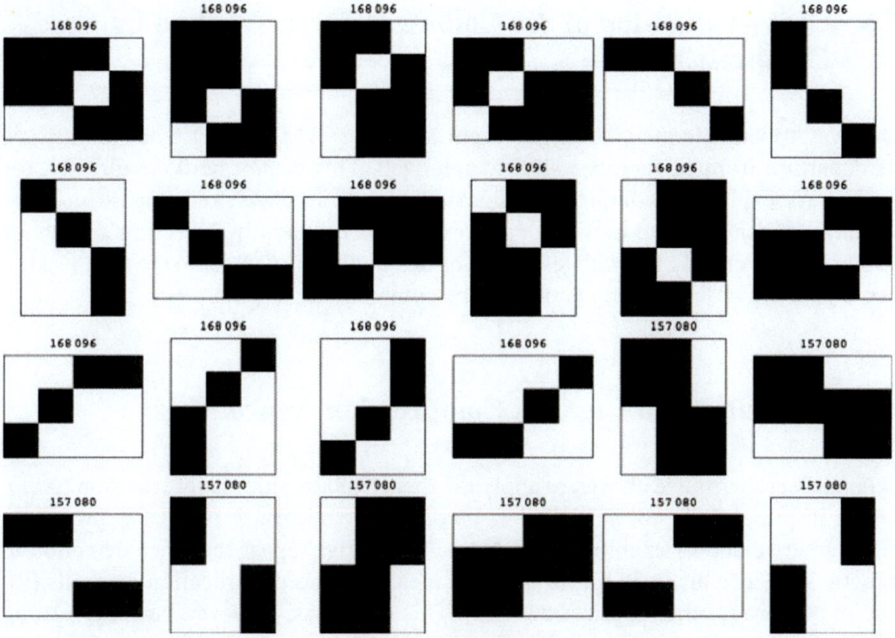

Fig. 5.9 Two "climbers" (and symmetric cases) in $D(4, 2)_{2D}$

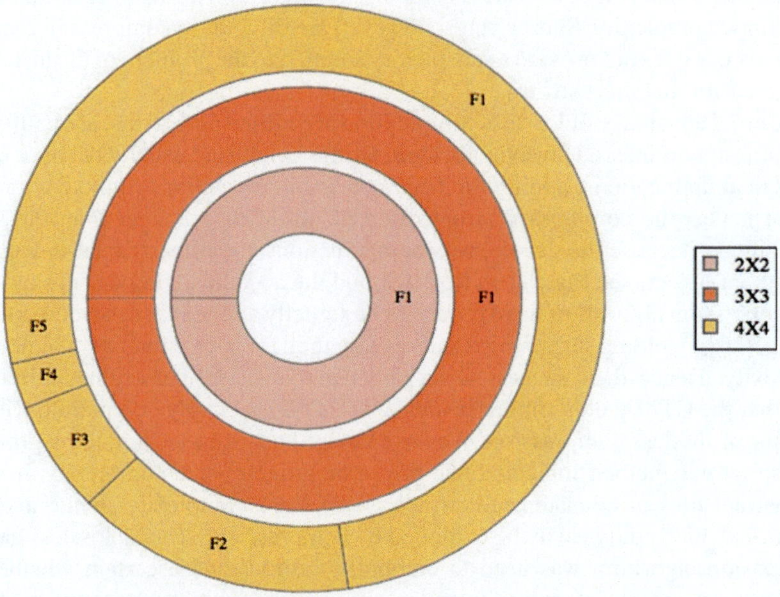

Fig. 5.10 Distribution of square binary arrays. $(4, 2)_{2D}$ was divided into 5 parts with F1 the first fifth, F2 the second fifth, and so on

5.5 Cross-Validation of the Coding Theorem Method by Compressibility

One way to validate the Coding Theorem Method (CTM) is to attempt to measure its departure from the compressibility approach. This cannot be done directly, for as we have explained, compression algorithms perform poorly on short strings, but we have found a way to partially circumvent this problem by selecting subsets of strings for which CTM calculates a high or low complexity which were then used to generate a file of length long enough to be compressed.

5.6 Limitations of Lossless Compression Approaches

When researchers have chosen to apply the theory of algorithmic information (AIT), which in principle is not supposed to be of any practical use [7], it has proven to be of great value, for example, for DNA false positive repeat sequence detection in genetic sequence analysis [8], in distance measures and classification methods [9], and in numerous other applications [10]. This effort has, however, been hamstrung by the limitations of compression algorithms (currently the only method used to approximate the Kolmogorov complexity of a string) given that this measure is not computable. The result of a lossless compression algorithm is an upper bound of the algorithmic complexity. Short strings, however, are difficult to compress in practice, and the theory does not provide a satisfactory solution to the problem of the instability of the measure for short strings.

The chief advantage of lossless compression algorithms is that they are a sufficient test of non-randomness. However, for short strings, which are usually the ones useful in practical applications, adding the decompression instructions to the compressed version makes the compressed string often, if not always, longer than the string itself, simply because the decompression instructions are at least equal in length to the original object (see Fig. 3.1 in Sect. 3.1 and the accompanying discussion).

Compression algorithms have proven to be signally applicable in several domains (see e.g. [10]), yielding surprising results as a method for approximating Kolmogorov complexity. Hence their success is in part a matter of their usefulness. Here we show that the CTM yields similar results, and that it is actually compatible with the behavior of lossless compression. We have devised an artful technique by grouping strings that our method indicated had the same program-size complexity, in order to construct files of concatenated strings of the same complexity (while avoiding repetition, which could easily be exploited by compression). Then a lossless general compression algorithm was used to compress the files and ascertain whether the files that were more compressed were the ones created with highly complex strings according to our method. The statistical evidence is strong enough to suggest that the CTM is sound and capable of producing satisfactory results. It is also the only currently available method for dealing with very short strings and in a sense is an expensive but powerful "microscope" for capturing the information content of very small objects.

5.7 CTM and Compression on Strings of Length 10 to 15

For this experiment we have selected the strings in $D(5)$ (Sect. 3.5) with lengths ranging from 10 to 15. Table 5.1 shows the number of $D(5)$ strings with these lengths. Up to length 13 we have almost all possible strings. For length 14 we have a considerable number and for length 15 there are less than 50% of the 2^{15} possible strings.

The distribution of complexities for these strings is shown in Fig. 5.12. Here, K_m, the used approximation to K, refers to $K_{D(5)}$ in Sect. 3.5. As expected, the longer the strings, the greater their average complexity. The overlapping of strings with different lengths that have the same complexity corresponds to climbers. The experiment consisted in creating files with strings of different K_m-complexity but equal length. Files with more complex (random) strings are expected to be less compressible than files with less complex (random) strings. This was done in the following way. For each l $(10 \leq l \leq 15)$, we let $S(l)$ denote the list of strings of length l, sorted by increasing BDM complexity. For each $S(l)$ we made a partition of 10 sets with the same number of consecutive strings. Let's call these partitions $P(l, p)$, $1 \leq p \leq 10$.

Then for each $P(l, p)$ we have created 100 files, each with 100 random strings in $P(l, p)$ in random order. We called these files $F(l, p, f)$, $1 \leq f \leq 100$. Summarizing, we now have:

- 6 different string lengths l, from 10 to 15, and for each length
- 10 partitions (sorted by increasing complexity) of the strings with length l, and
- 100 files with 100 random strings in each partition.

This makes for a total of 6000 different files. Each file contains 100 different binary strings, hence with length of $100 \times l$ symbols.

Table 5.1 Number of strings of length 10 to 15 found in $D(5)$

Length (l)	Strings
10	1024
11	2048
12	4094
13	8056
14	13068
15	14634

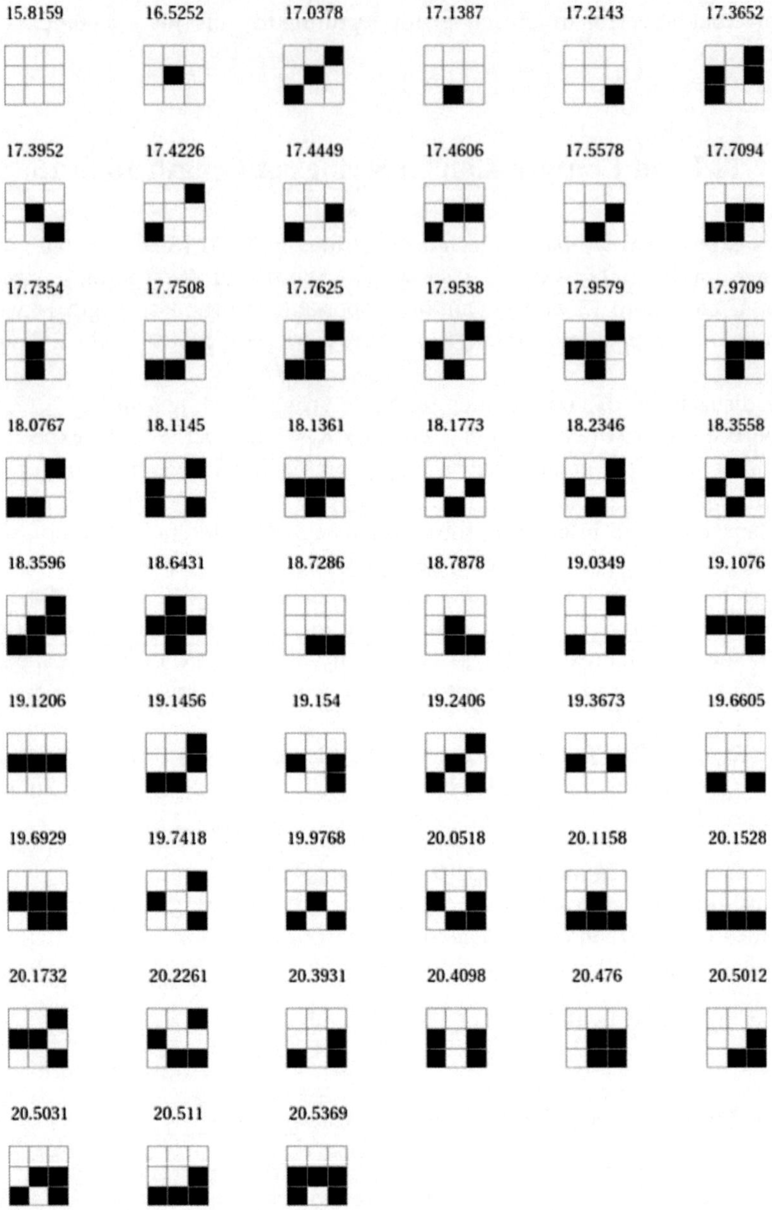

Fig. 5.11 Complete reduced set (non-symmetrical cases under reversion and complementation) of square arrays of size 3×3 in K_{2D} sorted from lowest to greatest Kolmogorov complexity after application of the coding theorem (1.11) to the output frequency of 2-D Turing machines. Observe that the 2 glider configurations in the Game of Life [6] come with high Kolmogorov complexity (with approximated values of 20.2261 and 20.5031)

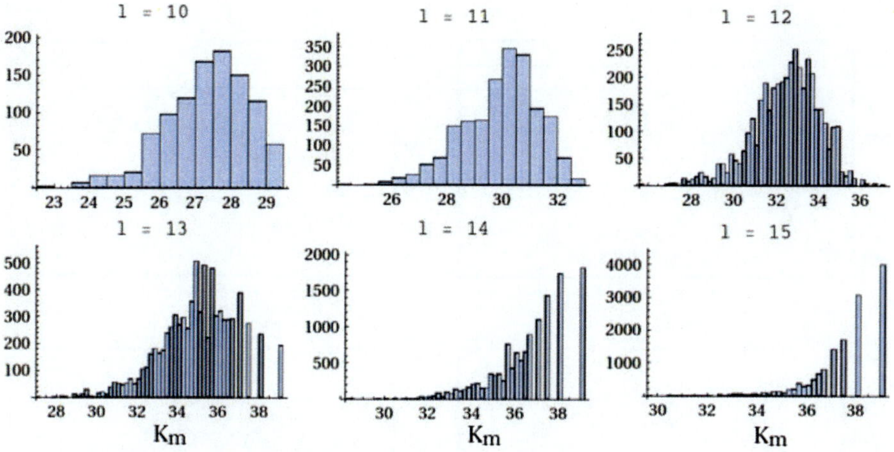

Fig. 5.12 Distribution of complexities for different string lengths (l)

A crucial step is to replace the binary encoding of the files by a larger alphabet, retaining the internal structure of each string. If we compressed the files $F(l, p, f)$ by using binary encoding then the final size of the resulting compressed files would depend not only on the complexity of the separate strings but on the patterns that the compressor discovers along the whole file. To circumvent this we chose two different symbols to represent the '0' and '1' in each one of the 100 different strings in each file. The same set of 200 symbols was used for all files. We were interested in using the most standard symbols we possibly could, so we created all pairs of characters from 'a' to 'p' (256 different pairs) and from this set we selected 200 two-character symbols that were the same for all files. This way, though we do not completely avoid the possibility of the compressor finding patterns in whole files due to the repetition of the same single character in different strings, we considerably reduce the impact of this phenomenon.

The files were compressed using the *Mathematics* function `Compress`, which is an implementation of the Deflate algorithm (Lempel-Ziv plus Huffman coding). Figure 5.13 shows the distributions of lengths of the compressed files for the different string lengths. The horizontal axis shows the 10 groups of files in increasing K_m. As the complexity of the strings grows (right part of the diagrams), the compressed files are larger, so they are harder to compress. The relevant exception is length 15, but this is probably related to the low number of strings of that length that we have found, which are surely not the most complex strings of this length.

We have used other compressors such as GZIP (which uses Lempel-Zip algorithm LZ77) and BZIP2 (Burrows-Wheeler block sorting text compression algorithm and Huffman coding), with several compression levels. The results are similar to those shown in Fig. 5.13.

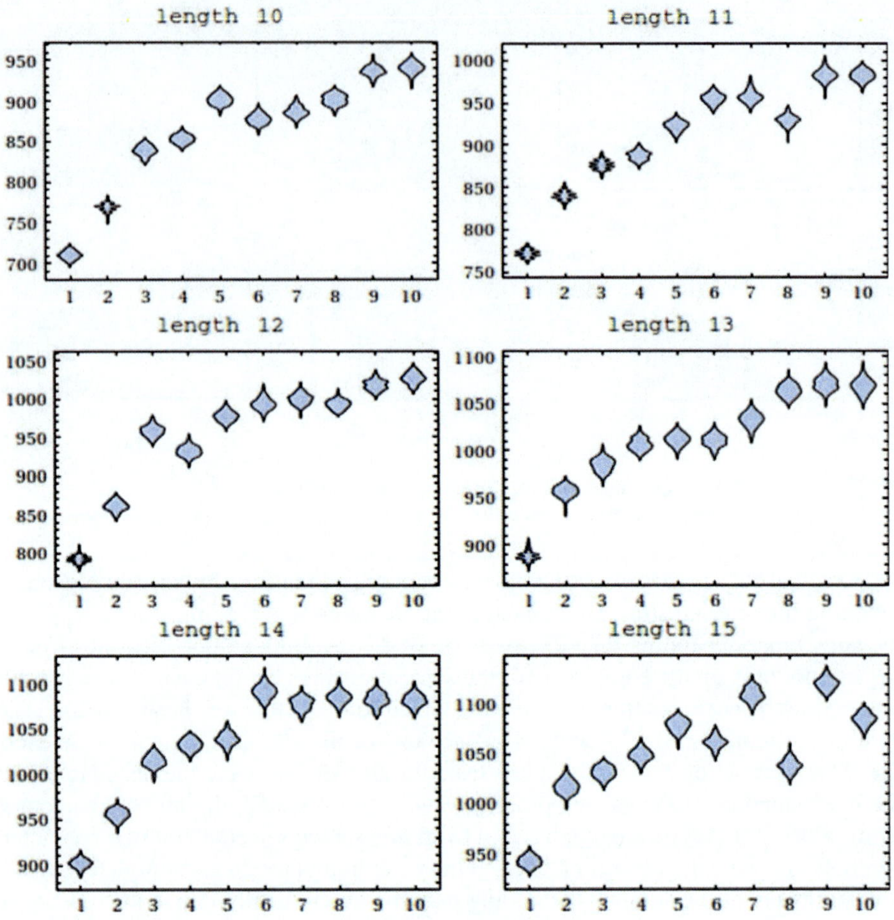

Fig. 5.13 Distribution of the compressed lengths of the files

5.8 Comparing 1-Dimensional Strings in $(4, 2)_{2D}$ to $(4, 2)$

We shall now look at how the approximation to K for 2D Turing machines in $(4, 2)_{2D}$ (Sect. 5.1) correlates with the approximation for one-dimensional machines (Chap. 3). In a sense this is like changing the Turing machine formalism to see whether the new distribution resembles distributions following other Turing machine formalisms, and whether it is robust enough.

All Turing machines in $(4, 2)$ are included in $(4, 2)_{2D}$ because these are just the machines that do not move up or down. We first compared the values of the 1832 output strings in $(4, 2)$ to the 1-dimensional arrays found in $(4, 2)_{2D}$. We are also interested in the relation between the ranks of these 1832 strings in both $(4, 2)$ and $(4, 2)_{2D}$.

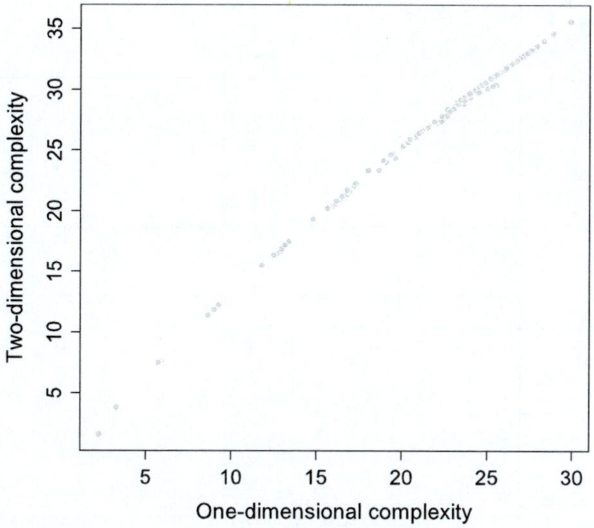

Fig. 5.14 Scatterplot of K_{2D} with 2-dimensional Turing machines as a function of K_m with 1-dimensional Turing machines

Figure 5.14 shows the link between K_{2D} with 2D Turing machines as a function of K_m. It suggests a strong almost-linear overall association. The correlation coefficient $r = 0.9982$ confirms the linear association, and the Spearman correlation coefficient $r_s = 0.9998$ proves a tight and increasing functional relation.

The length l of strings is a possible confounding factor. However Fig. 5.15 suggests that the link between one and 2-dimensional complexities is not explainable by l. Indeed, the partial correlation $r_{K_m K_{2D}.l} = 0.9936$ still denotes a tight association.

Figure 5.15 also suggests that complexities are more strongly linked with longer strings. This is in fact the case, as Table 5.2 shows: the strength of the link increases with the length of the resulting strings. One and 2-dimensional complexities are remarkably correlated and may be considered two measures of the same underlying feature of the strings. How these measures vary is another matter. The regression of K_{2D} on K_m gives the following approximate relation: $K_{2D} \approx 2.64 + 1.11 K_m$. Note that this subtle departure from identity may be a consequence of a slight non-linearity, a feature visible in Fig. 5.14.

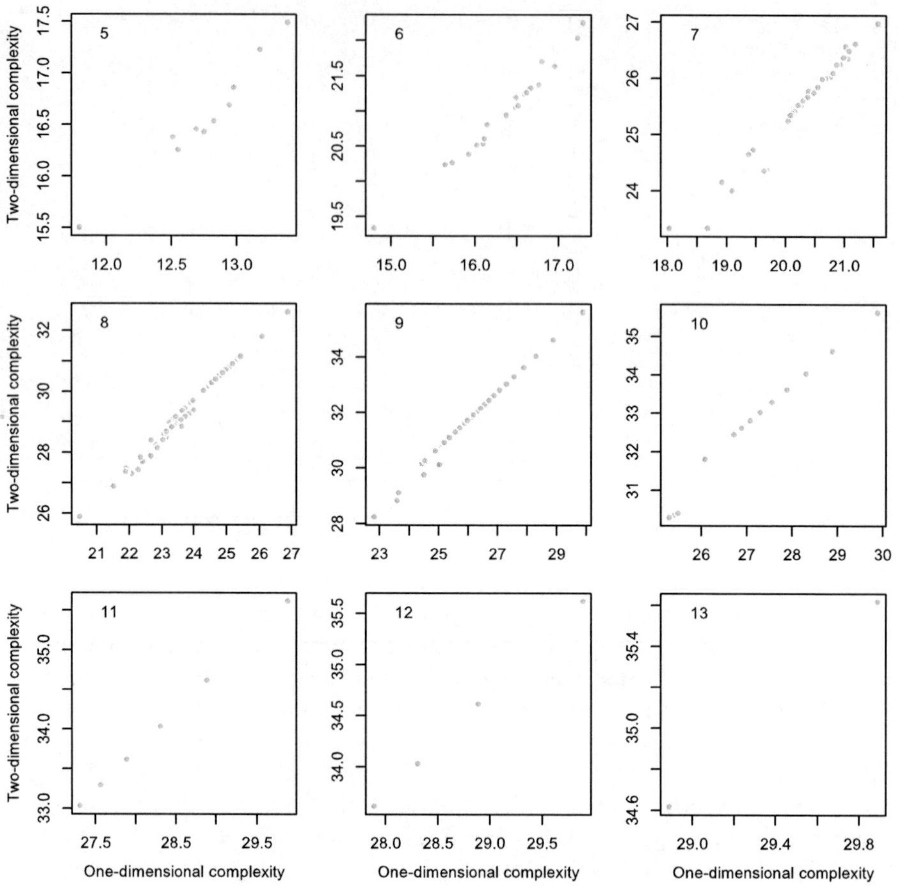

Fig. 5.15 Scatterplot of the complexity of 1-dimensional arrays (strings) produced by 2-dimensional Turing machines as a function of the complexity of the same arrays produced by 1-dimensional Turing machines. The analysis is done by length of strings, from 5 to 13

Table 5.2 Correlation coefficients between one and 2-dimensional complexities by length of strings

Length (l)	Correlation
5	0.9724
6	0.9863
7	0.9845
8	0.9944
9	0.9977
10	0.9952
11	1
12	1

5.9 Comparison of CTM and Compression of Cellular Automata

We have seen that CTM with associated measure K_m (or K_{2D} for 2D Kolmogorov complexity) is in agreement with compressibility, as we have reported in Sect. 5.7.

Now we can classify Elementary Cellular Automata (ECA) by using $D(4,2)_{2D}$ (Sect. 5.1). Classification of ECA by compressibility has been done before in [11] with results that are in complete agreement with our intuition and knowledge of the complexity of certain ECA rules (and related to Wolfram's classification [1]). In [11] both classifications by simplest initial condition and random initial condition were undertaken, leading to a stable compressibility classification of ECAs. Here we followed the same procedure for both simplest initial condition (single black cell) and random initial condition in order to compare the classification to the one that can be approximated by using $D(4,2)_{2D}$, as follows.

We will say that the space-time diagram (or evolution) of an Elementary Cellular Automaton c after time t has complexity:

$$K_{2D_{d\times d}}(c^t) = \sum_{r\in\{c^t\}_{d\times d}} K_{2D}(r) \tag{5.2}$$

That is, the complexity of a cellular automaton c is the sum of the complexities of the r arrays in the partition matrix $\{c^t\}_{d\times d}$ from breaking c^t into square arrays of length d produced by the ECA after t steps. An example of a partition matrix of an ECA evolution is shown in Fig. 5.16 for ECA Rule 30 and $d = 3$ where $r = 10$. Notice that the boundary conditions for a partition matrix may require the addition of at most $d-1$ empty rows or $d-1$ empty columns to the boundary as shown in Fig. 5.16 (or alternatively the dismissal of at most $d-1$ rows or $d-1$ columns) if the dimensions (height and width) are not multiples of d, in this case $d = 3$.

If the classification of all rules in ECA by $K_{2D_{d\times d}}$ (or simply K_{2D}) yields the same classification obtained by compressibility, one would be persuaded that K_{2D} is a good alternative to compressibility as a method for approximating the Kolmogorov complexity of objects, with the signal advantage that K_{2D} can be applied to very short strings and very short arrays such as images. Because all possible 2^9 arrays of size 3×3 are present in $(4,2)_{2D}$ we can use this distribution to try to classify all ECAs by Kolmogorov complexity using the Coding Theorem method. Figure 5.11 (Sect. 5.1) shows all relevant (non-symmetric) arrays.

Figure 5.18 displays the scatterplot of $K_{2D_{3\times 3}}$ against compression complexity calculated for every cellular automaton. It shows a positive link between the two measures. The Pearson correlation amounts to $r = 0.8278$, so the determination coefficient is $r^2 = 0.6853$. These values correspond to a strong correlation, although smaller than the correlation between 1- and 2-dimensional complexities calculated in Sect. 5.7.

Concerning orders arising from these measures of complexity, they too are strongly linked, with a Spearman correlation of $r_s = 0.92$. The scatterplots (Fig. 5.18)

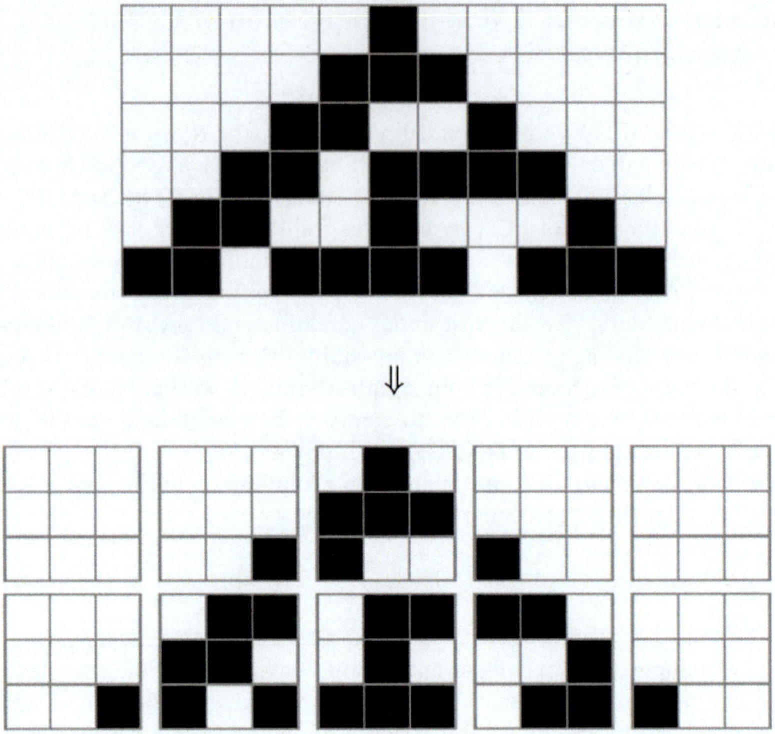

Fig. 5.16 Decomposing the evolution of Rule 30 (top) ECA after $t = 6$ steps into 10 subarrays of length 3×3 (bottom) in order to calculate $K_{2D_{3\times3}}$ to approximate its Kolmogorov complexity

show a strong agreement between the coding theorem method and the traditional compression method when both are used to classify ECAs by their approximation to Kolmogorov complexity.

The anomalies found in the classification of Elementary Cellular Automata (e.g. Rule 77 being placed among ECA with high complexity according to $K_{2D_{3\times3}}$) is a limitation of $K_{2D_{3\times3}}$ itself and not of the CTM which for $d = 3$ is unable to "see" beyond 3-bit squares. And yet the degree of agreement with compressibility is surprising (as well as with intuition, as a glance at Fig. 5.17 shows, and as the distribution of ECAs starting from random initial conditions in Fig. 5.20 confirms). In fact an average ECA has a complexity of about 20K bits, which is quite a large program-size when compared to what we intuitively gauge to be the complexity of each ECA, which may suggest that they should have smaller programs.

Adding the complexity approximation of each array in the partition matrix of a space-time diagram of an ECA provides an upper bound on the ECA Kolmogorov complexity, as it shows that there is a program that generates the ECA evolution picture with the length equal to the sum of the programs generating all the sub-arrays (plus a small length corresponding to the code length to join the sub-arrays). So if a

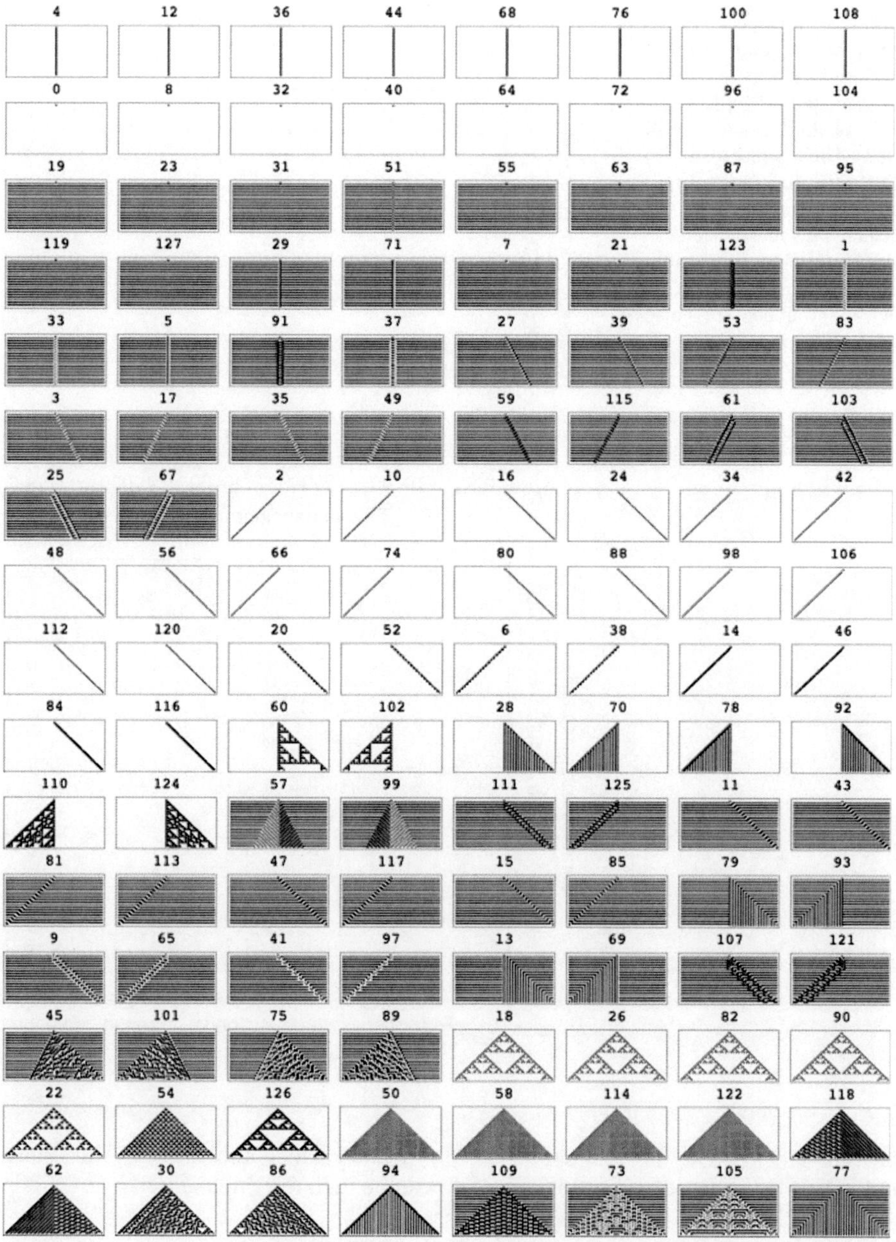

Fig. 5.17 All the first 128 ECAs (the other 128 are 0-1 reverted rules) starting from the simplest (black cell) initial configuration running for $t = 36$ steps, sorted from lowest to highest complexity according to $K_{2D_{3\times3}}$. Notice that the same procedure can be extended for its use on arbitrary images

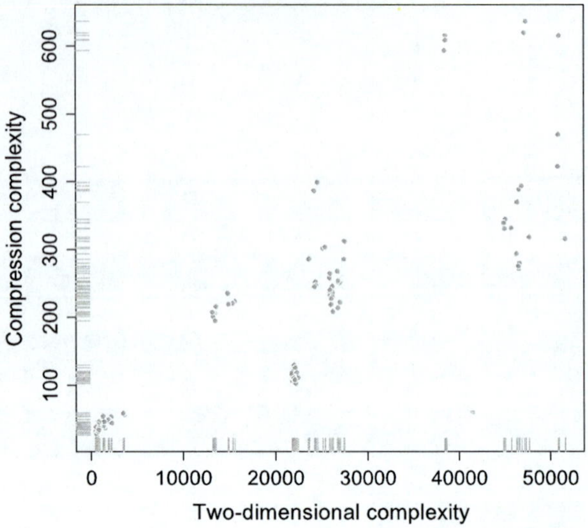

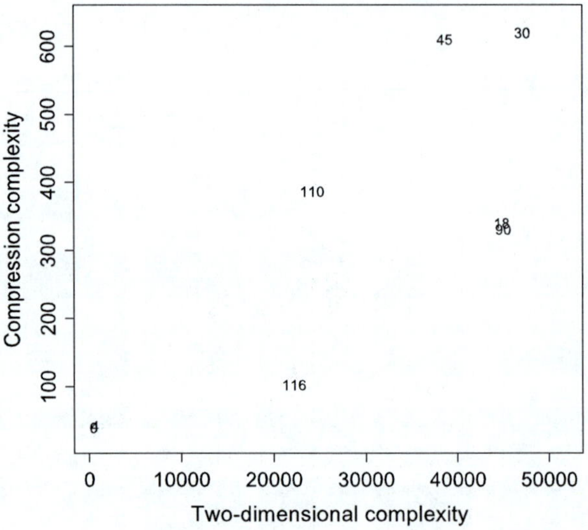

Fig. 5.18 Scatterplots of compression complexity against $K_{2D_{3\times3}}$ complexity as evaluated on the 128 first ECA evolutions after $t = 90$ steps. The top plot also shows the distribution of points along the axes displaying some clusters. The bottom plot shows a few of the ECA rules used in Fig. 5.20 (but here for a black cell initial condition)

sub-array appears n times we do not need to consider its complexity several times, but $\log_2(n)$. Taking into account this, (5.2) can be then rewritten as:

$$K\log_{2D_{d\times d}}(c^t) = \sum_{r_u\in\{c^t\}_{d\times d}} \log_2(n)K_{2D}(r_u) + K_{2D}(r_u) \tag{5.3}$$

where r_u are the different square arrays in the partition matrix $\{c^t\}_{d\times d}$ and n in $\log_2(n)$ the number of square arrays in $\{c^t\}_{d\times d}$ equal to r_u.

The classification of ECA, according to (5.3), is presented in Fig. 5.19. There is an almost perfect agreement with a classification by lossless compression length (see Figs. 5.19 and 5.20) which makes even one wonder whether the CTM is actually providing more accurate approximations to Kolmogorov complexity than lossless compressibility itself. Notice that the same procedure can be extended for its use on arbitrary images. In Chap. 6, measured based on variants of this idea are called the *Block Decomposition Method* (BDM).

Also worth notice that the fact that ECA can be successfully classified by $Klog_{2D_{3\times3}}$ suggests that output frequency distributions of ECA and Turing machines are strongly correlated, something that we had found and reported before in [12, 13].

We have shown in this chapter that the results of K_{2D} are in agreement with compressibility when applied to a concatenation of strings of the same complexity, and to classifications of Elementary Cellular Automata (ECA) by compressibility. However, K_{2D} is finer-grained compared to the number of different classes provided by compressibility. As a consequence, the Coding Theorem Method seems to have the advantage over lossless compression algorithms to better distinguish complexity of small objects. For example, among the ECA , $K_{2D_{3\times3}}$ provided 51 different complexity values, while compression retrieved 45 different values. These are not very different from each other but the smaller the object the more important it is to have a greater number of complexity values in order to distinguish the complexity of one string from another. For short enough objects compression algorithms collapse. This implies that the CTM is a complement to the compression method, especially where compression fails (small strings).

These results are extremely surprising, although what we would have expected in the best possible scenario, because while compression algorithms *see* global regularities, our CTM can only *see* local regularities, yet the agreement is shocking. In some deep way their agreement validates as much one method than the other and only remains asking which one may be slightly more accurate and whether for example the logarithmic penalization actually helps to tell apart some cases as it seems to do. The logarithmic penalization procedure is only necessary because 3×3 blocks may be too small compared to the object for which an approximation of its K value is needed. Chapter 6 presents several variations of this method.

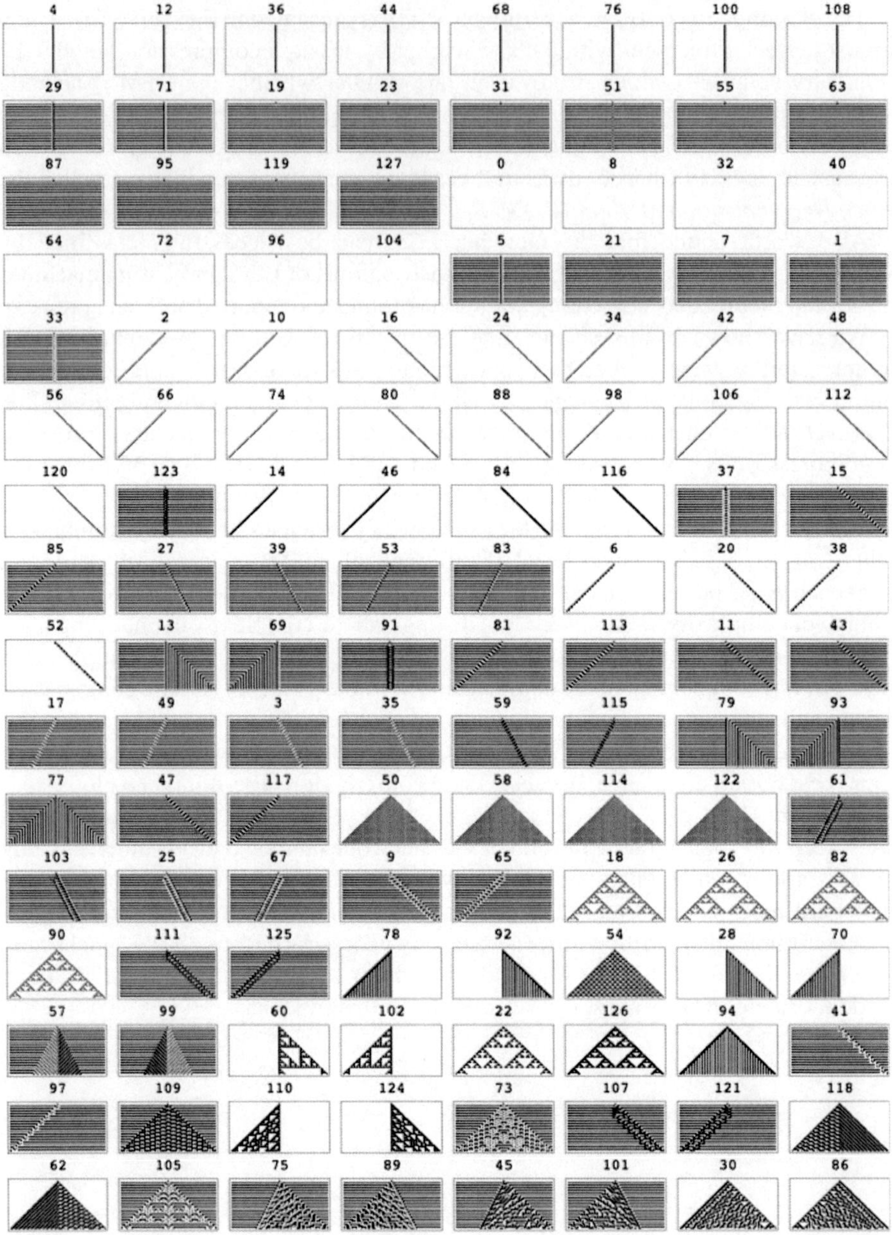

Fig. 5.19 All the first 128 ECAs (the other 128 are 0-1 reverted rules) starting from the simplest (black cell) initial configuration running for $t = 36$ steps, sorted from lowest to highest complexity according to $Klog_{2D_{3\times3}}$

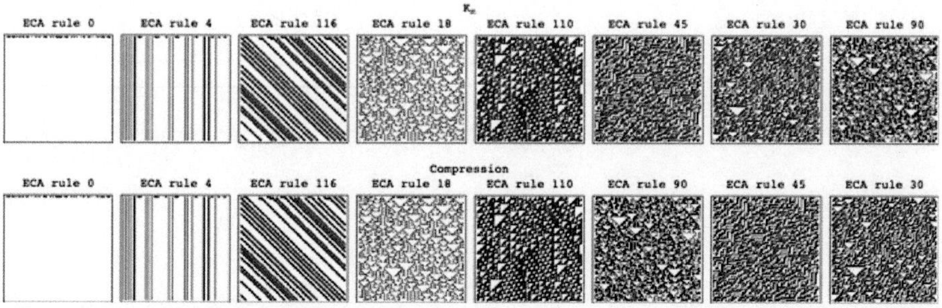

Fig. 5.20 Side by side comparison of 8 evolutions of representative ECAs, starting from a random initial configuration, sorted from lowest to highest $Klog_{2D_{3\times3}}$ values (top) and smallest to largest compression lengths using the Deflate algorithm as a method to approximate Kolmogorov complexity [11]

References

1. Wolfram, S.: A New Kind of Science. Wolfram Media, Champaign (2002)
2. Langton, C.G.: Studying artificial life with cellular automata. Phys. D: Nonlinear Phenomen. **22**(1), 120–149 (1986)
3. Zenil, H.: Une approche exprimentale la thorie algorithmique de la complexit, dissertation in fulfilment of the degree of Doctor in Computer Science (committee: J.-P. Highest honours. PhD thesis. Universit de Lille 1 (2011)
4. Delahaye, J.-P., Zenil, H.: Numerical evaluation of algorithmic complexity for short strings: A glance into the in- nermost structure of randomness. Appl. Math. Comput. **219**(1), 63–77 (2012)
5. Soler-Toscano, F., et al.: Calculating Kolmogorov complexity from the output frequency distributions of small turing machines. PLoS ONE **9**(5), 1–18 (2014)
6. Gardner, M.: Mathematical games - the fantastic combinations of john Conway's new solitaire game "life". Sci. Am. **23**, 120–123 (1970)
7. Chaitin, G.J.: From philosophy to program size (2003)
8. Rivals, E., et al.: Compression and genetic sequence analysis. Biochimie **78**(5), 315–322 (1996)
9. Cilibrasi, R., Vitanyi, P.: Clustering by compression. IEEE Trans. Inf. Theory **51**(4), 1523–1545 (2005)
10. Li, M., Vitnyi, P.: An Introduction to Kolmogorov Complexity and Its Applications. Springer, Berlin (2008)
11. Zenil, H.: Compression-based investigation of the dynamical properties of cellular automata and other systems. Complex Syst. **19**(1), 1–28 (2010)
12. Zenil, H., Delahaye, J-P.: On the algorithmic nature of the world. In: Information and Computation, vol. 10 (2010). ISSBN:97898142954820017
13. Delahaye, J-P., Zenil, H.: On the Kolmogorov-Chaitin complexity for short sequences. In: Calude, C., (ed.), Randomness and Complexity: From Leibniz to Chaitin. World Scientific (2007)

Chapter 6
The Block Decomposition Method

6.1 The Use and Misuse of Lossless Compression

Lossless compression algorithms have traditionally been used to approximate the Kolmogorov complexity of an object (e.g. a string). Data compression can be viewed as a function that maps data onto other data using the same units or alphabet (if the translation is into different units or a larger or smaller alphabet, then the process is called an encoding). Compression is successful if the resulting data are shorter than the original data plus the decompression instructions needed to fully reconstruct the original data. For a compression algorithm to be lossless, there must be a reverse mapping from compressed data to the original data. That is to say, the compression method must encapsulate a bijection between "plain" and "compressed" data, because the original data and the compressed data should be in the same units. By a simple counting argument, lossless data compression algorithms cannot guarantee compression for all input data sets, because there will be some inputs that do not get smaller when processed by the compression algorithm, and for any lossless data compression algorithm that makes at least one file smaller, there will be at least one file that it makes larger. Strings of data of length N or shorter are clearly a strict superset of the sequences of length $N-1$ or shorter. It follows therefore that there are more data strings of length N or shorter than data strings of length $N-1$ or shorter. And it follows from the *pigeonhole principle* that it is not possible to map every sequence of length N or shorter to a unique sequence of length $N-1$ or shorter. Therefore there is no single algorithm that reduces the size of all data.

One of the most time consuming steps of implementations of, for example, LZ77 compression (one of the most popular) is the search for the longest string match. Most lossless compression implementations are based upon the LZ algorithm. The classical LZ77 and LZ78 algorithms enact a greedy parsing of the input data. That is, at each step, they take the longest dictionary phrase which is a prefix of the currently unparsed string suffix. LZ algorithms are said to be 'universal' because,

H. Zenil et al., *Methods and Applications of Algorithmic Complexity*,
Emergence, Complexity and Computation 44,
https://doi.org/10.1007/978-3-662-64985-5_6

assuming unbounded memory (arbitrary sliding window length), they asymptotically approximate the entropy rate of the generating source [1]. Not only does lossless compression fail to provide any estimation of the algorithmic complexity of small objects [2, 3], it is also not more closely related to algorithmic complexity than Shannon entropy by itself, being only capable of exploiting statistical regularities.

The greatest limitation of lossless compression algorithms, in the light of algorithmic complexity, is that their implementations only exploit statistical regularities (repetitions up to the size of the sliding window length). Thus in effect no general lossless compression algorithm does better than provide the Shannon entropy rate (c.f. Sect. 6.1.1) of the objects it compresses. It is then obvious that an exploration of other possible methods for approximating K is not only desirable but needed, especially methods that can, at least in principle, and more crucially in practice, detect some truly algorithmic content.

6.1.1 Building Upon Entropy Rate

It is natural to ask how random a string appears when blocks of finite length are considered. That is, how do finite approximations of the entropy density converge.

For example, the string 01010101 ... is periodic, but for the smallest granularity (1 bit) or 1-symbol block, the sequence has maximal entropy, because the number of 0s and 1s is the same. Only for longer blocks can the string be found to be regular, and period 2 will be found by identifying the smallest entropy value for which the granularity is at its minimum.

Definition 6.1 Let $Pr(s_i, s_{i+1}, \ldots, s_{i+L}) = Pr(s)$ with $|s| = L$ denote the joint probability over blocks of L consecutive symbols. Let the *Shannon entropy rate* [4] (aka granular entropy, n-gram entropy) of a block of L consecutive symbols — denoted by $H_L(s)$— be:

$$H_L(s) = -\sum_{s_1 \in A} \ldots \sum_{s_L \in A} Pr(s_1, \ldots, s_L) \log_2 Pr(s_1, \ldots, s_L) \tag{6.1}$$

To determine the entropy rate of the sequence, we estimate the limit when $L \to \infty$. It is not hard to see, however, that $H_L(s)$ will diverge as L tends to infinity if the number of symbols increases, but if applied to a binary string $H_L(s)$, it will reach a minimum for the granularity in which a statistical regularity is revealed.

The Shannon Entropy of an object s is simply $H_L(s)$ for fixed block size $L = i$.

One way to solve the problem of divergence for sequences of strings with additional symbols is to take the so-called *entropy rate/density*:

Definition 6.2 The *entropy density* of s is given by:

$$H_{rate}(s) = \lim_{L \to \infty} \frac{H_L(s)}{|s|} \tag{6.2}$$

By summing up the overestimates of block entropy one obtains another interesting measure introduced by Crutchfield and Packard [5].

Definition 6.3 The *total excess entropy* of s is given by:

$$E(s) = \sum_{L=1}^{\infty} H_L(s) \tag{6.3}$$

If $E(s) = 0$ then the string is either of minimal entropy or maximal entropy (Borel normal). $E(s)$ is the amount (in bits), above and beyond $H(s)$, of apparent statistical randomness that is eventually "explained" by considering increasingly longer blocks.

The Block Decomposition Method (BDM) is based upon a similar decomposition approach and we will combine the strengths (and limitations) of both Shannon entropy and algorithmic complexity. We first introduce the algorithm upon which BDM is based for the local estimations of algorithmic complexity.

6.2 The Block Decomposition Method

Because finding the shortest program that reproduces a large object is computationally very expensive and ultimately uncomputable, one can aim at finding smaller programs that reproduce smaller parts of the original object, parts that together compose the larger object. And this is what the BDM does. It decomposes the original data into fragments, the Kolmogorov complexity of which we have good approximations of. The sum of the programs that reconstruct the original data is an approximation of its Kolmogorov complexity. This also means that the method is local. BDM cannot, however, work alone. It requires CTM (Chap. 3). In Sect. 6.5 we study a method to extend its range of application, thereby effectively extending its power by combining it with Shannon entropy.

The BDM is a hybrid complexity measure that combines Shannon Entropy in the long range but provides local estimations of algorithmic complexity. It is meant to improve the properties of Shannon Entropy that in practice are reduced to finding statistical regularities and to extend the power of CTM. It consists in decomposing objects into smaller pieces for which algorithmic complexity approximations have been numerically estimated using CTM, then reconstructing an approximation of the Kolmogorov complexity for the larger object by adding the complexity of the individual components of the object, according to the rules of information theory. For example, if s is an object and s^{10} is the repetition of ten times s, upper bounds for $K(s^{10})$ can be best achieved by $K(s^{10}) \simeq K(s) + \log_2(10)$ rather than $K(s^{10}) \simeq 10K(s)$, because we know that repetitions have a very low Kolmogorov complexity, given that one can describe repetitions with a short algorithm.

Here we introduce and study the properties of this *Block Decomposition Method* based on a method advanced in [2, 3] that takes advantage of the powerful relationship established by algorithmic probability between the frequency of a string produced by a random program running on a (*prefix-free*) universal Turing machine (UTM) and the string's Kolmogorov complexity. The chief advantage of this method is that it deals with small objects with ease, and it has shown stability in the face of changes of formalism, producing reasonable Kolmogorov complexity approximations. BDM must be combined with CTM if it is to scale up properly and behave optimally for upper bounded estimations of K. BDM + CTM is universal in the sense that it is guaranteed to converge to K due to the invariance theorem, and as we will prove later, if CTM no longer runs, then BDM alone approximates Shannon entropy.

Like compression algorithms, BDM is subject to a trade-off. Compression algorithms deal with the trade-off of compression power and compression/decompression speed.

6.2.1 *l*-Overlapping String Block Decomposition

Let us fix values for n and k and let $D(n, k)$ be the frequency distribution constructed from running all the Turing machines with n states and k symbols (Chap. 3). We define the BDM of a string X as follows,

Definition 6.4 Given a finite string X,

$$BDM(X, l, m) = \sum_{i}^{t} CTM(x_i, n, k) + \log_k(s_i) \tag{6.4}$$

where $\{x_1, \ldots, x_t\}$ is the set obtained from the decomposition of X into subsequences of length l. These subsequences are extracted by using a parameter m controlling the shift between them. It goes from 1 (then it takes all subsequences of length l in X) to l (no overlapping between subsequences). The value s_i is the multiplicity of x_i in the indicated decomposition of X. Depending on $|X|$, l and m there will be a remainder of X in the decomposition. $CTM(x_i, n, k)$ is the complexity of x_i as estimated by CTM from $D(n, k)$.

The choice of n and k depend only on the available resources for running CTM, which involves running the entire (n, k) space of Turing machines with n states and k symbols.

For example, for binary strings we can use $n = 5$ and $k = 2$ to produce the empirical output distribution $D(5, 2)$ of all machines with 5 states and 2 symbols by which all strings of size $l = 12$ are produced, except two (one string and its complement). But we assign them values $\max\{CTM(y, 5, 2) : |y| = 12\} + r$ where r is different from zero because the missing strings were not generated in $(5, 2)$ and therefore have a greater algorithmic random complexity than any other string produced in $(5, 2)$ of the same length. Then, for $l = 12$ and $m = 1$, $BDM(X, 12, 1)$ decomposes $X = 010101010101010101$ of length $|X| = 18$ into the following subsequences:

010101010101
101010101010
010101010101
101010101010
010101010101
101010101010
010101010101

with 010101010101 having multiplicity 4 and 101010101010 multiplicity 3.

We then get the CTM values for these sequences:

$$CTM(010101010101, 2, 2) = 26.99073$$
$$CTM(101010101010, 2, 2) = 26.99073$$

To calculate BDM, we then take the sum of the CTM values plus the sum of the $\log_k$ of the multiplicities, with $k = 2$ because the string alphabet is {0, 1}, the same of the Turing machines producing the strings. Thus:

$$\log_2(3) + \log_2(4) + 26.99073 + 26.99073 = 57.56642$$

6.2.2 2- and w-Dimensional Complexity

To look after the likelihood of an array, we can consider a 2-dimensional Turing machine. The Block Decomposition Method can then be extended to objects beyond the unidimensionality of strings, e.g. arrays representing bitmaps such as images, or graphs (by way of their adjacency matrices). We would first need CTM values for 2- and w-dimensional objects that we call base objects (e.g. base strings or base matrices).

A popular example of a 2-dimensional tape Turing machine is Langton's ant [6]. Another way to see this approach is to take the BDM as a way of deploying all possible 2-dimensional deterministic Turing machines of a small size in order to reconstruct the adjacency matrix of a graph from scratch (or smaller pieces that fully reconstruct it). Then, as with CTM, the algorithmic complexity of the adjacency matrix of the graph can be estimated via the frequency with which it is produced from running random programs on the (prefix-free) 2-dimensional Turing machine. More specifically,

$$BDM(M, \{x_i\}) = \sum_{(r_i,n_i)\in Adj(X)_{\{x_i\}}} \log(n_i) + CTM(r_i), \qquad (6.5)$$

where the set $Adj(X)_{\{x_i\}}$ is composed of the pairs (r, n), r is an element of the decomposition of M (as specified by a *partition* $\{x_i\}$) in different sub-arrays of size up to $d_1 \times \cdots \times d_w$ (where w is the dimension of the object) that we call *base matrices* (because CTM values were obtained for them) and s is the multiplicity

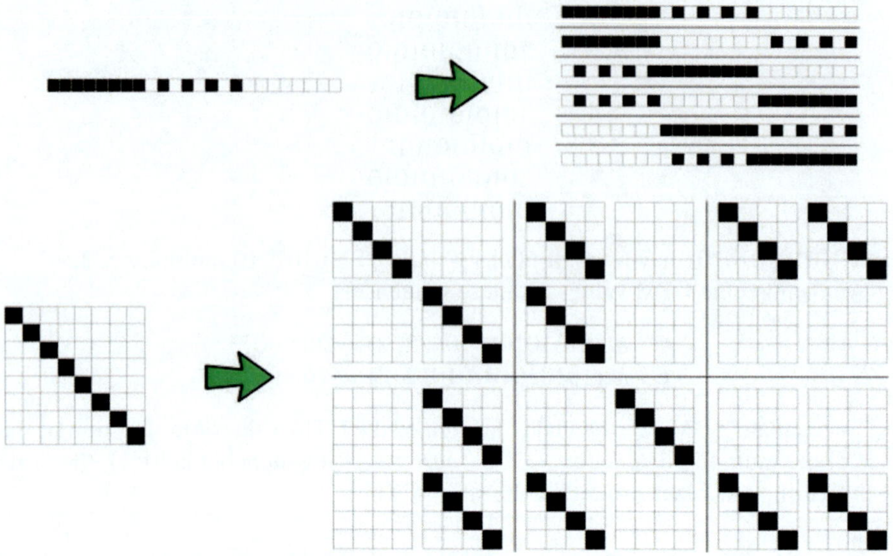

Fig. 6.1 Non-overlapping BDM calculations are invariant to block permutations (reshuffling base strings and matrices), even when these permutations may have different complexities due to the reorganization of the blocks that can produce statistical or algorithmic patterns. For example, starting from a string of size 24 (top) or an array of size 8×8 (bottom), with decomposition length $l = 8$ for strings and decomposition $l = 4 \times 4$ block size for the array, all 6 permutations for the string and all 6 permutations for the array have the same BDM value regardless of the shuffling procedure

of each component. $CTM(r)$ is a computable approximation from below to the algorithmic information complexity of r, $K(r)$, as obtained by applying the coding theorem method to w-dimensional Turing machines. In other words, $\{x_i\}$ is the set of *base objects*.

Because string block decomposition is a special case of matrix block decomposition, and square matrix block decomposition is a special case of w-block decomposition for objects of w dimensions, let us describe the way in which BDM deals with boundaries on square matrices, for which we can assume CTM values are known, and that we call base strings or base matrices.

Figure 6.1 shows that the number of permutations is a function of the complexity of the original object, with the number of permutations growing in proportion to the original object's entropy–because the number of different resulting blocks determines the number of different n objects to distribute among the size of the original object (e.g. 3 among 3 in Fig. 6.1 (top) or only 2 different 4×4 blocks in Fig. 6.1 (bottom)). This means that the non-overlapping version of BDM is not invariant vis-à-vis the variation of the entropy of the object, on account of which it has a different impact on the error introduced in the estimation of the algorithmic complexity of the object. Thus, non-overlapping objects of low complexity will have little impact, but with random objects non-overlapping increases inaccuracy. Overlapping decomposition solves this particular permutation issue by decreasing the number of possible permutations, in order to avoid trivial assignment of the same BDM values. However, overlapping has the undesired effect of systematically overestimating values of algorithmic complexity by counting almost every object of size n, $n-1$ times, hence overestimating at a rate of about $n(n-1)$ for high complexity objects of which the block multiplicity will be low, and by $n\log(n)$ for low complexity objects.

6.2.3 BDM Upper and Lower Absolute Bounds

In what follows we show the hybrid nature of the measure. We do this by setting lower and upper bounds to BDM in terms of the algorithmic complexity $K(X)$, the partition size and the approximation error of CTM, such that these bounds are tighter in direct relation to smaller partitions and more accurate approximations of K. These bounds are independent of the partition strategy defined by $\{x_i\}$.

Proposition 6.5 *Let BDM be the function defined in (6.5) and let X be an array of dimension w. Then $K(X) \le BDM(X, \{x_i\}) + O(\log^2|A|) + \epsilon$ and $BDM(X, \{x_i\}) \le |Adj(X)_{\{x_i\}}|K(X) + O(|Adj(X)_{\{x_i\}}|\log|Adj(X)_{\{x_i\}}|) - \epsilon$, where A is a set composed of all possible ways of accommodating the elements of $Adj(X)_{\{x_i\}}$ in an array of dimension w, and ϵ is the sum of errors for the approximation CTM over all the sub-arrays used.*

Proof Let $Adj(X)_{\{x_i\}} = \{(r_1, n_1), \ldots, (r_k, n_k)\}$ and $\{p_j\}$, $\{t_j\}$ be the sequences of programs for the reference prefix-free UTM U such that, for each $(r_j, n_j) \in Adj(X)_{\{x_i\}}$, we have $U(p_j) = r_j$, $U(t_j) = n_j$, $K(r_j) = |p_j|$ and $|t_j| \le 2\log(n_j) + c$. Let ϵ_j be a positive constant such that $CTM(r_j) + \epsilon_j = K(X)$; this is the error for each sub-array. Let ϵ be the sum of all the errors.

For the first inequality we can construct a program q_w, whose description only depends on w, such that, given a description of the set $Adj(X)_{\{b_i\}}$ and an index l, it enumerates all the ways of accommodating the elements in the set and returns the array corresponding to the position given by l.

Note that $|l|$, $|Adj(X)_{\{b_i\}}|$ and all n_j's are of the order of $\log|A|$. Therefore $U(q_w q_1 p_1 t_1 \ldots p_j t_j l) = X$ and

$$\begin{aligned}
K(X) \le& |q_w p_1 t_1 \ldots p_j t_j l| \\
\le& |q_w| + \sum_1^k (|q_j| + |p_j|) + |l| \\
\le& BDM(X, \{x_i\}) + \epsilon + |q_w| + (\log|A| + c)|Adj(X)_{\{b_i\}}| \\
&+ O(\log|A|) \\
\le& BDM(X, \{x_i\}) + O(\log^2|A|) + \epsilon,
\end{aligned}$$

which gives us the inequality.

Now, let q_X be the smallest program that generates X. For the second inequality we can describe a program $q_{\{x_i\}}$ which, given a description of M and the index j, constructs the set $Adj(X)_{\{x_i\}}$ and returns r_j, i.e. $U(q_{\{x_i\}}q_X j) = r_j$. Note that each $|j|$ is of the order of $\log |Adj(X)_{\{x_i\}}|$. Therefore, for each j we have $K(r_j) + \epsilon_j = |p_j| \leq |q_{\{x_i\}}| + |q_X| + O(\log |Adj(X)_{\{x_i\}}|)$ and $K(r_j) + \epsilon_j + \log(n_i) \leq |q_{\{x_i\}}| + |q_X| + O(\log |Adj(X)_{\{x_i\}}|) + \log(n_i)$. Finally, by adding all the terms over the j's we find the second inequality:

$$\begin{aligned} BDM(X, \{x_i\}) + \epsilon \leq & |Adj(X)_{\{x_i\}}|(|q_X| + |q_{\{x_i\}}| + \log(n_j) + \\ & O(\log |Adj(X)_{\{x_i\}}|)) \\ \leq & |Adj(X)_{\{x_i\}}|K(X) + O(|Adj(X)_{\{x_i\}}| \log |Adj(X)_{\{x_i\}}|). \end{aligned}$$

Corollary 6.6 *If the partition defined by* $\{x_i\}$ *is small, that is, if* $|Adj(X)_{\{x_i\}}|$ *is close to 1, then* $BDM(X, \{x_i\}) \approx K(X)$.

Proof Given the inequalities presented in Proposition 6.5, we have it that $K(X) - O(\log^2 |A|) - \epsilon \leq BDM(M, \{x_i\})$ and

$$BDM(M, \{x_i\}) \leq |Adj(X)_{\{x_i\}}|K(X) + O(|Adj(X)_{\{x_i\}}| \log |Adj(X)_{\{x_i\}}|) + \epsilon,$$

which at the limit leads to $K(X) - \epsilon \leq BDM(X) \leq K(X) - \epsilon$ and $BDM(X) = K(X) - \epsilon$. From [3], we can say that the error rate ϵ is small, and that by the invariance theorem it will converge towards a constant value.

6.3 Dealing with Object Boundaries

Strategies to cleverly partition and deal with boundary conditions can be conceived. Here we introduce a strategy for partition minimization and *base object* size maximization that we will illustrate for 2-dimensionality. The strategies are intended to overcome under- or over-fitting complexity estimations that are due to conventions, not just technical limitations (due to, e.g., uncomputability and intractability).

In Sect. 6.2.3, we have shown that using smaller partitions for BDM yields more accurate approximations to the algorithmic complexity K. However, the computational costs for calculating CTM are high. We have compiled an exhaustive database for square matrices of size up to 4×4. Therefore it is in our best interest to find a method to minimize the partition of a given matrix into squares of size up to $d \times d$ for a given d.

6.3.1 Recursive BDM

The strategy consists in taking the biggest base matrix multiple of $d \times d$ on one corner and dividing it into adjacent square submatrices of the given size. Then we group the remaining cells into 2 submatrices and apply the same procedure, but now for $(d-1) \times (d-1)$. We continue dividing into submatrices of size 1×1.

Let X be a matrix of size $m \times n$ with $m, n \geq d$. Let's denote by quad $= \{UL, LL, DR, LR\}$ the set of quadrants on a matrix and by quad^d the set of vectors of quadrants of dimension d. We define a function $part(X, d, q)$, where d is a natural number and $\langle q_1, \ldots, q_d \rangle \in$ quad^d, as follows:

$$\begin{aligned} part(X, d, q_i) =& max(M, d, q) \cup part(resL(X, d, q), d-1, q_{i+1}) \\ & \cup part(resR(X, d, q), d-1, q_{i+1}) \\ & \cup part(resLR(X, d, q), d-1, q_{i+1})\}, \end{aligned}$$

where $max(X, q, d)$ is the largest set of adjacent submatrices of size $d \times d$ that can be clustered in the corner corresponding to the quadrant q_i, $resR(X, d-1, q_i)$ is the submatrix composed of all the adjacent rightmost cells that could not fit on $max(X, q, d_i)$ and are not part of the leftmost cells, $resL(X, d-1, q_i)$ is an analogue for the leftmost cells and $resLR(X, d-1, q)$ is the submatrix composed of the cells belonging to the rightmost and leftmost cells. We call the last three submatrices *residual matrices*.

By symmetry, the number of matrices generated by the function is invariant with respect to any vector of quadrants $\langle q_1, \ldots, q_d \rangle$. However, the final BDM value can (and will) vary according to the partition chosen. Nevertheless, with this strategy we can evaluate all the possible BDM values for a given partition size and choose the partition that yields the minimum value, the maximum value, or compute the average for all possible partitions.

The partition strategy described can easily be generalized and applied to strings (1 dimension) and tensors (objects of n-dimensions).

6.3.2 Periodic Boundary Conditions

One way to avoid having remaining matrices (from strings to tensors) of different sizes is to embed a matrix in a topological torus (see Fig. 6.2 bottom) such that no more object borders are found. Then let X be a square matrix of arbitrary size m. We screen the matrix M for all possible combinations to minimize the number of partitions maximizing block size. We then take the combination of smallest BDM for fixed *base matrix* size d and we repeat for $d-1$ until we have added all the components of the decomposed X. This procedure will, however, overestimate the complexity values of all objects (in unequal fashion along the complexity spectra) but will remain bounded, as we will show in Sect. 6.4.

Fig. 6.2 One way to deal with the decomposition of n-dimensional tensors is to embed them in an n-dimensional torus($n = 2$ in the case of the one depicted here), making the borders cyclic or periodic by joining the borders of the object. Depicted here are three examples of graph canonical adjacency matrices embedded in a 2-dimensional torus that preserves the object complexity on the surface, a complete graph, a cycle graph and an Erdös-Rényi graph with edge density 0.5, all of size 20 nodes and free of self-loops. Avoiding borders has the desired effect of producing no residual matrices after the block decomposition with overlapping

Without loss of generality the strategy can be applied to strings (1 dimension) and tensors (any larger number of dimensions, e.g. greater than 2), the former embedded in a cylinder while tensors can be embedded in n-dimensional tori (see Fig. 6.2).

6.4 Error Estimation

One can estimate the error in different calculations of BDM, regardless of the error estimations of CTM (quantified in [2, 3]), in order to calculate their departure and deviation both from granular entropy and algorithmic complexity, for which we know lower and upper bounds. For example, a maximum upper bound for binary strings is the length of the strings themselves. This is because no string can have an algorithmic complexity greater than its length, simply because the shortest computer program (in bits) to produce the string may be the string itself.

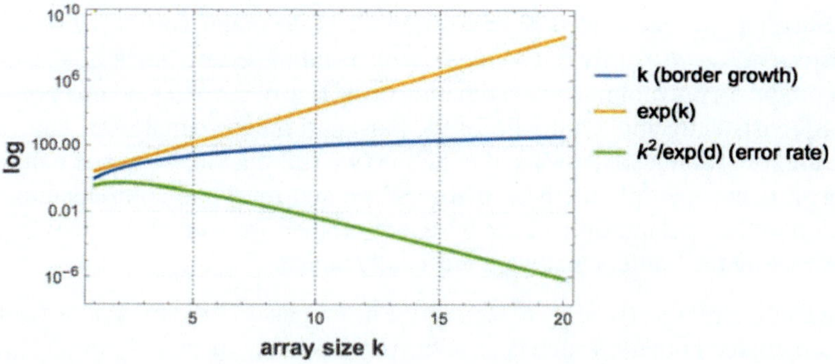

Fig. 6.3 Error rate for 2-dimensional arrays. With no loss of generalization, the error rate for n-dimensional tensors $\lim_{d \to \infty} \frac{k^n}{n^k} = 0$ is convergent and thus negligible, even for the discontinuities disregarded in this plot which are introduced by some BDM versions, such as non-overlapping and trimming boundary condition related discontinuities

In the calculation of BDM, when an object's size is not a multiple of the base object of size d, boundaries of size $< d$ will be produced, and there are various ways of dealing with them to arrive at a more accurate calculation of an object that is not a multiple of the base. First we will estimate the error introduced by ignoring the boundaries or dealing with them in various ways, and then we will offer alternatives to take into consideration in the final estimation of their complexity.

If a matrix X of size k is not a multiple of the base matrix of size d, it can be divided into a set of decomposed blocks of size $d \times d$, and R, L, T and B residual matrices on the right, left, top and bottom boundaries of M, all of smaller size than d.

Then boundaries R, L, T and B can be dealt with in the following way:

- Trimming boundary condition: R, L, T and B are ignored, so $BDM(X) = BDM(X, R, L, T, B)$, with the undesired effect of general underestimation for objects not multiples of d. The error introduced (see Fig. 6.3) is bounded between 0 (for matrices divisible by d) and $k^2/exp(k)$, where k is the size of X. The error is thus convergent ($exp(k)$ grows much faster than k^2) and can therefore be corrected, and is negligible as a function of array size as shown in Fig. 6.3.
- Cyclic boundary condition (Fig. 6.2 bottom): The matrix is mapped onto the surface of a torus such that there are no more boundaries and the application of the overlapping BDM version takes into consideration every part of the object. This will produce an over-estimation of the complexity of the object but will for the most part respect the ranking order of estimations if the same overlapping values are used with maximum overestimation $d - 1 \times \max\{CTM(b)|b \in M\}$, where $K(b)$ is the maximum CTM value among all base matrices b in M after thedecomposition of M.

- Full overlapping recursive decomposition: X is decomposed into $(d-1)^2$ base matrices of size $d \times d$ by traversing X with a sliding square block of size d. This will produce a polynomial overestimation in the size of the object of up to $(d-1)^2$, but if consistently applied it will for the most part preserve ranking.
- Adding low complexity rows and columns (we call this 'add col'): If a matrix of interest is not multiple the base matrices, we add rows and columns until completion to the next multiple of the base matrix, then we correct the final result by substracting the borders that were artificially added.

The BDM error rate (see Fig. 6.3 top) is the discrepancy of the sum of the complexity of the missed borders, which is an additive value of, at most, polynomial growth. The error is not even for a different complexity. For a tensor of d dimensions, with all 1s as entries, the error is bounded by $\log(k^d)$ for objects with low algorithmic randomness and by $\frac{k^d}{d^k}$ for objects with high algorithmic randomness.

Ultimately there is no optimal strategy for making the error disappear, but in some cases the error can be better estimated and corrected (Fig. 6.3) and all cases are convergent, hence asymptotically negligible, and in all cases complexity ranking is preserved and under- and over-estimations bounded.

6.4.1 BDM Convergence Towards Shannon Entropy

Let $\{x_i\}$ be a partition of X defined as in the previous section for a fixed d. Then the Shannon entropy of X for the partition $\{x_i\}$ is given by:

$$H_{\{x_i\}}(X) = - \sum_{(r_j,n_j)\in Adj(X)_{\{b_i\}}} \frac{n_j}{|\{x_i\}|} \log\left(\frac{n_j}{|\{x_i\}|}\right), \tag{6.6}$$

where $Pr(r_j) = \frac{n_j}{|\{x_i\}|}$ and the array r_j is taken as a symbol itself.

The following proposition establishes the asymptotic relationship between $H_{\{x_i\}}$ and BDM.

Proposition 6.7 *Let M be a 2-dimensional matrix and $\{x_i\}$ a partition strategy with elements of maximum size $d \times d$. Then:*

$$|BDM_{\{x_i\}}(X) - H_{\{x_i\}}(X)| \leq O(\log(|M|))$$

Proof First we note that $\sum n_j = |\{x_i\}|$ and, given that the set of matrices of size $d \times d$ is finite, there exists a constant c_d such that $|Adj(X)_{\{b_i\}}| CTM(n_j) < c_d$. Therefore:

$$
\begin{aligned}
BDM_{\{x_i\}}(X) - H_{\{x_i\}}(X) &= \sum \left(CTM(r_j) + \log(n_j) + \frac{n_j}{|\{x_i\}|} \log\left(\frac{n_j}{|\{x_i\}|}\right)\right) \\
&\leq c_d + \sum \left(\log(n_j) + \frac{n_j}{|\{x_i\}|} \log\left(\frac{n_j}{|\{x_i\}|}\right)\right) \\
&= c_d + \sum \left(\log(n_j) - \frac{n_j}{|\{x_i\}|} \log\left(\frac{|\{x_i\}|}{n_j}\right)\right) \\
&= c_d + \frac{1}{|\{x_i\}|} \sum \left(|\{x_i\}| \log(n_j) - n_j \log\left(\frac{|\{x_i\}|}{n_j}\right)\right) \\
&= c_d + \frac{1}{|\{x_i\}|} \sum \log\left(\frac{n_j^{|\{x_i\}|+n_j}}{|\{x_i\}|^{n_j}}\right)
\end{aligned}
$$

Now, let's recall that the number of n_j's is bounded by c_d. Therefore c'_d exists such that

$$
\begin{aligned}
\frac{1}{|\{x_i\}|} \sum \log\left(\frac{n_j^{|\{x_i\}|+n_j}}{|\{x_i\}|^{n_j}}\right) &\leq \frac{c_d}{|\{x_i\}|} \log\left(\frac{|\{x_i\}|^{|\{x_i\}|+c'_d|\{x_i\}|}}{|\{x_i\}|^{c'_d|\{x_i\}|}}\right) \\
&= \frac{c_d}{|\{x_i\}|} \log(|\{x_i\}|^{|\{x_i\}|}) \\
&= c_d \log(|\{x_i\}|)
\end{aligned}
$$

Finally, given that $|\{x_i\}|$ is of the order of $|X|$, we have the result.

6.5 Normalized BDM

A normalized version of BDM is useful for applications in which a maximal value of complexity is known or desired for comparison purposes. The chief advantage of a normalized measure is that it enables a comparison among objects of different sizes, without allowing size to dominate the measure. This will be useful in comparing arrays and objects of different sizes. First, for a square array of size $n \times n$, we define:

$$
MinBDM(n)_{d\times d} = \lfloor n/d \rfloor + \min_{x \in M_d(\{0,1\})} CTM(x) \tag{6.7}
$$

where $M_d(\{0, 1\})$ is the set of binary matrices of size $d \times d$ and $CTM(x)$ is the complexity of x using the Coding Theorem Method. For any n, $MinBDM(n)_{d\times d}$ returns the minimum value of (6.5) for square matrices of size n, so it is the minimum BDM value for matrices with n nodes. It corresponds to an adjacency matrix composed of repetitions of the least complex $d \times d$ square. It is the all-1 or all-0 entries matrix, because $0_{d,d}$ and $1_{d,d}$ are the least complex square base matrices (hence the most compressible) of size d.

Secondly, for the maximum complexity, (6.5) returns the highest value when the result of dividing the adjacency matrix into the $d \times d$ base matrices contains the highest possible number of different matrices (to increase the sum of the right terms in (6.5)) and the repetitions (if necessary) are homogeneously distributed along those squares (to increase the sum of the left terms in (6.5)) which should be the most complex ones in $M_d(\{0, 1\})$. For $n, d \in \mathbb{N}$, we define a function

$$f_{n,d} : M_d(\{0, 1\}) \longmapsto \mathbb{N}$$

that verifies:

$$\sum_{r \in M_d(\{0,1\})} f_{n,d}(r) = \lfloor n/d \rfloor^2 \tag{6.8}$$

$$\max_{r \in M_d(\{0,1\})} f_{n,d}(r) \leq 1 + \min_{r \in M_d(\{0,1\})} f_{n,d}(r) \tag{6.9}$$

$$CTM(r_i) > CTM(r_j) \Rightarrow f_{n,d}(r_i) \geq f_{n,d}(r_j) \tag{6.10}$$

The value $f_{n,d}(r)$ indicates the number of occurrences of $r \in M_d(\{0, 1\})$ in the decomposition into $d \times d$ squares of the most complex square array of size $n \times n$. Condition (6.8) establishes that the total number of component squares is $\lfloor n/d \rfloor^2$. Condition (6.9) reduces the square repetitions as much as possible, so as to increase the number of differently composed squares as far as possible and distribute them homogeneously. Finally, (6.10) ensures that the most complex squares are the best represented. Then, we define:

$$MaxBDM(n)_{d \times d} = \sum_{\substack{r \in M_d(\{0,1\}),\\ f_{n,d}(r) > 0}} \log_2(f_{n,d}(r)) + CTM(r)$$

Finally, the normalized BDM value of an array M is:

Definition 6.8 Given a matrix M of size n, if the size of the squares in $\{x_i\}$ is $d \times d$, then $NBDM(M)_d$ is defined as

$$NBDM(M)_d = \frac{BDM(M, \{x_i\}) - MinBDM(n)_{d \times d}}{MaxBDM(n)_{d \times d} - MinBDM(n)_{d \times d}} \tag{6.11}$$

This way we take the complexity of an array M to have a normalized value which is not dependent on the size of M but rather on the relative complexity of M with respect to other arrays of the same size. Figure 6.4, provides an example of high complexity for illustration purposes. The use of $MinBDM(n)_{d \times d}$ in the normalization is relevant. Note that the growth of $MinBDM(n)_{d \times d}$ is linear with n, and the growth of $MaxBDM(n)_{d \times d}$ exponential. This means that for high complexity matrices, the result of normalizing by using just $CTM(M)/MaxBDM(n)_{d \times d}$ would be similar to $NBDM(M)_d$. But it would not work for low complexity arrays, as when the com-

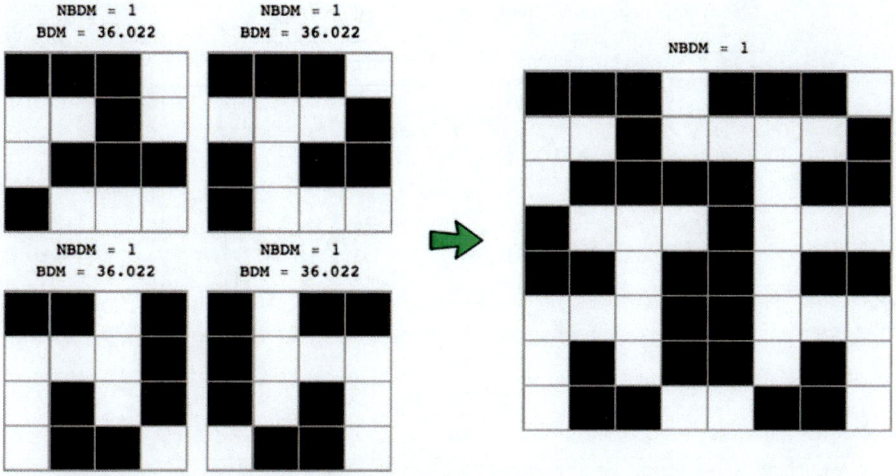

Fig. 6.4 NBDM assigns maximum value 1 to any base matrix with highest CTM or any matrix constructed out of base matrices. In this case, the 4 base matrices on the left are those with the highest CTM in the space of all base matrices of the same size, while the matrix to the left is assigned the highest value because it is built out of the maximum complexity base matrices. The CTM complexity of individual 4×4 squares has been estimated from a random sample of 1291×10^9 random Turing machines in $(5, 2)_{2D}$ (Sect. 5.1), with a runtime bound of 2000 steps

plexity of M is close to the minimum, the value of $CTM(M)/MaxBDM(n)_{d\times d}$ drops exponentially with n. For example, the normalized complexity of an empty array (all 0s) would drop exponentially in size. To avoid this, Eq. (6.11) considers not only the maximum but also the minimum.

Notice the heuristic character of $f_{n,d}$. It is designed to ensure a quick computation of $MaxBDM(n)_{d\times d}$, and the distribution of complexities of squares ensures that $MaxBDM(n)_{d\times d}$ is actually the maximum complexity of a square matrix of size n, but for other distributions it could work in a different way. For example, condition (6.9) assumes that the complexities of the elements in $M_d(\{0, 1\})$ are similar. This is the case for $d \in \{3, 4\}$ in $D(5, 2)_{2D}$, but it may not be true for other distributions. But at any rate it offers a way of comparing the complexities of different arrays independent of their size.

6.6 Behaviour of CTM to BDM Transition

How BDM scales CTM remains a question, as does the rate at which BDM loses the algorithmic estimations provided by CTM. Also unknown is what the transition between CTM and CTM+BDM looks like, especially in the face of applications involving objects of medium size between the range of application of CTM (e.g. 10 to 20 bit strings) and larger objects (e.g. longer sequences in the hundreds of bits).

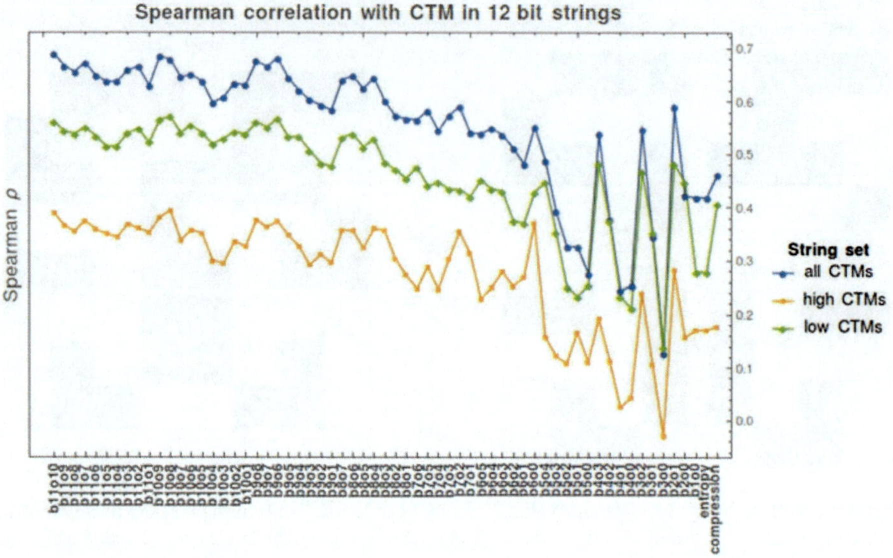

Fig. 6.5 Spearman correlation coefficients (ρ) between CTM and BDM of all possible block sizes and overlap lengths for 12 bit strings, compared with the correlation between CTM and Shannon entropy, and the correlation between CTM and compression length (shown at the rightmost edge of the plot) in blue. ρ coefficients for the 2048 strings below and above the median CTM value are shown in green and orange, respectively. BDM block size and overlap increases to the left. Compression length was obtained using *Mathematica*'s `Compress[]` function. All values were normalized as described in Sect. 6.5

We perform a Spearman correlation analysis to test the strength of a monotonic relationship between CTM values and BDM values with various block sizes and block overlap configurations in all 12 bit strings. We also test the strength of this relationship with CTM on Shannon entropy and compression length.

Figure 6.5 shows the agreement between BDM and CTM for strings for which we have exact CTM values, against which BDM was tested. The results indicate an agreement between CTM and BDM in a variety of configurations, thereby justifying BDM as an extension of the range of application of CTM to longer strings (and to longer objects in general).

In the set of all 12 bit strings, the correlation is maximal when block size is 11 and the overlap is 10 (b11o10, $\rho = 0.69$); Shannon entropy has $\rho = 1$ with BDM when strings are divided in blocks of size 1 and overlap 0 (b1o0, $\rho = 0.42$), as is expected from what is described in Sect. 6.4.1.

The Spearman rank test performed on the first 4096 binary strings has p-values $<1 \times 10^{-15}$, while the Spearman rank test on the 2048 strings with CTM below the median has p-values $<1 \times 10^{-9}$. Finally the Spearman rank test on the 2048 strings with CTM value above the median has p-values $<1 \times 10^{-5}$, in all cases except those corresponding to b4o1, b4o0, and b3o0, where $\rho < 0.03$ and $0.045 \leq$ p-value ≥ 0.25. The lower ρ coefficients in above median CTM strings indicates that there is a greater difficulty in estimating the algorithmic complexity of highly irregular strings through either BDM, entropy, or compression length than in detecting their regularity. Figure 6.5 shows that for block size greater than 6 the Spearman ρ of BDM is always higher than the correlation of CTM with either Shannon entropy or compression length. Some block configurations of size smaller than 6 (e.g., b2o1) also have higher ρ than both Shannon entropy and compression.

While BDM approximates the descriptive power of CTM and extends it over a larger range, we prove in Sect. 6.2.3 that BDM approximates Shannon entropy if base objects are no longer generated with CTM, but if CTM approximates algorithmic complexity, then BDM does.

6.6.1 Smooth BDM (and 'Add Col')

As an alternative method for increasing accuracy while decreasing computational cost, is the use of a weighted function as penalization parameter in BDM. Let the base matrix size be 4×4. We first partition the matrix into sub matrices of the matrix base size 4×4. If the matrix size is not divisible by 4 we (1) use a smooth BDM with full overlap boundary condition (we call this method 'smooth' BDM) or (2) add an artificial low complexity boundary to 'complete' the matrix to the next multiple of 4 and apply 'smooth' (we call this approach 'add col' in future sections).

When using the BDM full overlap boundary condition, we screen the entire matrix by moving a sliding square of size 4×4 over it (as it is done for 'recursive BDM'). When adding artificial low complexity boundaries we only calculate non overlapping sub-matrices of size 4×4 because the expanded matrix of interest is of multiple of 4. These artificial low complexity boundary are columns and rows of one symbols (zeroes or ones). We then correct the final result by subtracting the information added to the boundaries from $\log(|R|) + \log(|C|)$.

To prevent the undesired introduction of false patterns in the 'completion' process (add col), we use the minimum BDM of the extended matrix for both cases (column and rows of zeroes and ones denoted by $BDM_1(X)$ and $BDM_0(X)$ respectively).

In both cases, to distinguish the occurrence of rare and thus highly complex patterns, we assign weights to each base matrix based on the probability of seeing each pattern, denoted by W_i, where i is the index of the base matrix. We thereby effectively 'smooth' the transition to decide matrix similarity, unlike the previous versions of BDM which counts multiplicity of equal matrices. Thus the main difference introduced in the 'smooth' version of BDM is the penalization by base matrix (statistical) similarity rather than only perfect base matrix match.

To simplify notation, in what follows let us denote the adjacency matrix $Adj(X)$ of a matrix M simply as M. The *smooth* version of BDM is then calculated as follows:

$$BDM_0(X) = \sum_{i=1}^{n} (BDM(B_i) \times W_i + \log(n_i)) - 2 \times BDM(0) - \log(|R|) - \log(|C|)$$

$$BDM_1(X) = \sum_{i=1}^{n} (BDM(B_i) \times W_i + \log(n_i)) - 2 \times BDM(1) - \log(|R|) - \log(|C|)$$

Then,

$$BDM(X) = \min(BDM_0, BDM_1)$$

$$BDM_f(X) = \sum_{i=1}^{n} BDM(B_i) \times W_i + \log(n_i)$$

6.6.2 Weighted Smooth BDM with Mutual Information

The Smooth BDM version assigns a weight to each base matrix depending on its statistical likelihood, which is equivalent to assigning a weight based on the entropy of the base matrix over the distribution of all base matrices of 4×4. An equivalent version that is computationally more expensive is the use of classical mutual information. This is arrived at by measuring the statistical similarity between base matrices precomputed by mutual information.

Mutual information is a measure of the statistical dependence of a random variable X on another random variable Y in the joint distribution of X and Y relative to the joint distribution of X and Y under an assumption of independence. If $MI(X, Y) = 0$, then X and Y are statistically independent, but if the knowledge of X fully determines Y, $MI(X, Y) = 1$, then X and Y are not independent. Because MI is symmetric $MI(X, Y) = MI(Y, X)$; if $MI(X, Y) = 1$, then knowing all about Y also implies knowing all about X. In one of its multiple versions MI of X and Y can be defined as:

$$MI(X, Y) = H(X) - H(X|Y) \tag{6.12}$$

where $H(X)$ is the Shannon entropy of X and $H(X|Y)$ the conditional Shannon entropy of X given Y.

In this way, statistically similar base matrices are not counted as requiring 2 completely different computer programs, one for each base matrix, but rather a slightly modified computer program producing 2 similar matrices accounting mostly for one and for the statistical difference of the other. More precisely, BDM can be defined by:

$$BDM(X) = \sum_{i}^{n} MIBDM(B_i) + \log_b n_{x_i} \tag{6.13}$$

where $MIBDM$ is defined by:

$$MIBDM(B_i) = \min\{MI(B_i, B_j)CTM(X_i) + (1 - MI(B_j, B_i))CTM(X_j), \\ MI(B_i, B_j)CTM(X_j) + (1 - MI(B_j, B_i))CTM(X_i)\} \tag{6.14}$$

and where $MI(B_i, B_j)$ is a weight for each CTM value of each base matrix such that j is the index of the matrix that maximizes MI (or maximizes statistical similarity) over the distribution of all the base matrices such that $MI(B_i, B_j) \geq MI(B_i, B_k)$ for all $k \in \{1, \ldots, N\}$.

However, this approach requires $N \times N$ comparisons $MI(B_i, B_j)$ between all base matrices B with indexes $i \in \{1, \ldots, N\}$ and $j \in \{1, \ldots, N\}$.

Notice that because MI is symmetric, then $MI(B_i, B_j) = MI(B_j, B_i)$, but the min in (6.14) is because we look for the minimum CTM value (i.e. the length of the shortest program) for the 2 cases in which one base matrix is the one helping define the statistical similarities of the other and vice versa.

6.7 A Logical Depth-Based Measure

Logical Depth is a measure that distinguishes "structure" from algorithmic randomness (i.e. high algorithmic complexity) through the introduction of computational time. CTM can also provide estimations based upon this logical depth. According to the concept of logical depth [7], the complexity of a string is best defined by the time that an unfolding process takes to reproduce the string from its shortest description. The longer it takes, the more complex. Hence complex objects are those that can be seen as "containing internal evidence of a nontrivial causal history" [7]. It is thus more in keeping with our intuitive sense of the complexity of physical objects, because trivial and random objects are intuitively easy to produce, decompressing quickly in both cases. Bennett provides a careful development [7] of the notion of logical depth, taking into account near-shortest programs rather than the shortest one, in order to arrive at a more robust measure. The simplest version is defined as follows.

Definition 6.9 Let s be a string and d a significance parameter. A string's depth at significance d, is given by [7]:

$$LD_d(s) = \min\{T(p) : (|p| - |p'| < d) \text{ } and \text{ } (U(p) = s)\}, \tag{6.15}$$

where $T(p)$ is the number of steps in the computation $U(p) = s$, $|p'|$ is the length of the shortest program that produces s, thus effectively the algorithmic complexity $K(s)$ of s. In other words, $LD_d(s)$ is the fastest computing time $T(p)$ required to compute s from a d-incompressible program p on a Turing machine U; that is, a program that cannot be compressed by more than a fixed (small) number of d bits.

For algorithmic complexity the choice of UTM is bounded by an additive constant, as shown by the *invariance theorem*, but for logical depth it is bounded by a multiplicative factor [7]. We will denote by $LD_{CTM}(s)$ the measure based upon the concept of logical depth for a string s as explained in Sect. 3.11.1).

Definition 6.10 We define a BDM version for logical depth LD_{BDM} that applies a multiplicative log penalization instead of an additive one, as we did for previously studied versions of BDM.

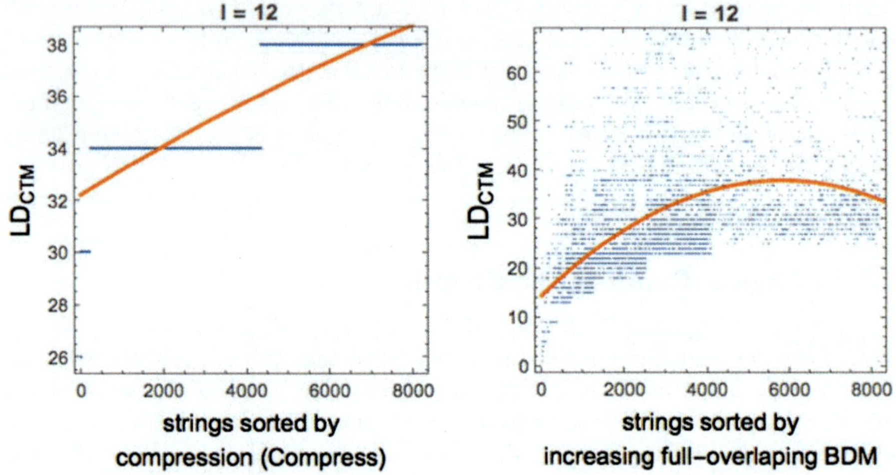

Fig. 6.6 Unlike approximations to algorithmic complexity by lossless compression, LD-based values (versus CTM and BDM) conform to theoretical expectation regarding LD behaviour

$$LD_{BDM}(s, l, m) = \sum_{i}^{t} (\log_2 n_i + 1) \times LD_{CTM}(x_i) \tag{6.16}$$

where n_i is the multiplicity of the base string x_i.

In Fig. 6.6, a numerical experiment is undertaken between BDM and LD for CTM and lossless compression, with LD_{CTM} and LD_{BDM} performing as theoretically expected, i.e., assigning low logical depth to high complexity strings as estimated by BDM. LD_{CTM} and LD_{BDM} can also be defined in higher dimensions, as was done for CTM and BDM separately, based on the replacement of the unidimensional Turing machine tape by an n-dimensional tape.

6.8 BDM Asymptotic Time Complexity and Scalability

Consider the complexity of each step involved in the Block Decomposition Method for approximating the Kolmogorov complexity of a graph. The calculation of BDM for the Kolmogorov complexity approximation of graphs grows in proportion to the growth of the adjacency matrix of M, $Adj(M)$. That is, $O(n^2)$ in the number of vertices n. For every element in an array with floating values between zero and one, one can first precompute the output of the expensive function, as we did with $D(5, 2)$, and store it in a minimal perfect hash table. The index of the array is associated with the value stored there, allowing us to look up the value in constant time. It is only the size of the adjacency matrix that registers an increase of n^2 in the number of vertices n that dominates the computational time required to compute K ($K \log_m$ is also bounded by a small polynomial as it only involves multiplication, which is computationally efficient). The algorithm for tuples calculation involves either knowledge of the length of the object (to calculate the subsets) or traversing the object to perform the partition.

CTM, BDM and LD as functions of the object's size (and therefore the size of the Turing machine rule space that has to be explored) have the following time complexity:

- CTM is uncomputable but for decidable cases runs in $O(exp)$ time.
- Non-overlapping string BDM and LD runs in $O(1)$ linear time and $O(n^d)$ polynomial time for d-dimensional objects.
- Overlapping BDM runs in $O(nx)$ time with m the overlapping offset.
- Full overlapping with $m = 1$ runs in $O(2^n)$ polynomial time as a function of the number of overlapping elements n.
- Smooth BDM runs in $O(1)$ linear time.
- Mutual Information BDM runs in $O(exp)$ time for strings and d exponential for dimension d.

Table 6.1 Summary of ranges of application and scalability of CTM and all versions of BDM. d stands for the dimension of the object

	Short strings <100 bits	Long strings >100 bits	Scalability
Lossless compression	×	✓	$O(n)$
Coding Theorem Method (CTM)	✓	×	$O(exp)$ to $O(\infty)$
Non-overlapping BDM	✓	✓	$O(n)$
Full-overlapping Recursive BDM	✓	✓	$O(n^{(d-1)})$
Full-overlapping Smooth BDM	✓	✓	$O(n^{(d-1)})$
Smooth add col BDM	✓	✓	$O(n)$

Table 6.1, summarizes the range of applications, with CTM and BDM preeminent in that they can more efficiently deal with short, medium and long sequences and other objects such as graphs, images, networks and higher dimensionality objects.

6.9 Algorithmic Complexity of Integer Sequences

Here we introduce an application of BDM to integer sequences. We show that it constitutes an alternative or complementary tool to lossless compression algorithms, widely used to find estimations of algorithmic complexity.

We use BDM by decomposing long strings into strings of maximum length 12 that can be found in $D(5)$. The final estimation of the complexity of a string longer than 12 bits is then the result of the sum of the complexities of the different substrings of length not exceeding 12 in $D(5)$ if they are different, but the sum of only $\log_2(n)$ if n substrings are the same. The BDM version used here is,

$$BDM(X) = \sum_{i}^{t} m_5(x_i) + \log_2(s_i)$$

where s_i is the multiplicity of x_i, and x_i the subsequences from the decomposition of X into t subsequences, with a possible remainder sequence $y < x$ if $|X|$ is not a multiple of the decomposition length l.

The On-Line Encyclopedia of Integer Sequences (OEIS) is a database with the largest collection of integer sequences. It is created and maintained by Neil Sloane and the OEIS Foundation.

Widely cited, the OEIS stores information on integer sequences of interest to both professional mathematicians and amateurs. As of 30 December 2016 it contained nearly 280 000 sequences, making it the largest database of its kind.

We found 875 binary sequences in the OEIS database, accessed through the knowledge engine WolframAlpha Pro and downloaded with the Wolfram Language.

Examples of descriptions found to have the greatest algorithmic probability include the sequence "A maximally unpredictable sequence" with associated sequence 01001101011100010000111101100101001001111; or A068426, the "Expansion of ln_2 in base 2" and associated sequence 01000110110000010100-1110010111011100000. This contrasts with sequences of high entropy such as sequence A130198, the single paradiddle, a four-note drumming pattern consisting of two alternating notes followed by two notes with the same hand, with sequence 01001011010010110100101101001011010010111, or sequence A108737, found to be among the less compressible, with the description "Start with $S =$. For $m = 0, 1, 2, 3, \ldots$ let u be the binary expansion of m. If u is not a substring of S, append the minimal number of 0s and 1s to S to remedy this. The sequence gives S" and sequence 0101100111000101011011110000100101001101. We found that the measure most driven by description length was compressibility.

The longest description of a binary sequence in the OEIS, identified as A123594, reads "Unique sequence of 0s and 1s which are either repeated or not repeated with the following property: when the sequence is 'coded' in writing down a 1 when an element is repeated and a 0 when it is not repeated and by putting the initial element in front of the sequence thus obtained, the above sequence appears".

We found that the textual description length as derived from the database is, as illustrated above, best correlated with the AP-based (BDM) measure, with Spearman test statistic 0.193, followed by compression (only the sequence is compressed, not the description) with 0.17, followed by entropy, with 0.09 (Fig. 6.7). Spearman rank correlation values among complexity measures reveal how these measures are related to each other with BDM v Compress: 0.21, BDM v Entropy: 0.029 and Compress v Entropy: −0.01 from 875 binary sequences in the OEIS database.

We noticed that the descriptions of some sequences referred to other sequences to produce a new one (e.g. "A051066 read mod 2"). This artificially made some sequence descriptions look shorter than they should be. When avoiding all sequences referencing others, all Spearman rank values increased significantly, with values 0.25, 0.22 and 0.12 for BDM, compression and entropy respectively.

To test whether the AP-based (BDM) measure captures some algorithmic content that the best statistical measures (Compress and entropy) may be missing, we compressed the sequence description and compared again against the sequence complexity. The correlation between the compressed description and the sequence compression came closer to that of the AP-estimation by BDM, and BDM itself was even better. The Spearman values after compressing textual descriptions were 0.27, 0.24 and 0.13 for BDM, Compress and entropy respectively.

We then looked at 139 546 integer sequences from the OEIS database, avoiding other non-integer sequences in the database. Those considered represent more than half of the database. Every integer was converted into binary, and for each binary

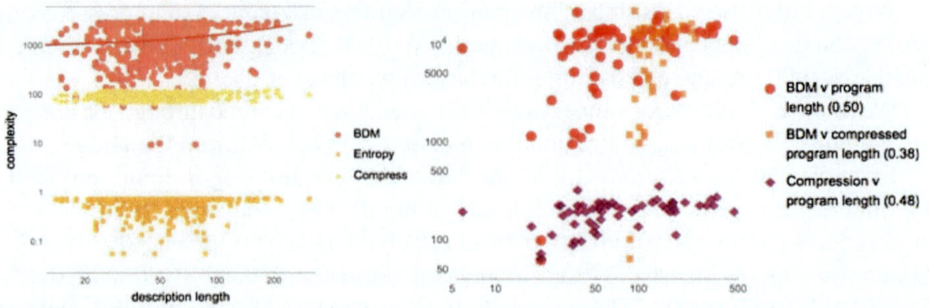

Fig. 6.7 Left: Correlation between the estimated algorithmic complexity (log) by the AP-based measure (BDM) an the length of the text description of each sequence from the OEIS. Fitted line for highest correlation (BDM) is given by $1064.84 + 7.29x$ using least squares. Right: Algorithmic complexity estimation by BDM (log) and of compression on program length (in the Wolfram Language/Mathematica) as coming from the OEIS. In parenthesis the Spearman rank correlation values for each case. Further compressing the program length using Compress resulted in a lower correlation value and BDM outperformed lossless compression

sequence representing an integer an estimation of its algorithmic complexity was calculated. We compared the total sum of the complexity of the sequence (first 40 terms) against its text description length (both compressed and uncompressed) by converting every character into its ASCII code, program length and function lengths, these latter in the Wolfram Language (using Mathematica). While none of those descriptions can be considered the shortest possible, their lengths are upper bounds of the maximum possible lengths of the shortest versions. As shown in Fig. 6.7, we found that the AP-based measure (BDM) performed best when comparing program size and estimated complexity from the program-generated sequence.

6.10 Summary

We have introduced a well-grounded, theoretically sound and robust measure of complexity that beautifully connects two of the main branches of information theory, classical and algorithmic. We have shown that the methods are scalable in various ways, including native n-dimensional variations of the same measure. The properties and numerical experiments are in alignment with theoretical expectations and represent the only truly different alternative and more accurate measure of algorithmic complexity currently available. We have also shown that BDM is computationally efficient, hence complementing the effective use of lossless compression algorithms for calculation of upper bounds of Kolmogorov complexity.

There are thus three methods available today for approximating K (two of which have been advanced by us, one being completely novel: BDM; and one that was known but had never been calculated before: CTM). Here they are described by their range of application:

- CTM deals with 1–12 bits (perhaps a bit longer with 6 states, at most perhaps 20 bits, and 30 if one were being excessively optimistic)
- BDM deals with 12 bits to hundreds of bits (with a cumulative error that grows by the length of the strings–if not applied in conjunction with CTM). The worst case occurs when substrings share information content with other decomposed substrings and BDM just keeps adding their K values individually.
- CTM+BDM (deals with any string length but it is computationally extremely expensive)
- Lossless compression deals with no less than 100 bits and is unstable up to about 1K bits.

Because BDM locally estimates algorithmic complexity based on algorithmic probability and the invariance theorem, it is more independent of object description than other computable measures such as Shannon entropy, though in the 'worst case' it behaves like Shannon entropy. We have also shown that the various flavours of BDM are extremely robust, both by calculating theoretical errors on tensors and by numerical investigation, establishing that any BDM version is fit for use in most cases. Hence the most basic and efficient one can be used without much concern as to the possible alternative methods that could have been used in its calculation, as we have exhaustively and systematically tested most, if not all, of them.

References

1. Lempel, A., Ziv, J.: On the complexity of finite sequences. IEEE Trans. Inf. Theory **22**(1), 75–81 (1976)
2. Delahaye, J.P., Zenil, H.: Numerical evaluation of algorithmic complexity for short strings: a glance into the in- nermost structure of randomness. Appl. Math. Comput. **219**(1), 63–77 (2012)
3. Soler-Toscano, F., et al.: Calculating kolmogorov complexity from the output frequency distributions of small turing machines. PLoS ONE **9**(5), 1–18 (2014)
4. Shannon, C.E.: A mathematical theory of communication. Bell Syst. Tech. J. **27**, 379–423, 623–656 (1948)
5. Crutchfield, J.P., Packard, N.H.: Symbolic dynamics of onedimensional maps: entropies, finite precision, and noise. Int. J. Theor. Phys. **21**(6), 433–466 (1 June 1982). issn: 1572-9575. https://doi.org/10.1007/BF02650178
6. Langton, C.G.: Studying artificial life with cellular automata. Phys. D: Nonlinear Phenom. **22**(1), 120–149 (1986)
7. Bennett, C.H.: Logical depth and physical complexity. In: Herken, R. (ed.) The Universal Turing Machine-a Half-Century Survey, pp. 227–257. Oxford University Press, Oxford (1988)

Chapter 7
Conditional BDM

7.1 Introduction

One key challenge is how to extend and/or combine classical mutual information to mutual algorithmic information using CTM and BDM as introduced in [1–3]. This version of BDM has been key in the introduction of algorithmic complexity in machine learning to give raise to what we call algorithmic machine learning.

Let X and Y be tensors of the same dimension and $\{\alpha_i\}$ a computable *partition strategy* into objects for which the CTM values are known. The BDM value for X with respect of $\{\alpha_i\}$ is defined as

$$BDM(X) = \sum_{(r_i, n_i) \in Adj(X)} CTM(r_i) + \log(n_i),$$

where Adj is the set of pairs (r_i, n_i) resulting from applying the partition $\{\alpha_i\}$ to X, r_i is a sub-tensor for which the CTM value is known and n_i is the respective multiplicity within the tensor. We will call the blocks for CTM values are known *base blocks* or *CTM blocks*.

Definition 7.1 Given the definition stated above we can define the *conditional BDM of X with respect of the tensor Y with respect of $\{\alpha_i\}$* as

$$BDM(X|Y) = \sum_{(r_i, n_i) \in Adj(X) - Adj(Y)} (CTM(r_i) + \log(n_i)) + \sum_{Adj(X) \cap Adj(Y)} f(n_j^x, n_j^y)$$

where n_j^x and n_j^y are the multiplicity of the sub-tensor within X and Y respectively and f is the function defined as

$$f(n_j^x, n_j^y) = \begin{cases} 0 & \text{if } n_j^x = n_j^y \\ \log(n_j^x) & \text{otherwise.} \end{cases}$$

H. Zenil et al., *Methods and Applications of Algorithmic Complexity*,
Emergence, Complexity and Computation 44,
https://doi.org/10.1007/978-3-662-64985-5_7

Intuitively, the Definition 7.1 approximates the algorithmic information within the tensor X that is not present in Y: If we assume knowledge of Y, then to describe the elements of the decomposition of X using the strategy partition strategy $\{\alpha_i\}$ we need the description of the subtensors that are not Y. In the case of common subtensors, if the multiplicity is the same then we can assume that X does not contain addition information, but that it does if the multiplicity differs.

7.1.1 *On the Second Term*

As stated in the previous section, the term

$$\sum_{Adj(X) \cap Adj(Y)} f(n_j^x, n_j^y)$$

quantifies the additional information contained within X when the multiplicity of the sub-tensors differs between X and Y. This term is important in cases when such multiplicity dominates the complexity of the objects. An alternative definition for the term is

$$\sum_{Adj(X) \cap Adj(Y)} |n_j^x - n_j^y|,$$

however we consider that this would introduce ambiguity given that we do not have information of whose multiplicity was higher. Moreover, the stated definition will show to be *well behaved* in the Sect. 7.3.

7.2 Joint and Mutual BDM

Of joint BDM harder to extend from classical information theory. On classical information theory, we can think on joint entropy as the information contained on two or more events occurring concurrently, and the joint entropy of the channels as the average over all combinations of events. Within AIT, we interpret this concept as the "*amount of information contained within two or more objects*".

In contrast with classic information theory, we started by defining conditional BDM, therefore we think that the best way to define joint BDM is from the *chain rule*.

Definition 7.2 The *joint BDM of X and Y with respect of* $\{\alpha_i\}$ is defined as

$$JointBDM(X, Y) = BDM(Y|X) + BDM(X).$$

Following the same path, we could define *mutual BDM*:

Definition 7.3 The *mutual BDM of X and Y with respect of* $\{\alpha_i\}$ is defined as

$$MutualBDM(X, Y) = BDM(X) - BDM(X|Y).$$

7.3 The Relationship Between Conditional, Joint and Mutual Information

The results shown in this section are evidence that my definition for conditional BDM is *well behaved*, as they are analogue to important properties for conditional, joint and mutual entropy.

Proposition 7.1 *If* $X = Y$ *then* $BDM(X|Y) = 0$.

Proof Is direct consequence of the Definition 7.1.

Is important to note that $BDM(X|Y) = 0$ does not imply that $X = Y$. However, it implies that $Adj(X) = Adj(Y)$. This is consequence that BDM does not measure the information encoded in the position of the subtensors.

Proposition 7.2 $BDM(X) \geq BDM(X|Y)$.

Proof As we consider subsets of $Adj(X)$, is direct consequence of the Definition 7.1.

Proposition 7.3 *If X and Y are independent with respect to the partition* $\{\alpha_i\}$, *this is equivalent to* $Adj(X) \cap Adj(Y) = \emptyset$, *then* $BDM(X|Y) = BDM(X)$.

Proof Is direct consequence of the Definition 7.1, given that we have that

$$Adj(X) - Adj(Y) = Adj(X).$$

Proposition 7.4 $MutualBDM(X, Y) = MutualBDM(Y, X)$.

Proof First, consider the equation

$$\begin{aligned} MutualBDM(X, Y) =& BDM(X) - BDM(X|Y) \\ =& \sum_{(r_i, n_i) \in Adj(X)} CTM(r_i) + \log(n_i) \\ &- \sum_{(r_i, n_i) \in Adj(X) - Adj(Y)} (CTM(r_i) + \log(n_i)) \\ &- \sum_{Adj(X) \cap Adj(Y)} f(n_k^x, n_k^y). \end{aligned}$$

While the other hand we have that

$$\begin{aligned} MutualBDM(Y,X) =& BDM(Y) - BDM(Y|X) \\ =& \sum_{(r_j,n_j)\in Adj(Y)} CTM(r_j) + \log(n_j) \\ &- \sum_{(r_j,n_j)\in Adj(Y)-Adj(X)} (CTM(r_j) + \log(n_j)) \\ &- \sum_{Adj(Y)\cap Adj(X)} f(n_k^y, n_k^x). \end{aligned}$$

Notice that on both equations we have the sum over all the pairs that are in both sets, $Adj(X)$ and $Adj(Y)$, with the difference being on the terms corresponding to the *multiplicity*. Now we have to consider two cases. If $n_i^x = n_i^y$ we have the equality. Otherwise, on the first equation we have terms of the form $\log(n_j^x) - f(n_k^x, n_k^y)$, which, by definition of f, is 0; analogously for the second equation. Therefore, we have the equality.

Proposition 7.5 *The relation between mutual and joint BDM is given by,*

$$MutualBDM(X,Y) = BDM(X) + BDM(Y) - JointBDM(X,Y).$$

Proof

$$\begin{aligned} MutualBDM(X,Y) &= MutualBDM(Y,X) \\ &= BDM(Y) - BDM(Y|X) \\ &= BDM(Y) + BDM(X) - (BDM(Y|X) + BDM(X)) \\ &= BDM(X) + BDM(Y) - JointBDM(X,Y) \end{aligned}$$

7.4 Properties and Relationship with Entropy

As mentioned before, the goal behind the definition of Conditional BDM, $BDM(X|Y)$, is to measure the amount of information contained X not present in Y. Ideally, this is measured by the conditional algorithmic information $K(X|Y)$. However, this is value is not computable and approximating it by means of CTM would require extensive computation resources in exponential relationship with the ones required for the values used in the current iterations of BDM. The definition of conditional BDM presented in this text relies heavily on the entropy-like behaviour of BDM, perhaps having relating closer to the latter than BDM itself on most use cases. However, unlike conditional entropy, conditional BDM extends *naturally* to tensors of larger size.

7.4.1 The Impact of Partition Strategy

As shown in previous results, BDM better approximates the universal measure $K(X)$ as the number of elements resulting from applying the partition strategy $\{\alpha_i\}$ to X. This is not the case for conditional BDM, instead $BDM(X|Y)$ is a good approximation to $K(X|Y)$ when the $Adj(X)$ and $Adj(Y)$ share a high number of *base tensors* r_i, and the probability of this occurring is lower in inverse proportion to the number of elements of the partition. For this reason we have empathised through this text that conditional BDM is dependant on the chosen partition strategy $\{\alpha_i\}$.

For a simple example, think of the binary string $X = 11110000$ and consider its inverse $Y = 00001111$. Since we have the CTM approximation for strings of size 8, the best BDM value for each string is found when $Adj(X) = \{(11110000, 1)\}$ and $Adj(Y) = \{(00001111, 1)\}$. However, given that the elements of the partitions are different we have that

$$BDM(11110000|00001111) = BDM(11110000) = 25.1899,$$

even thought we intuitively know that, algorithmic information wise, they should be very close. However, conditional BDM is able to capture this with partitions of size 1–4 with no overlapping, assigning a value of 0 to $BDM(X|Y)$.

In general, it seems that there is no general strategy for finding a *best partition strategy*. This is an issue shared with conditional block entropy, and just like the original BDM definition, at its worse conditional BDM will behave like conditional entropy when comparable, while maintaining best cases close to the ideal of conditional algorithmic complexity.

7.4.2 Conditional BDM as an Extension of Conditional Entropy

Conditional entropy $H(X|Y)$ is defined as

$$\begin{aligned} H(X|Y) &= \sum_{y \in Y} p(y)\, H(X|Y = y) \\ &= -\sum_{y \in Y} p(y) \sum_{x \in X} p(x|y) \log p(x|y), \end{aligned}$$

which can be interpreted as "*the amount of information introduced by X when Y us assumed to be known*" [scholarpedia] and is defined for two variables over the same probability space, on other words, where the probability $p(x, y)$ is defined. In this section I will show how restrictions creates problems when will try to use this mathematical construct over the probability distribution suggested by finite binary strings.

Two important properties for conditional entropy are that $H(X|Y) \leq H(X)$ and $H(X|Y) = H(Y, X) - H(Y)$, where $H(X, Y)$ can be considered a "vector valued variable" [cite Elements of IT], formally

$$H(X, Y) = -\sum_{x\in X}\sum_{y\in Y} p(x, y) \log p(x, y).$$

Now, lets consider two binary strings $X = 1111$ and $Y = 1010$. The suggested probability by X is $P(X = 1) = 1$ while for Y is $P(Y = 1) = 0.5$. Therefore, by the second relation we have that

$$\begin{aligned} H(X, Y) &= -\sum_{x\in\{1\}}\sum_{y\in\{0,1\}} p(x, y) \log p(x, y) \\ &= -p(1, 0) \log p(1, 0) - p(1, 1) \log p(1, 1). \end{aligned}$$

Now, given that $H(1010) = 1$ and $H(1111) = 0$, if we assume that the distributions are independent, therefore $p(x, y) = p(x)p(y)$, we have that $H(X|Y) = H(X, Y) - H(Y) = 1 - 1 = H(X)$. But if the distributions are always assumed to be independent, then we will always have the case of $H(X|Y) = H(X)$ given that Y will have nothing to say about X and $\sum_{x\in X} p(x|y) \log p(x|y)$, and therefore $H(X|Y)$, will always be equal to $H(X)$. However, if we assume that the variables are not independent then we face an impossible task given that $H(Y, X) - H(Y)$ must be equal or less than zero, the value of 1 can only be reach when both probabilities are $\frac{1}{2}$ and entropy cannot be a negative value. This issue is heightened when using block entropy since, with very high probability, the resulting *alphabets* will be unequal.

The problem exposed lies in that Shannon's entropy was originally defined for communication channels for which a finite string would only give us a glimpse of the underlying distribution. For instance, if we assume that there is an underlying common distribution for the strings 1010 and 1111 then conditional entropy will behave as expected.

In contrast, conditional BDM implicitly assumes the universal distribution as the underlying probability distribution for the elements of a finite strings and uses algorithmic information techniques to compute an approximation to the information content of a string as a whole with respect to other. For this reason, as shown in Sect. 7.3, conditional BDM is always well behaved across different finite strings **when assuming the same partition strategy**. Furthermore, as we will show empirically in the next section, conditional BDM is able to capture the conditional information content of strings that come from the same distribution as well as conditional entropy.

7.5 Numerical Results

In this section we explore the behaviour of conditional BDM and contrast it with conditional entropy over a series of numerical experiments.

7.5.1 *Conditional BDM Compared to Conditional Entropy over Biased Distributions*

For the first experiment we generated a sample of 19,000 pseudo-random binary strings of length 20 that are pairwise related by coming for one of 19 *biased* distributions where the expected number of 1's varies from 1 to 19. For each pair we computed the conditional BDM with partitions of size 1 and divided it by the conditional BDM of the first string with respect of a random string coming from an uniform distribution. To both, the divisor and the dividend, we added 1 to avoid divisions by zero. We repeated the experiment for conditional entropy. Both results where normalized by diving the obtained quotients by the maximum obtained value for each distribution. In the plot Fig. 7.1 we show the average obtained for each biased distribution.

From the plot Fig. 7.1 we can see that as the underlying distribution associated for string is increasingly biased, the expected shared information content of two related

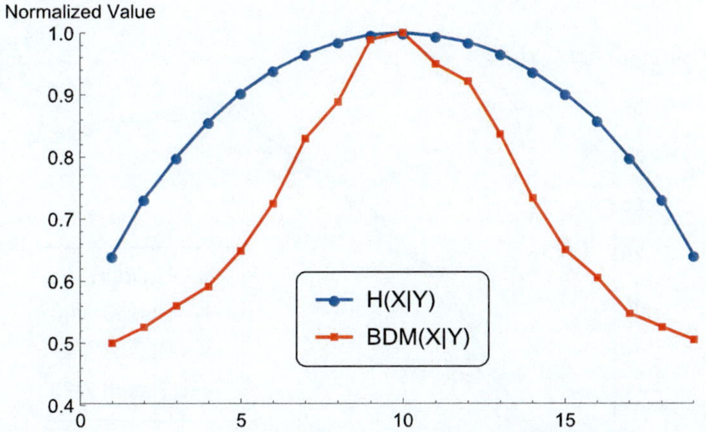

Fig. 7.1 Each point represents the normalized average of the conditional BDM ($BDM(X|Y)$) and conditional entropy ($H(X|Y)$), respectively, corresponding to 5000 pairs of strings randomly chosen from a distribution where the expected number of 1's is the value shown in the *x axis* divided between the conditional BDM or conditional entropy of the first element of the pair and an unrelated randomly chosen binary string. All strings are of length 20. The partition strategy used for BDM is that of sets of size 1. From this plot we can see that conditional BDM manages to capture the statistical relationship of finite strings generated from the same distribution

strings is higher (conditional BDM is lower) when compared to the conditional BDM of two unrelated strings. This behaviour is congruent with what we expect and observe for conditional entropy. The area under the normalized cube being smaller is expected given that BDM is a finer-graded information content measure than entropy and is not perfectly symmetric as BDM and CTM are computational approximations to an uncomputable function and are also inherently more sensible to the fundamental limits of computable random number generators.

7.5.2 Conditional BDM over Different Partition Sizes

The experiment shown in the previous section used BDM with partitions of size 1 since is that under such partition strategy where we expect the behaviour to be closer to entropy. In this section we will briefly explore the effect that different partition sizes have in the conditional BDM of random strings.

For this experiment we generated 2,400,000 random binary strings of size 20 with groups of 600,000 strings belonging to one four different distributions: *uniform* (ten 1's expected), *biased 3/20* (three 1's expected), *biased 1/4* (five 1's expected) and *biased 7/20* (seven 1's expected). Then, we formed pairs of strings belonging to the same distribution and computed the conditional BDM using different partition sizes from 1 to 20, for a total of 30,000 pairs per data point, normalising the result by diving it by the partition size to avoid this factor being the dominant one. In the plot Fig. 7.2 we show the average obtained for each data point.

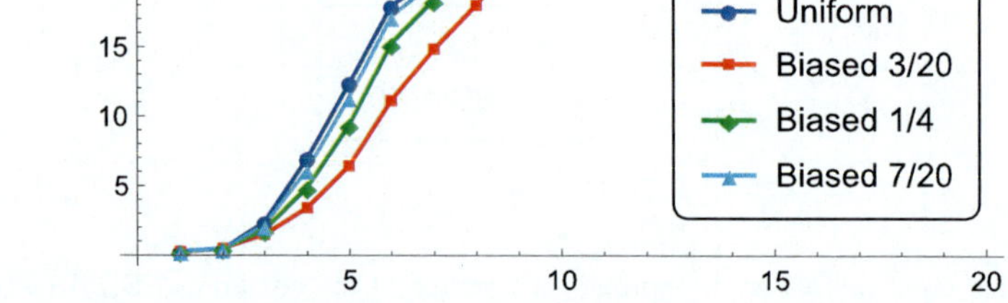

Fig. 7.2 Each point represents the average of the conditional BDM from 30,000 pairs of binary strings of size 20 randomly generated from four different distributions: *uniform* (ten 1's expected), *biased 3/20* (three 1's expected), *biased 1/4* (five 1's expected) and *biased 7/20* (seven 1's expected). The *y axis* indicates the partition size used to compute the respective conditional BDM value, which was normalized by dividing it by the partition size

From the plot Fig. 7.2 we can observe two main behaviours. The first one is that as the partition size increases so does the conditional BDM value. This is because bigger partitions take into account more information from the position of each bit and we do not expect that randomly generated strings to share positional information. The drop observed after partitions of size 12 is result of CTM values being available up to strings of size 12, the point where the program start to rely on BDM for the computation, additionally the partition strategy ignores smaller partitions than the ones stated, therefore reducing the overall amount of information taken into account.

The second one is that not only conditional BDM is able to capture the discrepancies expected from the different distributions for partition sizes where there is no loss of statistical information (this being from size 1–10), but seems to improve on its ability to do so with larger partition sizes up to 10, therefore improving upon the results presented at Sect. 7.5.1.

It is important to note that an important reduction of accuracy for partitions of size larger than 10 was expected given that the partition strategy used discards substrings of smaller sizes than the ones stated. For instance, the partition of size 3 of the string 10111 is just $\{101\}$, therefore losing information. Furthermore, for big partitions sizes with respect of the string length, the statistically similarity banishes, given that now each substring is considered a *different symbol of an alphabet*. Therefore, the abrupt change of behaviour observed beyond partitions of size 15 is expect and product of causality.

7.6 Distance to Conditional Kolmogorov Complexity

The ideal and universal measure of conditional information content is the conditional algorithmic information complexity $K(X|Y)$, which is defined as

$$K(X|Y) = \min\{|p| : p(Y) = X\},$$

where p is a program of length $|p|$ for a reference Universal Turing Machine U that receives Y as an input. This conditional complexity function measures the total amount of information of X that can be recovered from Y using any computable method.

Although theoretically sound, the conditional information content is an incomputable function, therefore it represents a theoretical ideal that cannot be reached in practice. By construction, conditional BDM is an approximation to this measure, however it differs by not taking into account two information sources: the information content shared between base blocks and the position of each block.

As an example of the first limitation, consider the string 1010 and its *negation* 0101. We know that they are algorithmically close, but for a partition strategy of size 2 with no overlapping, the Adj sets $\{("10", 2)\}$ and $\{("01", 2)\}$ are disjoint, therefore conditional BDM assigns the maximum BDM value to the shared information content. Within this limitation, conditional BDM represents a better approximation to K in comparison to entropy mainly because BDM uses the CTM approximation value for each block, rather than just its distribution, and the information content of its multiplicity, therefore representing a more accurate approximation to the overall information content of the non-shared base blocks.

As for the second limitation, can only become significant when the size of the base blocks is *small* when compared to that of the objects we are analysing, given that the positional information can become the dominant factor of the information content within an object. This is an issue shared with entropy: when we have this case BDM, and therefore conditional BDM, approaches entropy. However, conditional BDM has the added benefit that conditional BDM is defined for finite tensors generated from different distributions.

7.7 Hybrid Classical and Algorithmic Conditional BDM

As explained in the previous section, one weakness of the previous definition of conditional BDM in comparison to conditional entropy, is its inability to discern the shared information between the different CTM blocks on which an object is partitioned. The *ideal* way to deal with this issue is to compute a set of **conditional CTM blocks**:

Definition 7.4 Given the objects X and Y, the *conditional CTM* of X in relation to Y, denoted by $\mathrm{CTM}(X|Y)$ value is the number of Turing machines that given the input Y produce Y as an output.

This set data can be computed in a similar way that current CTM data sets have been produced. However, it would require an exponentially higher number of computations, so is not feasible to use this definition for the short to medium term.

Another way to address this issue is to define an hybrid information measure just in the spirit of the original BDM definition between classical Shannon entropy combined with local approximations of algorithmic complexity.

Definition 7.5 The *entropy-weighted conditional BDM of X with respect of the tensor Y with respect of* $\{\alpha_i\}$ as

$$HBDM(X|Y) = \sum_{(r_i,n_i)\in Adj(X)-Adj(Y)} \frac{H(r_i|Y)}{H(r_i)}(CTM(r_i) + \log(n_i)) + \sum_{Adj(X)\cap Adj(Y)} f(n_j^x, n_j^y)$$

where n_j^x and n_j^y are the multiplicity of the subtensor within X and Y respectively and f is the function defined as

$$f(n_j^x, n_j^y) = \begin{cases} 0 & \text{if } n_j^x = n_j^y \\ \log(n_j^x) & \text{otherwise,} \end{cases}$$

and $H(r_i|Y)$ is the conditional entropy of the subtensor r_i with respect of Y. When $H(r_i) = 0$ then the factor $\frac{H(r_i|Y)}{H(r_i)}$ is assumed to be 0.

The idea behind the previous definition is simple: we weight the amount of algorithmic information that the subtensor r_i contains of X by means of the *quantity of classic information* that Y has over r_i.

References

1. Hernndez-Orozco, S. et al.: Algorithmic information regularization and weighting. In: submitted (2019)
2. Hernndez-Orozco, S. et al.: Algorithmic probability classification. In: submitted (2019)
3. Hernndez-Orozco, S. et al.: Algorithmic complexity and integrated information. In: submitted (2019)

Part II
Applications

Chapter 8
Applications to Graph and Network Complexity

We show that numerical approximations of Kolmogorov complexity of graphs and networks capture some group-theoretic and topological properties of empirical networks, ranging from metabolic to social networks, and of small synthetic networks that we have produced. That K and the size of the group of automorphisms of a graph are correlated opens up interesting connections to problems in computational geometry, and thus connects several measures and concepts from complexity science. We derive these results via two different Kolmogorov complexity approximation methods applied to the adjacency matrices of the graphs and networks. The methods used are the traditional lossless compression approach to Kolmogorov complexity, and the normalised version of the Block Decomposition Method (Chap. 6) based on algorithmic probability theory.

8.1 Introduction

Graphs are an important tool for mathematically analysing many systems, from interactions of chemical agents, to ecological networks, to representing data objects in computer science [15, 22]. An interesting perspective regarding such graphs is to investigate the complexity [11, 17] or information content of a graph [28]. While Shannon information theory [21, 24, 28] and counting symmetries [19, 20] have been applied to measure information content/complexity of graphs, little has been done, by contrast, to demonstrate the utility of Kolmogorov complexity as a numerical

The material in this section was adapted from **Correlation of automorphism group size and topological properties with program-size complexity evaluations of graphs and complex networks**, Authors: Hector Zenil, Fernando Soler-Toscano, Kamaludin Dingle and Ard A. Louis, Published in *Physica A: Statistical Mechanics and its Applications*, Pages 341–358, Copyright 2014, with permission from Elsevier.

H. Zenil et al., *Methods and Applications of Algorithmic Complexity*,
Emergence, Complexity and Computation 44,
https://doi.org/10.1007/978-3-662-64985-5_8

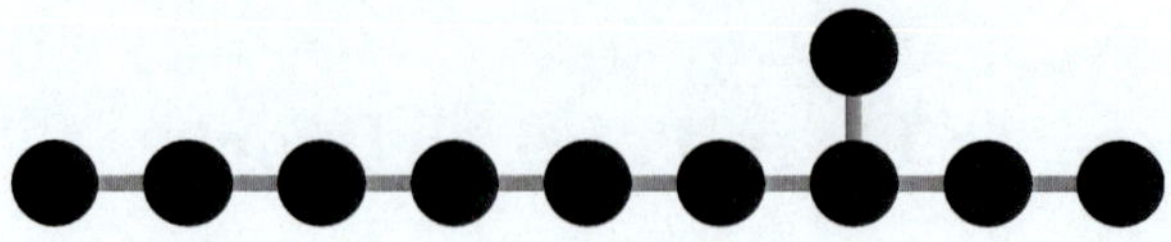

Fig. 8.1 An example of a graph which has no symmetries (i.e. automorphism group size of 1, cf. Definition 8.2), despite being intuitively simple, i.e. just a string of nodes with a single side node

tool for graph and real-world network investigations. Some theoretical connections between graphs and algorithmic randomness have been explored (e.g. [7]), but these are mostly related to formal properties of random graphs.

Here we computationally study some of these numerical and real-world directions and show how Kolmogorov complexity can capture group-theoretic and topological properties of abstract and empirical graphs and networks. We do this by introducing a measure of graph complexity based on approximating the Kolmogorov complexity of the adjacency matrix representation of a graph, which we achieve by applying the *Block Decomposition Method* (Chap. 6). A theoretical advantage of using Kolmogorov complexity is that it captures *all* structure (i.e. non-randomness) in an object, such as a graph. In contrast, only looking at symmetries, for example, can miss structure and potential simplicity in a graph. For example, Fig. 8.1 shows a graph with no symmetries, despite being far from random, and indeed is intuitively simple.

8.2 Preliminaries

A graph $g = (V, E)$ consists of a set of vertices V (also called nodes) and a set of edges E. Two vertices, i and j, form an edge of the graph if $(i, j) \in E$. Let the binary adjacency matrix of g be denoted by $Adj(A)$. A graph can be represented by its adjacency matrix. Assuming that the vertices are indices from 1 to n, that is, that $V = \{1, 2, \ldots, n\}$, then the adjacency matrix of g is an $n \times n$ matrix, with entries $a_{i,j} = 1$ if $(i, j) \in E$ and 0 otherwise. The distance $D(g)$ of a graph g is the maximum distance between any 2 nodes of g. The size $V(g)$ of a graph g is the vertex count of g; similarly $E(g)$ denotes the edge count of g.

Definition 8.1 Two graphs g and h are *isomorphic* if and only if there exists a permutation λ such that $\lambda(g) = h$. (That is, g and h are topologically equivalent).

The general problem of graph isomorphism appears, for example, in chemistry [1, 2] and biology [13, 16]. An automorphism of a graph g is an isomorphism of g's vertices that preserves g's edge relation. Every graph has a trivial symmetry (the identity) that maps each vertex to itself. More formally,

Definition 8.2 An *automorphism* of a graph g is a permutation λ of the vertex set V, such that the pair of vertices (i, j) forms an edge if and only if the pair $(\lambda(i), \lambda(j))$ also forms an edge.

The set of all automorphisms of an object forms a group, called the *automorphism group*. Intuitively, the size of the automorphism group $A(g)$ provides a direct measure of the abundance of symmetries in a graph or network. Every graph has a trivial symmetry (the identity) that maps each vertex to itself.

A clustering coefficient is a measure of the degree to which nodes in a graph tend to cluster together (for example, friends in social networks [10]). Let N_i be the neighbourhood of a node v_i, that is, the set of v_j such that $D(v_i, v_j) = 1$ and $n_i = |N_i|$. Then,

Definition 8.3 The *local clustering coefficient* of node v_i, $C(v_i)$, is given by,

$$C(v_i) = \frac{2\,|E(N_i)|}{n_i(n_i - 1)}$$

where $E(N_i)$ denotes the set of edges with both nodes in N_i

That is, the number of edges possessed by a node in the neighbourhood of another, divided by the maximum number of possible edges in the neighbourhood (the complete subgraph formed by the neighbouring nodes). The mean global clustering coefficient of the whole network (or a sub-graph) g is simply $C(g)$ averaged over its nodes. Global clustering coefficient values are always between 0 and 1.

8.3 Applying BDM to Graphs

By using the adjacency matrix of a graph as its representation, we can use BDM to estimate K for graphs. Given the 2-dimensional nature of adjacency matrices we can use a variation of a Turing machine that runs on a 2-dimensional tape in order to estimate upper bounds of K of the adjacency matrix of a graph. A popular example of a 2-dimensional tape Turing machine is Langton's ant [3]. Another way to see this approach is to take the BDM as a form of deploying all possible 2-dimensional deterministic Turing machines of a small size in order to reconstruct the adjacency matrix of a graph from scratch (or smaller pieces that fully reconstruct it). Then as with the Coding Theorem Method (above), the Kolmogorov complexity of the adjacency matrix of the graph can be estimated via the frequency that it is produced from running random programs on the (prefix-free) 2-dimensional Turing machine.

BDM will use the decomposition of the adjacency matrix into 4×4 submatrices, all of then found in a sample of 1291×10^9 random Turing machines in $(5, 2)_{2D}$ (Sect. 5.1), with a runtime bound of 2000 steps. Then, $D(5, 2)_{2D}(r)$ is the probability for a random Turing machine in $(5, 2)_{2D}$ of halting and producing r. The BDM version used for graph analysis is

$$K\log_m(g) = \sum_{(r_u,n_u)\in Adj(g)_{4\times4}} \log_2(n_u) - \log_2 D(5, 2)_{2D}(r_u) \tag{8.1}$$

where $Adj(g)_{4\times 4}$ represents the set with elements (r_u, n_u), obtained when decomposing the adjacency matrix of g into non-overlapping squares of size 4 by 4. In each (r_u, n_u) pair, r_u is one such square and n_u its multiplicity (number of occurrences). To abbreviate, K_m will be frequently used instead of $K \log_m$ to denote (8.1). The normalization of BDM, called NBDM, is given by (6.11) (Sect. 6.6).

8.3.1 Complexity and Graph Vertex Order

As a final note on our method, the order of the graph nodes in the adjacency matrix is relevant for the complexity retrieved by K_m. This is especially important in highly symmetrical graphs. Figure 8.3 shows five different adjacency matrices for the wheel graph of 18 nodes (Fig. 8.2), with the nodes ordered in different ways. The extreme left matrix of Fig. 8.3 represents the nodes in consecutive order, with the central one at the right. The other four matrices sort the nodes in random order. As we can see in this very regular graph, the lowest complexity corresponds to the organized matrix.

Hence, when studying the complexity of graphs, for several applications we will be interested not only in the K_m value of a particular adjacency matrix, but in several randomisations of the vertex order. Notice that it is just the adjacency matrix representation that changes; topologically the graph is the same in all permutations of the nodes. In estimating complexity, it is reasonable to consider that the complexity of a graph corresponds to the lowest K_m value of all permutations of the adjacency matrix, as the shortest program generating the simplest adjacency matrix

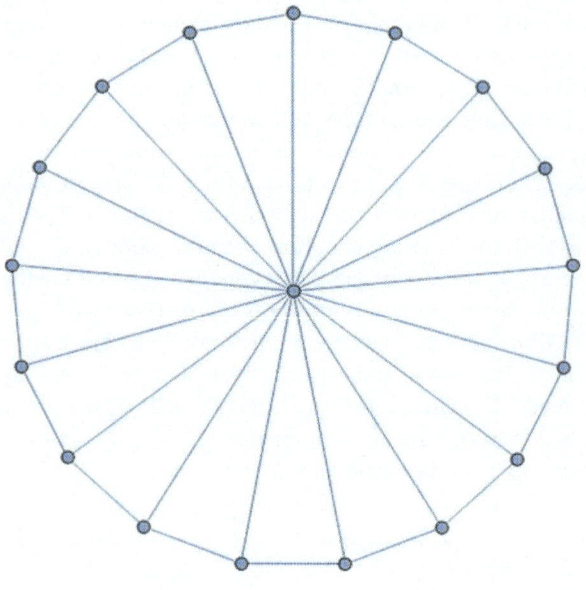

Fig. 8.2 Wheel-18 graph

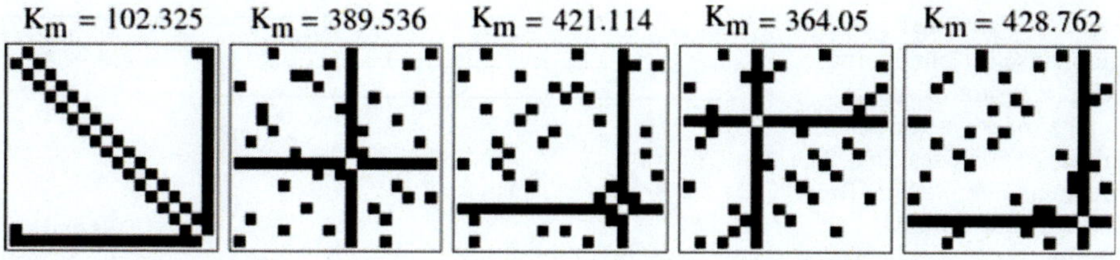

Fig. 8.3 Wheel-18 graph with the nodes in five different orders. Despite each adjacency matrix corresponding to the same topology (i.e. the Wheel-18 graphs), the different node orders affect the estimated complexity

is the shortest program generating the graph. Hence when estimating K_m we find a large number of programs producing the array from which a minimum complexity value is calculated by means of the frequency of the programs producing the array.

8.3.2 Complexity and Edge Density

There has been a protracted discussion in the literature as to whether the complexity of a structure increases with its connectivity, beginning with a disconnected graph with no edges, or whether instead it reaches a maximum before returning to zero for complete graphs [11]. In [4], for example, Gell-Mann asks about the algorithmic complexity (description length) of small graphs with eight vertices each and an increasing number of edges $E(g) = 0$ to $E(g) = V(g)(V(g) - 1)/2$ (complete graph). Gell-Mann reasonably argues that these two extreme cases should have roughly equal complexity. Graphs with 0.5 edge density should fall in between and the other cases are more difficult to tell apart. Here we provide an answer to the general question of the relation between description complexity and edge count, an answer which is in agreement with Gell-Mann while being at odds with several other tailor made measures of complexity [11, 12, 25, 26]. We have created a number of random graphs, all with 50 nodes but different numbers of edges, ranging from 1 to $\binom{50}{2} = 1225$, in intervals of $1225/20$ edges. For each interval, we created 20 random graphs and generated 100 random permutations of each (see Fig. 8.4). All the graphs (Fig. 8.4) in the same interval correspond to aligned points. There are always 20 aligned points, but in most cases they overlap. Points represent the minimum (Top) and standard deviations (Bottom) of the NBDM complexity of the 100 permutations of each group.

In fact, that the most complex binary strings (hence binary adjacency matrices) will have roughly equal zeros and ones is a standard result from Kolmogorov complexity theory. This can be shown straightforwardly by the following relation of K to Shannon entropy: If $x = x_1, \ldots, x_l$ is a bit string, then [14],

$$K(x) \leq l H_0 \left(\frac{1}{l} \sum_{i=1}^{l} x_i \right) + O(\log_2(l)) \tag{8.2}$$

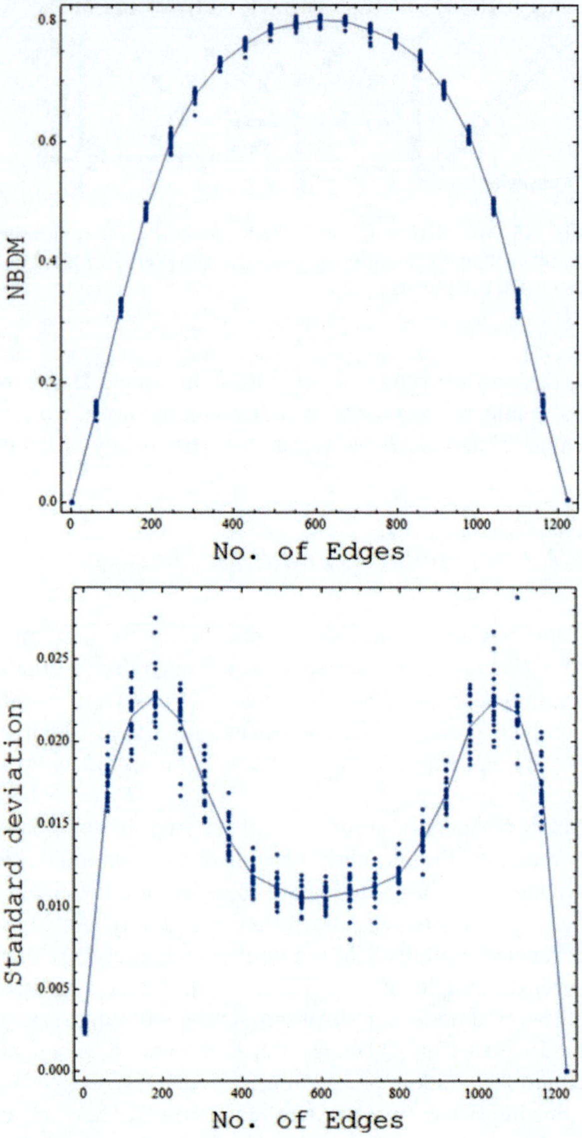

Fig. 8.4 Estimated (normalised) Kolmogorov complexity for increasing number of edges for random graphs of 50 nodes each. The minimum complexity (Top) and standard deviation (Bottom) is shown for 100 random permutations of 20 graphs in each group

where

$$H_0(p) = -p \log_2(p) - (1 - p) \log_2(1 - p) \tag{8.3}$$

is the Shannon entropy (in bits) of the string x. This then implies that if the number of edges diverges from roughly half of the maximum possible, then p must diverge from 0.5, and so the entropy H_0 decreases, implying that x cannot be algorithmically random. Hence, the most complex strings must have roughly equal zeros and ones, or in graph terms, they must have roughly half the number of possible edges. Additionally, (8.2) predicts that Fig. 8.4 (Top) should look roughly like the graph of $H_0(p)$ versus p, which it does: peaking at the centre, with gradual decay to zero at $p = 0, 1$. Also, note the clear symmetry of the plots in Fig. 8.4. This is because the value of K_m is the same for any given graph and its complement. Complexity is minimal for empty or complete graphs (the most homogeneous matrices, all 0 or all 1), and so the standard deviation is also minimal. These observations show that our measure is behaving as expected from theory.

8.3.3 Graph Duality

The dual graph of a planar graph g is a graph that has a vertex corresponding to each face of g, and an edge joining two neighbouring faces for each edge in g. If $g\prime$ is a dual graph of g, then $A(g\prime) = A(g)$, making the calculation of the Kolmogorov complexity of graphs and their dual graphs interesting because of the correlation between Kolmogorov complexity and $A(g\prime)$, which should be the same for $A(g)$. One should also expect the estimated complexity values of graphs to be the same as those of their dual graphs, because the description length of the dual graph generating program is $O(1)$.

Note that unlike the t-statistic, the value of the D statistic (and hence the P value) is not affected by scale changes like using log. The KS-test is a robust test that cares only about the relative distribution of the data.

We numerically approximated the Kolmogorov complexity of dual graphs using both lossless compression and the BDM applied to the adjacency matrices of 113 regular graphs with non-identical dual graphs found in *Mathematica*'s built-in repository GraphData[]. The values (Fig. 8.5) were normalised by a multiple of the size of the adjacency matrices $c|Adj(A)|$. Graphs (g) and their dual graphs ($g\prime$) were found to have estimated Kolmogorov complexity values that are close to each other. The Spearman coefficient r between $K_m(g)$ and $K_m(g\prime)$ NBDM estimations amounts to $r = 0.96$. Values approximated by lossless compression were calculated with Deflate (a standard compression algorithm) implemented in *Mathematica*'s Compress[] function. Notice that the BDM (and therefore the NBDM) accumulates errors for larger graphs if the complementary algorithm (CTM) is not run for a greater number of Turing machines. A more accurate analysis of divergence rates and accumulated errors should be investigated. Robustness of algorithmic probability approximations for different (Turing-universal) computational formalisms (e.g. deterministic cellular automata, Post tag systems, etc.) was investigated in [23], where frequency distributions were calculated and their ranking order correlation also quantified with the Spearman coefficient.

These results show that Kolmogorov complexity applied to adjacency matrices of graphs as a measure of complexity behaves as expected, agreeing both with theory and intuition. Having verified that our measure behaves as expected, we can now apply our measure to analysing automorphism in graphs.

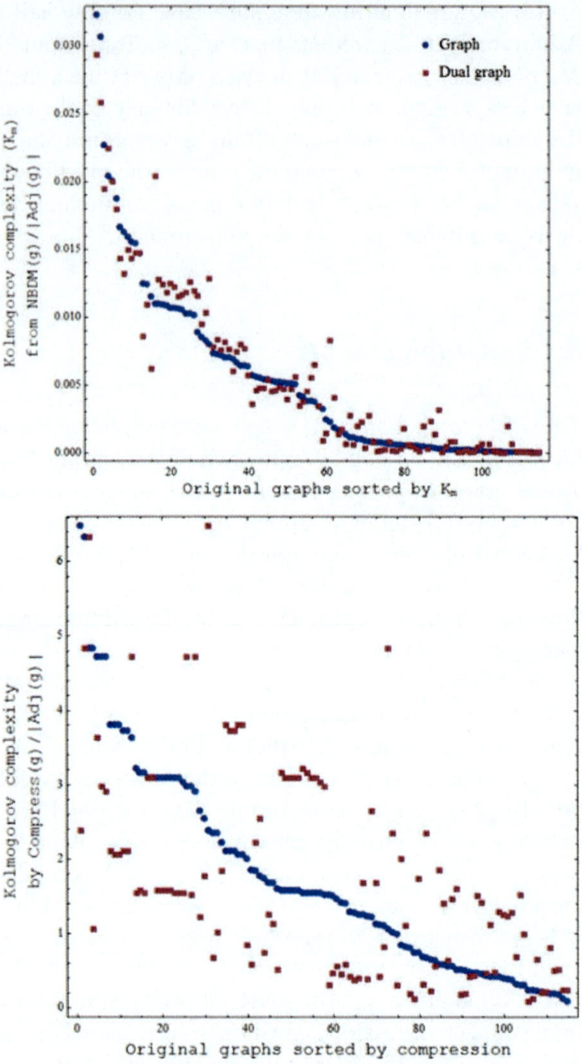

Fig. 8.5 Log scatter-plots of graphs ranked by Kolmogorov complexity approximated by two different methods. Top: Dual graphs not ranked by the BDM method closely follow the distribution of their respective duals as one would expect from a complexity measure given that an $O(1)$ program can build a graph from its dual and vice-versa. Bottom: Lossless compression in agreement with BDM and the expected complexity of dual graphs

8.4 Testing BDM and Boundary Condition Strategies

A test for both CTM and BDM can be carried out using objects that have different representations or may look very different but are in fact algorithmically related. First we will prove some theorems relating to the algorithmic complexity of dual and cospectral graphs and then we will perform numerical experiments to see if CTM and BDM perform as theoretically expected.

A dual graph of a planar graph G is a graph that has a vertex corresponding to each face of G, and an edge joining two neighbouring faces for each edge in G. If G' is

Table 8.1 Spearman ρ values of various BDM versions tested on dual and cospectral graphs that theoretically have the same algorithmic complexity up to a (small) constant

	Non-overlapping BDM	Fully overlapping recursive BDM	Smooth fully overlapping BDM	Smooth add row or column BDM
Duality test	0.874	0.783	0.935	0.931
Cospectrality test	0.943	0.933	0.9305	0.931

a dual graph of G, then $A(G') = A(G)$, making the calculation of the Kolmogorov complexity of graphs and their dual graphs interesting–because of the correlation between Kolmogorov complexity and $A(G')$, which should be the same for $A(G)$. One should also expect the estimated complexity values of graphs to be the same as those of their dual graphs, because the description length of the dual graph generating program is $O(1)$.

Cospectral graphs, also called *isospectral* graphs, are graphs that share the same graph spectrum. The set of graph eigenvalues of the adjacency matrix is called the spectrum $Spec(G)$ of the graph G. This cospectrality test for complexity estimations is interesting because two non-isomorphic graphs can share the same spectrum.

Compression lengths and BDM values in Table 8.1 and Fig. 8.6 are obtained from the adjacency matrices of 113 dual graphs and 193 cospectral graphs from *Mathematica*'s `GraphData[]` repository. Graphs and their dual graphs were found by BDM to have estimated algorithmic complexities close to each other. While Entropy and Entropy rate do not perform well in any test compared to the other measures, compression retrieves similar values for cospectral graphs as compared to BDM, but it is outperformed by BDM on the duality test (Fig. 8.7). The best BDM version for duals was different from that for cospectrals. For the duality test, the smooth, fully overlapping version of BDM outperforms all others, but for cospectrality, overlapping recursive BDM outperforms all others.

Let us address the task of quantifying how many strings with maximum entropy rate are actually algorithmically compressible, i.e., have low algorithmic complexity. That is, how many strings are actually algorithmically (as opposed to simply statistically) compressible but are not compressed by lossless compression algorithms, which are statistical (entropy rate) estimators. We know that most strings have both maximal entropy (most strings look equally statistically disordered, a fact that constitutes the foundation of thermodynamics) and maximal algorithmic complexity (according to a pigeonhole argument, most binary strings cannot be matched to shorter computer programs as these are also binary strings). But the gap between those with maximal entropy and low algorithmic randomness diverges and is infinite at the limit (for an unbounded string sequence). That is, there is an infinite number of sequences that have maximal entropy but low algorithmic complexity.

The promise of BDM is that, unlike compression, it does identify some cases of strings with maximum entropy that actually have low algorithmic complexity. Figure 8.8 shows that indeed BDM assigns lower complexity to more strings than

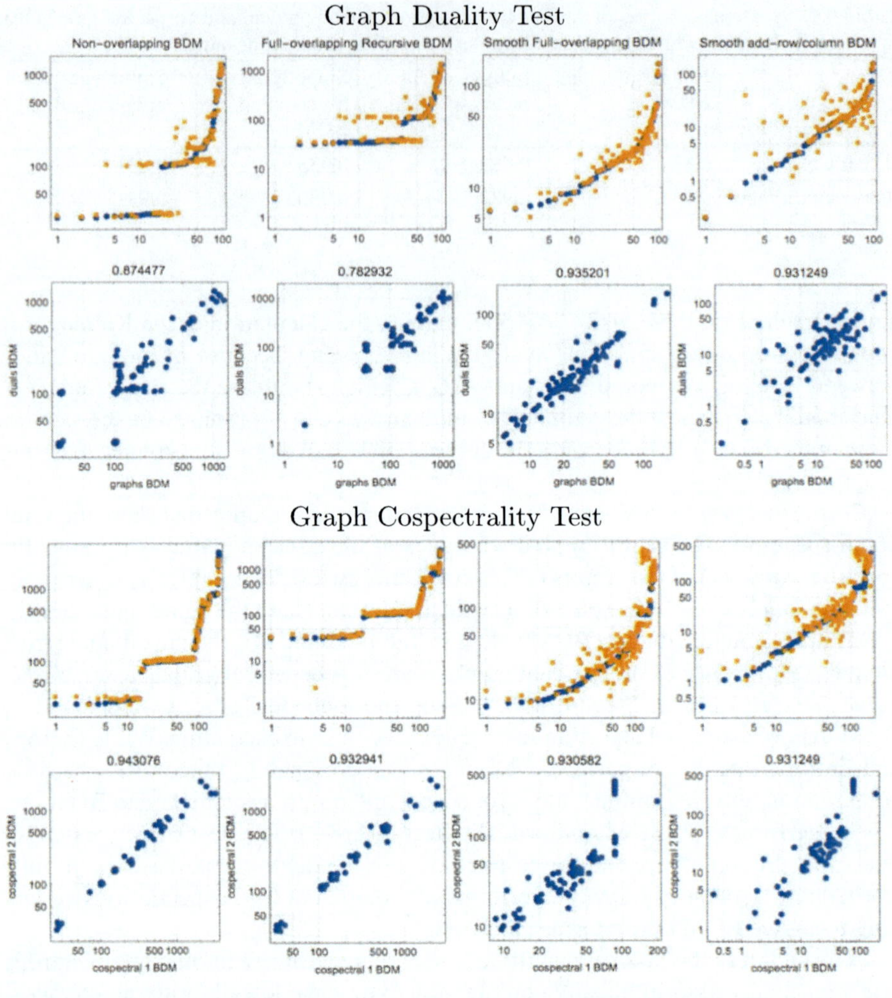

Fig. 8.6 Scatterplots comparing the various BDM versions tested on dual and cospectral graphs that theoretically have the same algorithmic complexity up to a (small) constant. x-axis values for each top row plot are sorted by BDM for one of the dual and for the cospectral graph series. Bottom rows: on top of each corresponding scatterplot are the Spearman ρ values

entropy, as expected. Unlike entropy, and implementations of lossless compression algorithms, BDM recognizes some strings that have no statistical regularities but have algorithmic content that makes them algorithmically compressible.

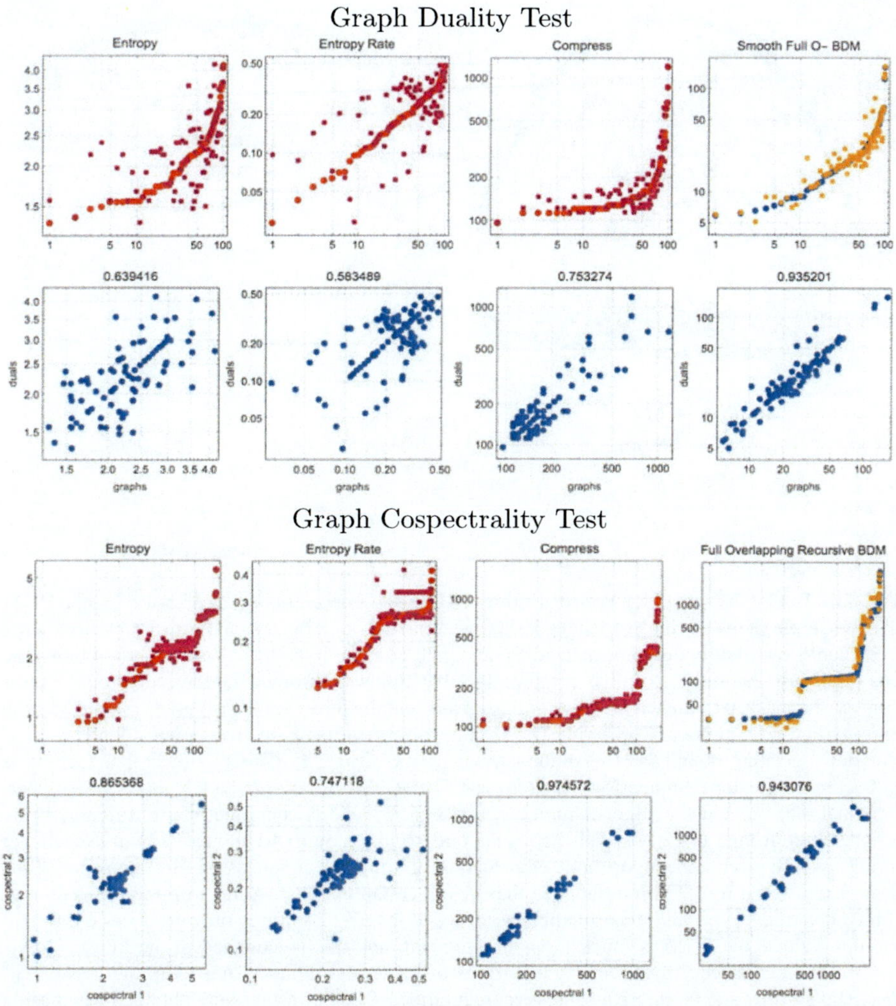

Fig. 8.7 Scatterplots comparing other measures against the best BDM performance. x-axis values for each top row plot are sorted by BDM for one of the dual and for the cospectral graph series. Bottom rows: on top of each corresponding scatterplot are the Spearman ρ values

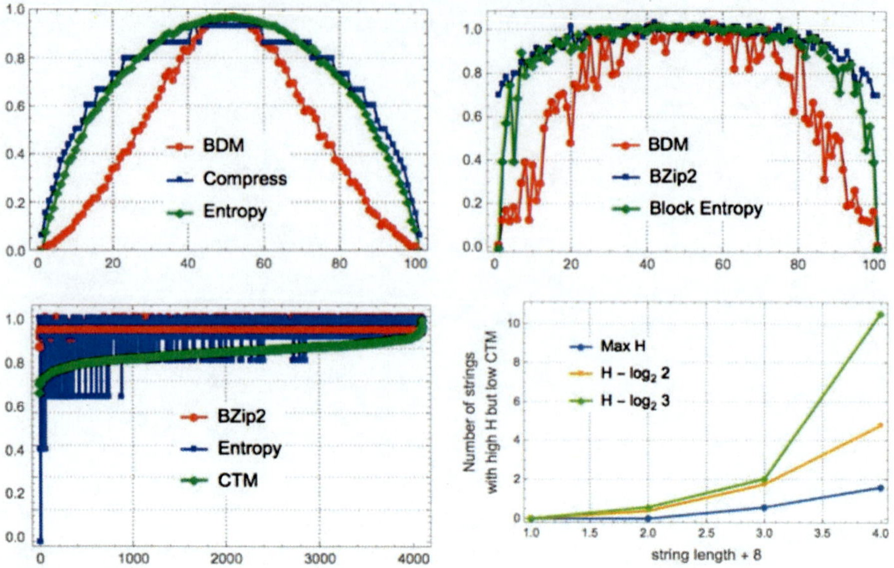

Fig. 8.8 Top left: Comparison between values of Entropy, compression (`Compress[]` and BDM over a sample of 100 strings of length 10 000 generated from a binary random variable following a Bernoulli distribution and normalized by maximal complexity values. Entropy just follows the Bernoulli distribution and, unlike compression that follows entropy, BDM values produce clear convex-shaped gaps on each side assigning lower complexity to some strings compared to both entropy and compression. Top right: The results confirmed using another popular lossless compression algorithm BZip2 on 100 random strings of 100 bits each (BZip2 is slower than Compress but achieves greater compression). Bottom left: The $CTM_{low}(s) - H_{high}(s)$ gap between near-maximal entropy and low algorithmic complexity grows and is consistent along different string lengths, here from 8 to 12 bits. This gap is the one exploited by BDM and carried out over longer strings, which gives it the algorithmic edge against entropy and compression. Bottom right: When strings are sorted by CTM, one notices that BZip2 collapses most strings to minimal compressibility. Over all $2^{12} = 4096$ possible binary strings of length 12, entropy only produces 6 different entropy values, but CTM is much more fine-grained, and this is extended to the longer strings by BDM, which succeeds in identifying strings of lower algorithmic complexity that have near-maximal entropy and therefore no statistical regularities. Similar results were obtained when using a third lossless compression algorithm, LZMA

8.5 Graph Automorphisms and Kolmogorov Complexity

Intuition tells us that a graph with more symmetries can be more compactly described and should therefore have a lower Kolmogorov complexity. For example, if an object is symmetrical a single symbol can be used for each repetition, in addition to an additive constant of information describing the type of transformation (e.g. rotation, reversion, translation). Specifically, any collection of graph nodes within the same orbit of a group transformation will require only the bonds of one node to be specified, with the other nodes then using the same information to specify their bonds. Hence, the number of bits required to specify the graph would be signifcantly low-

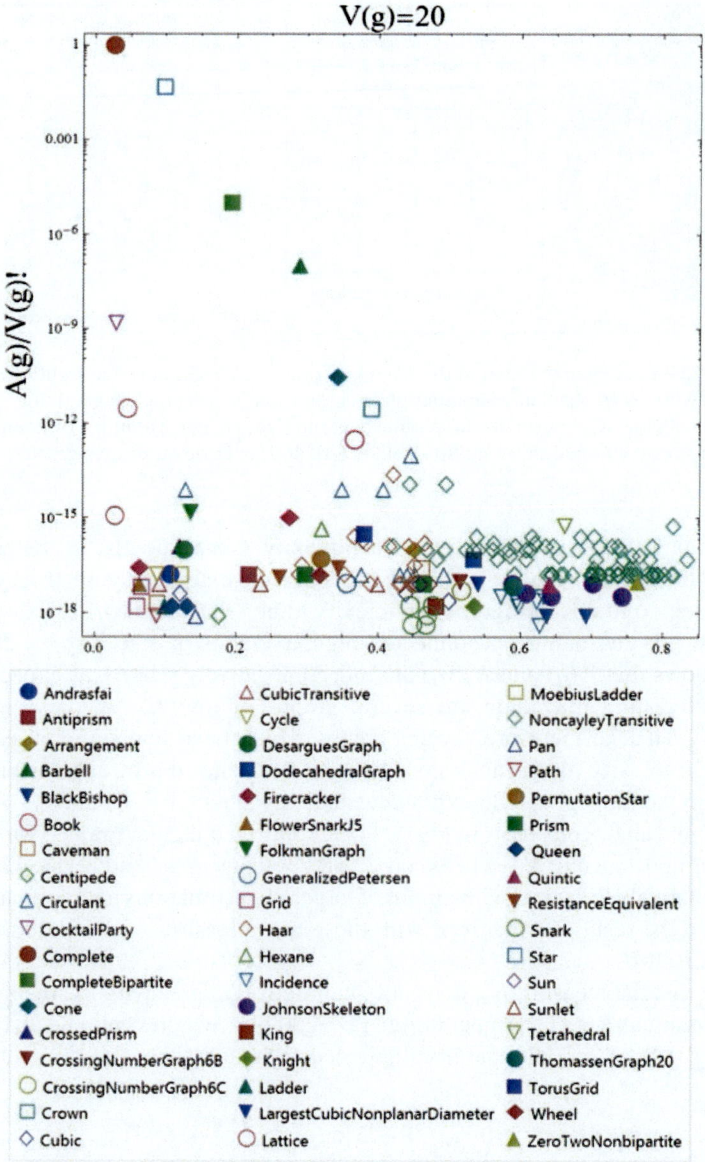

Fig. 8.9 Plot of graph automorphism group size $A(g)$ (y-axis) of all connected regular graphs of size $V(g) = 20$ available in *Mathematica*'s GraphData[] versus Kolmogorov complexity (x-axis) estimated by BDM. As theoretically expected, the larger automorphism group size $A(g)$ the smaller Kolmogorov complexity estimations

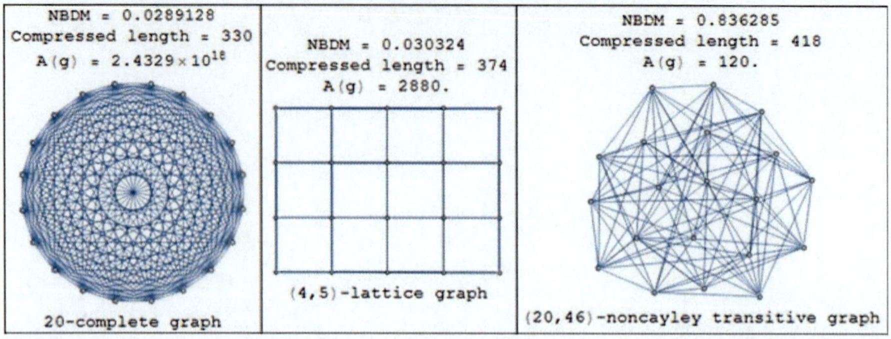

Fig. 8.10 The three graphs found in the boundaries of Fig. 8.9. From left to right: The graph at the top left, with low K_m and large automorphism group size is the complete graph for $V(g) = 20$. Bottom left with low K_m but small automorphism group size is the (4, 5)-lattice. Bottom right, with high K_m and small automorphism group size: the (20, 46)-noncayley transitive graph

ered, reducing the graph Kolmogorov complexity. Consequently, as the size of its automorphism group $A(g)$ of a graph g measures the extent of symmetries in g, one would expect to find Kolmogorov complexity to be related to $A(g)$.

We test this reasoning on connected regular graphs of size $V(g) = 20$ nodes; Fig. 8.9 shows that $A(g)$ and $K_m(g)$ are indeed negatively related, as expected.

It is interesting that there are several graphs of low K_m and also low $A(g)$ (Fig. 8.10). That is, there are several graphs which have few symmetries but yet also low complexity (this is analogous to Fig. 8.1). Hence our measure is picking up structure in the graphs which are not detectable by symmetry search.

We plotted analogous plots to Fig. 8.9 (i.e. connected regular graphs) for different number of nodes, using $V(g)$ between 22 and 36 (Figs. 8.17 and 8.18). The results were qualitatively the same, with graphs of larger K_m estimations having smaller $A(g)$ values, and the results also agreed with those using lossless compression (Deflate) instead of BDM.

Notice that the measure K_m does not quantitatively agree with the theoretical K, as the theoretical K has an upper bound of ~200 bits, which would be arrived at by specifying $Adj(g)$ literally and in full, i.e. using the fact that

$$K(g) \leq \binom{V(g)}{2} + 2\log_2 \binom{V(g)}{2} + O(1) \tag{8.4}$$

On the other hand, BDM retrieves values up to ~800. Nonetheless, BDM values (just like compression results) are consistent upper bounds. BDM can provide better approximations but it requires the calculation of a larger sample of random programs [29] with square matrices of larger size d, compared to the current $d = 4$ that the experiments here introduced used. But in order for BDM to *scale up* and provide better K_m approximations to the theoretical (and ultimately uncomputable) K, the full method to consider is both Coding Theorem Method (CTM) + BDM. CTM is, however, computationally very expensive, while BDM is computationally very cheap, so there is a trade-off in the application of the algorithm. BDM alone is limited by the data generated by CTM, and the range of application of BDM will be limited in this respect, specially for increasingly larger objects (where lossless compression can take over and behave better than for small objects, hence CTM + BDM and lossless compression are complementary tools). The chief advantage of the CTM + BDM approach, is that CTM needs to run only once and BDM can be then efficiently applied and used many times on all sorts and types of data.

8.6 Applying BDM to Real-World Natural and Social Networks

We now move on to applying NBDM to real world networks of different sizes. The 88 real-world networks range from metabolic to social networks and have between 200 and 1000 nodes, and were extracted from the function "[`ExampleData NetworkGraph`]" in Wolfram *Mathematica* v.9. A subset of 20 of the specific real-world network examples used in this experiment are in Table 8.2. Figure 8.11 shows complexity plotted against automorphism group size for these networks. We find that the same qualitative relationship between K and $A(g)$ as reported in Fig. 8.9 is obtained for these larger real-world networks. This provides further evidence that approximations of Kolmogorov complexity identify group-theoretic properties (as well as topological properties, as will be shown in Sect. 8.7) of graphs and networks. The automorphism group sizes $A(g)$ were calculated using the software Saucy 3.0 (http://vlsicad.eecs.umich.edu/BK/SAUCY/, accessed in April 2019), a scalable symmetry-finding tool [27]. Saucy only deals with undirected graphs hence only directed versions of the real-world sample of networks was used to calculate the automorphism group size $A(g)$ of each with Saucy. We made directed graphs into undirected graphs by setting an edge between two nodes i and j if in the original directed graph there was a directed edge from i to j, j to i, or both. Clearly undirected graphs are simpler than directed graphs, in general. In the case of labelled nodes, it is easy to see that the Kolmogorov complexity of a random undirected graph is typically half of a directed labelled graph, as the binary adjacency matrix for undirected graphs is symmetric. For unlabelled graphs, it is not so straightforward, due to complications of isomorphism (recall that there are typically many adjacency graphs representing a given unlabelled graph). Nonetheless directed graphs typically require more bits to specify their links, and are hence more complex. Exploring directed graphs will be left for future work.

Table 8.2 Random sample of 20 real-world networks [5, 8] from the 88 included in the study (and plotted in Fig. 8.11), sorted from smallest to largest estimated Kolmogorov complexity values (NBDM). While the (negative) correlation between K_m and $V(g)$ is almost perfect (Pearson coefficient −0.95) the (negative) correlation between K_m and $A(g)$ is significant (Pearson coefficient −0.178) after normalisation by $V(g)$ for the 88 elements

Network description (g)	$V(g)$	Normalised $K_m(g)$ (BDM)	$A(g)/V(g)$
Metabolic Network of Actinobacillus Actinomycetemcomitans	993	0.00336	4.42×10^{74}
Metabolic Network Neisseria Meningitidis	981	0.00344	2.86×10^{76}
Perl Module Authors Network	840	0.00350	4.63×10^{470}
Metabolic Network Campylobacter Jejuni	946	0.00370	6.97×10^{74}
Metabolic Network Emericella Nidulans	916	0.00378	3.43×10^{68}
Pyrococcus Horikoshii Network	953	0.00382	4.22×10^{70}
Pyrococcus Furiosus Network	931	0.00384	3.37×10^{68}
Metabolic Network Thermotoga Maritima	830	0.00477	2.05×10^{64}
Mycoplasma Genitalium Network	878	0.00480	6.75×10^{92}
Treponema Pallidum Network	899	0.00499	2.71×10^{84}
Chlamydia Trachomatis Network	822	0.00511	1.73×10^{75}
Metabolic Network Pyrococcus Furiosus	751	0.00511	2.86×10^{50}
Rickettsia Prowazekii Network	817	0.00523	1.39×10^{76}
Arabidopsis Thaliana Network	768	0.00535	1.93×10^{60}
Oryza Sativa Network	744	0.00569	3.45×10^{57}
Chlamydia Pneumoniae Network	744	0.00635	2.00×10^{70}
Metabolic Network Oryza Sativa	665	0.00640	9.49×10^{47}
Metabolic Network Rickettsia Prowazekii	456	0.01080	1.10×10^{34}
Metabolic Network Mycoplasma Pneumoniae	411	0.01280	1.85×10^{28}
Metabolic Network Borrelia Burgdorferi	409	0.01460	2.10×10^{36}

8.7 The Algorithmic Randomness of Synthetic Complex Networks

An objective and universal measure of complexity should take into account symmetries as a simplifying factor when it comes to description size. Here we explore how K_m can characterize topological properties of complex networks.

The study of complex networks is currently an active area of research [15]. The field has been driven largely by observations that many real-world networks (e.g. internet links or metabolic networks) have properties very different from both regular and random graphs; the latter having been extensively studied in foundational work by Paul Erdös and Alfréd Rényi. Specifically, two topological properties of many

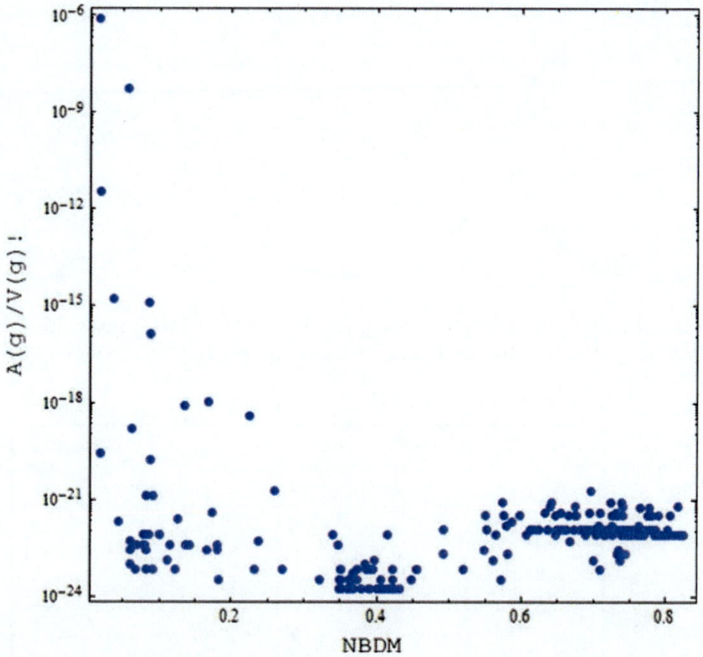

Fig. 8.11 Real-world networks also display the same correlation between Kolmogorov complexity and automorphism group size $A(g)$. Networks with more symmetries have lower estimated Kolmogorov complexity. Automorphisms count is normalised by network size

complex networks that have been a focus of interest are (a) a scale-free (or power law) distribution in node degree distributions, and (b) the "small-world" property where graphs have high clustering and the average graph distance D grows no faster than the log of the number of nodes: $D \sim \log(V(g))$.

Observations of these properties have motivated the development of many models to explain these features. Of these, The Barabási-Albert model [9] reproduces complex network features using a preferential attachment mechanism, and the Watts-Strogatz model [6] also provides a mechanism for constructing small-world networks with a rewiring algorithm (which involves starting from a regular network and randomly rewiring). See Fig. 8.12 for an example.

8.7.1 Network Connectedness and Complexity

We have theoretically substantiated and experimentally demonstrated how network size, both in terms of the number of nodes and the density of edges for a fixed number of nodes, can impact Kolmogorov complexity values (above). However, Fig. 8.13 demonstrates that node and edge count do not exclusively dominate K_m,

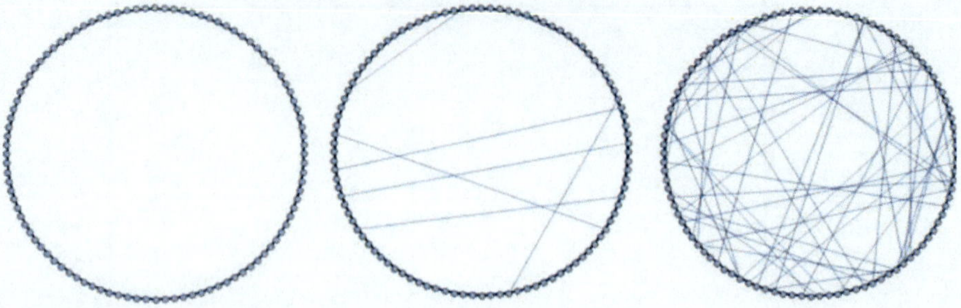

Fig. 8.12 Example of a Watts–Strogatz rewiring algorithm for $n = 100$-vertex graphs and rewiring probability $p = 0, 0.01$ and 0.1 starting from a $2n$-regular graph. The larger p the closer to a random graph (rewiring $p = 1$)

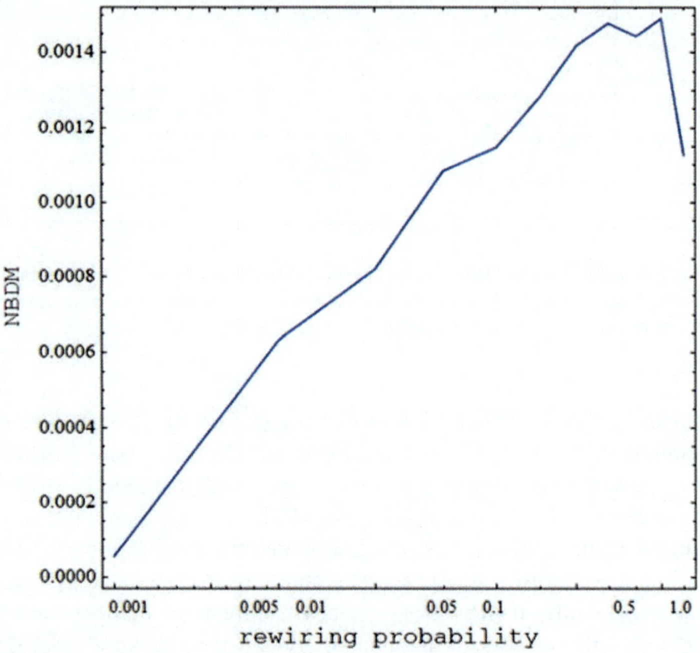

Fig. 8.13 Kolmogorov complexity of the Watts-Strogatz model as a function of the rewiring probability on a 1000-node network starting from a regular ring lattice. Both the number of nodes and the number of links are kept constant, while p varies; Kolmogorov complexity increases with p

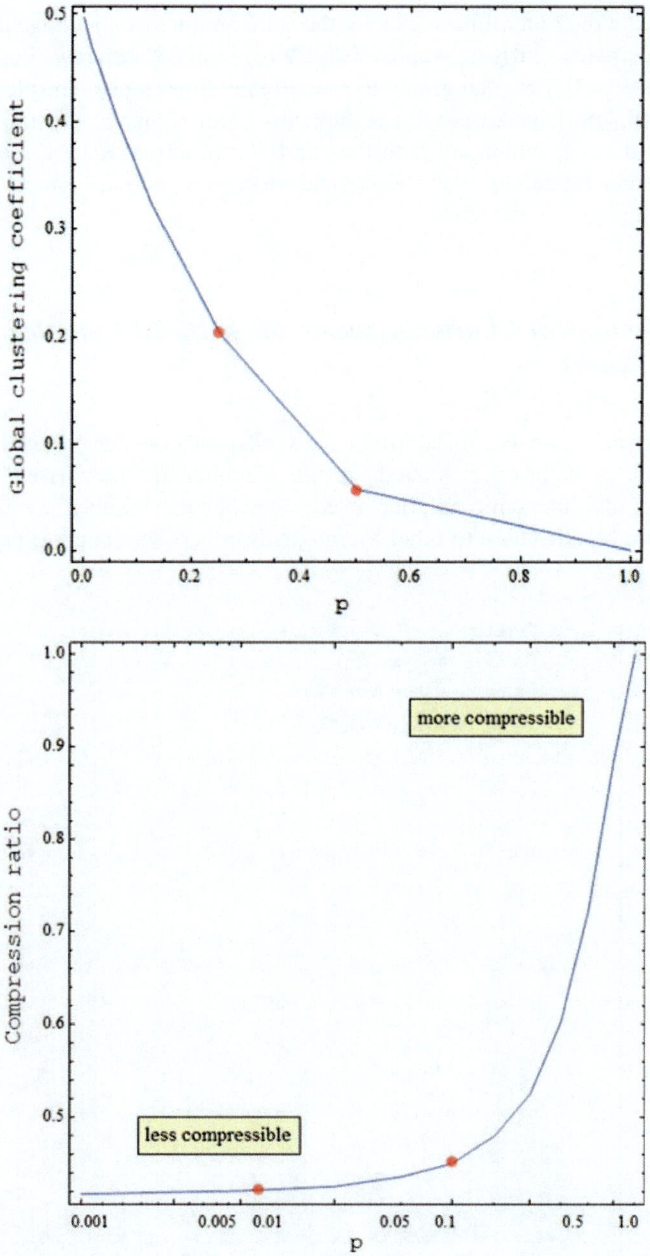

Fig. 8.14 The Watts-Strogatz model starts from a ring lattice, hence highly compressible, then with rewiring probability p (x-axis) the global clustering coefficient drops fast (Top) in this 15×10^3-node W-S network while approaching a random graph ($p = 1$) slowing down at points $p \sim 0.2$ and 0.5 (red dots in plots) where compression ratios (Bottom) also display slight slope variations

as the graphs in the plot all have exactly the same graph size and edge density. The plot considers Watts–Strogatz graphs of size $V(g) = 1000$ with rewiring probability ranging from $p = 0$ (a regular graph) to $p = 1$ (a random graph). The plot shows that the graph complexity increases with p, thus illustrating a change not subject to graph size or edge density, which are both the same for all cases. Rather, the increasing complexity must be due to other topological properties such as connectedness, link distribution and graph diameter.

8.7.2 *Topological Characterization of Artificial Complex Networks*

Random networks and Barabási-Albert networks not only have exactly the same vertex count $V(g)$ in the experiment the results of which are summarised in Figs. 8.15 and 8.16, but also the same number of edges on average. The K_m difference can only therefore be attributed to other topological properties related to each network

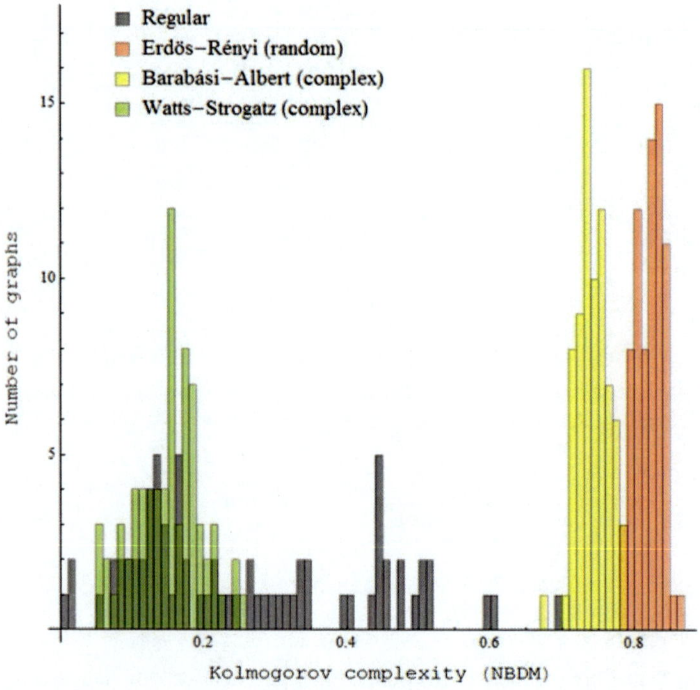

Fig. 8.15 Distribution of 292 regular, Watts-Strogatz, Barabási-Albert and Erdös-Rényi networks with $V(g) = 30$ (73 networks each) with W-S rewiring probability $p = 0.05$. The number 73 comes from the number of regular graphs of size $V(g) = 30$ in the used repository (*Mathematica*'s GraphData[])

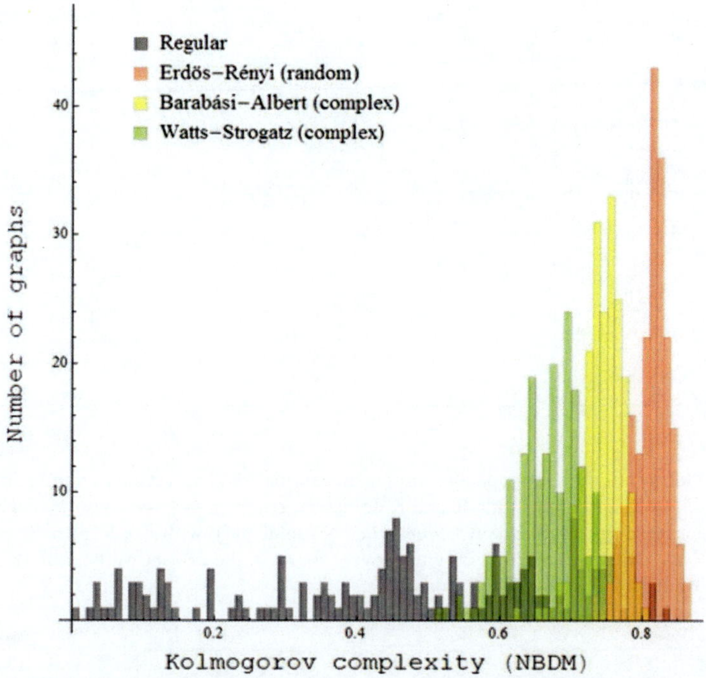

Fig. 8.16 Distribution of 792 regular, Watts-Strogatz, Barabási-Albert and Erdös-Rényi networks with $V(g) = 20$ (198 networks each) with W-S rewiring probability from $p = 0.05$ in Fig. 8.15 to $p = 0.5$ as an experiment introducing randomness to witness the shift of the W-S networks towards higher complexity values of random graphs. For combinatorial reasons, there are more regular networks of size 30 than 20, hence this time the number 198 comes from the number of regular graphs found in the (*Mathematica*'s GraphData[])

model. Observing that K can be affected by topological features such as clustering coefficient (as shown in Fig. 8.14 for Watts-Strogatz networks), we proceeded to examine other network models. As shown in Figs. 8.15 and 8.16, K_m approximated by the BDM assigns low Kolmogorov complexity to regular graphs and Watts-Strogatz networks and higher complexity to Barabási-Albert networks, with random networks displaying the greatest Kolmogorov complexity as expected. Indeed, that random graphs are the most algorithmically complex is clear from a theoretical point of view: nearly all long binary strings are algorithmically random, and so nearly all random unlabelled graphs are algorithmically random [18].

Barabási-Albert networks are often referred to as scale free, because the node distribution follows a power law. However for small graphs such as those we analyse, it is questionable whether the global scaling is meaningful. An advantage of the method is, however, that it still differentiates between different network models despite their small size. The theory of algorithmic information formally characterizes any object in terms of the properties specified in its description from which the object can be fully recovered. It is therefore *vis-à-vis* small objects that the theory presents

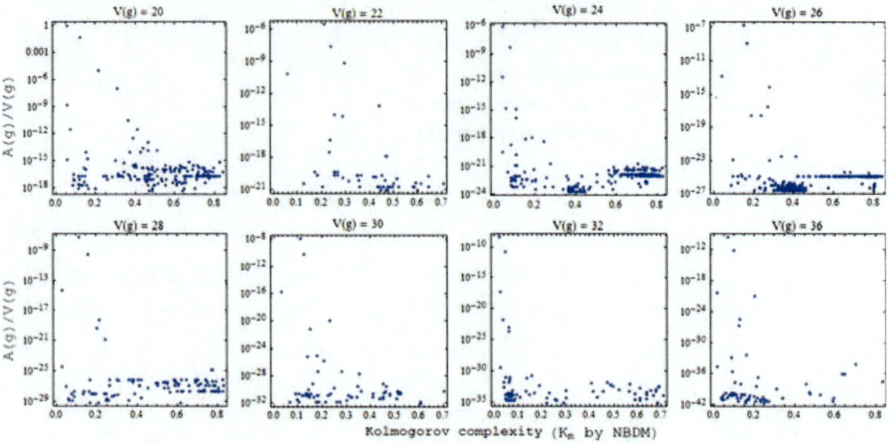

Fig. 8.17 Plots of number of graph automorphisms normalised by maximum number of edges of g, $A(g)/V(g)!$ (y-axis) versus (normalised) Kolmogorov complexity (x-axis) estimated by NBDM for connected regular graphs found in *Mathematica* (GraphData[]) with size $V(g) = 20$ to 36 nodes (only vertex sizes for which at least 20 graphs were found in the dataset were plotted). The decay can be seen, though the relationship is noisy

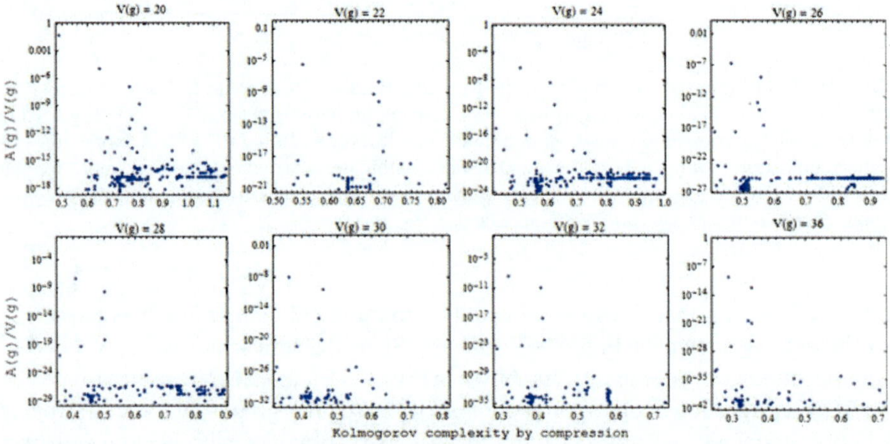

Fig. 8.18 Plots of number of graph automorphisms normalised by maximum number of edges of g, $A(g)/V(g)!$ (y-axis) versus Kolmogorov complexity (x-axis) estimated by lossless compressed length (Deflate) of connected regular graphs in *Mathematica*'s GraphData[] with size $V(g) = 20$ to 36 nodes (only vertex sizes for which at least 20 graphs were found in the dataset were plotted)

its greatest challenges, given that the invariance theorem [18] does not tell us anything about the rate of convergence in values (Fig. 8.17).

We also considered regular graphs (Figs. 8.18 and 8.19) for this comparison including Haars, circulants, noncayley transitives, snarks, cubics, books, lattices and suns among other types of regular networks. The average NBDM complexity value

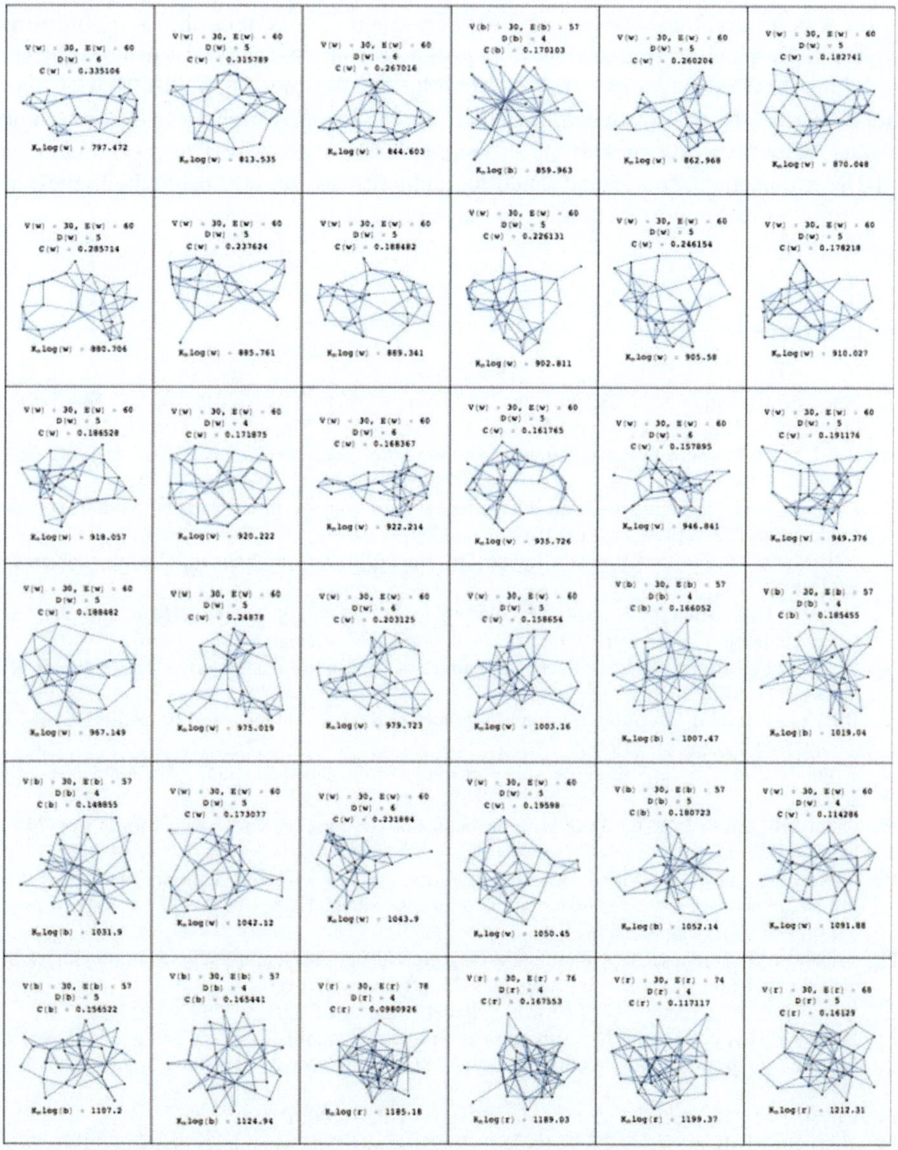

Fig. 8.19 Random (Erdös–Rényi) graphs (denoted by r) versus complex networks (Watts-Strogatz and Barabási-Albert) (denoted by w and b) sorted by Kolmogorov complexity (smallest to largest K_m log) as approximated by the BDM

of the regular graph group remained very low (and the distribution mostly uniform with a slight peak at around 0.5). The group of complex networks peak at different K_m values, with Watts-Strogatz networks ranking low for $p = 0.01$ when the small world effect is produced, and then moves towards high Kolmogorov complexity when rewiring probability p increases. Barabási-Albert networks peak at NBDM value equal to 0.75 and random graphs (Erdös-Rényi) ranked the highest at an estimated Kolmogorov complexity value close to 0.9 (Fig. 8.19).

References

1. Hopcroft, J., Tarjan, R.: Efficient Planarity Testing. J. ACM 21.4 (Oct 1974), 549–568. ISSN: 0004-5411
2. Read, R.C., Corneil, D.G.: The graph isomorphism disease. J. Graph Theory **1.1**, 339–363 (1977)
3. Langton, C.G.: Studying artificial life with cellular automata. In: Physica D: Nonlinear Phenomena, vol. 22.1, pp. 120–149 (1986)
4. Gell-Mann, M.: The Quark and the Jaguar: Adventures in the Simple and the Complex. Macmillan (1995)
5. Johnson, D.S., Trick, M.A.: Cliques, coloring, and satisfiability: second DIMACS implementation challenge, 11–13 Oct. 1993, vol. 26. American Mathematical Soc (1996)
6. Watts, D.J., Strogatz, S.H.: Collective dynamics of 'small-world' networks. Nature **393.6684**, 409–410 (1998)
7. Buhrman, H. et al.: Kolmogorov random graphs and the in-compressibility method. SIAM J Comput **29.2**, 590–599 (1999)
8. Jeong, H. et al.: The large-scale organization of metabolic networks. Nature **407.6804**, 651–654 (2000)
9. Albert, R., Barabsi, A.-L.: Statistical mechanics of complex networks. Rev. Mod. Phys. **74.1**, 47 (2002)
10. Girvan, M., Newman, M.E.J.: Community structure in social and biological networks. In: Proceedings of the National Academy of Sciences, vol. 99.12, pp. 7821–7826 (2002). https://doi.org/10.1073/pnas.122653799
11. Bonchev, D., Buck, G.A.: Quantitative measures of network complexity. In: Complexity in Chemistry, Biology, and Ecology, pp. 191–235. Springer (2005)
12. Standish, R.K.: Complexity of networks. Recent Adv. Artif. Life **3**, 253–263 (2005)
13. Baskerville, K., Paczuski, M.: Subgraph ensembles and motif discovery using an alternative heuristic for graph isomorphism. Phys. Rev. E Stat Nonlin Soft Matter Phys **74.5** Pt 1, 051903 (2006)
14. Cover, T.M., Thomas, J.A.: Information Theory. Wiley (2006)
15. Newman, M., Barabasi, A.-L., Watts, D.J.: The Structure and Dynamics of Networks. Princeton University Press (2006)
16. Baskerville, K., Grassberger, P., Paczuski, M.: Graph animals, subgraph sampling, and motif search in large networks. Phys. Rev. E **76.3**, 36107 (2007)
17. Kim, J., Wilhelm, T.: What is a complex graph? In: Physica A: Statistical Mechanics and its Applications, vol. 387.11, pp. 2637–2652 (2008)
18. Li, M., Vitnyi, P.: An Introduction to Kolmogorov Complexity and Its Applications. Springer (2008)
19. Xiao, Y. et al.: Emergence of symmetry in complex networks. Phys. Rev. E **77.6**, 066108 (2008)
20. Xiao, Y. et al.: Network quotients: structural skeletons of complex systems. Phys. Rev. E **78.4**, 046102 (2008)

21. Mowshowitz, A., Mitsou, V.: Entropy, orbits, and spectra of graphs. In: Analysis of Complex Networks: From Biology to Linguistics (2009)
22. Newman, M.: Networks: An Introduction. Oxford University Press (2010)
23. Zenil, H., Delahaye, J.-P.: On the algorithmic nature of the world. Inf. Comput. **10** (2010). 9789814295482 0017
24. Adami, C. et al.: Information content of colored motifs in complex networks. Artif. Life **17.4**:375–390 (2011)
25. Dehmer, M., Mowshowitz, A.: A history of graph entropy measures. Inform. Sci. **181.1**, 57–78 (2011)
26. Dehmer, M., Sivakumar, L.: Recent developments in quantitative graph theory: information inequalities for networks. PloS One **7.2**, e31395 (2012)
27. Katebi, H., Sakallah, K.A., Markov, I.L.: Conflict anticipation in the search for graph automorphisms. In: Logic for Programming, Artificial Intelligence, and Reasoning. Springer, pp. 243–257 (2012)
28. Mowshowitz, A., Dehmer, M.: Entropy and the complexity of graphs revisited. Entropy **14.3**:559–570 (2012)
29. Soler-Toscano, F. et al.: Calculating kolmogorov complexity from the output frequency distributions of small turning machines. PLoS One **9.5**, 1–18 (2014)

Chapter 9
Algorithmic Complexity in Cognition

In this chapter, we review a series of topics relevant to psychological science in which the Algorithmic Complexity of Short Strings (ACSS), as estimated using the methods described in the first part of this book, proved useful. These topics are remarkably diverse, including fields such as development, working memory, reasoning, aesthetic preferences, visual cognition, randomness perception and production, language evolution [1], and even belief in conspiracy theories [2].

9.1 Working Memory and Intelligence

Although there is no doubt that working memory (WM)—a system including short-term memory (STM) and cognitive resources to operate on the items stored in short-term memory—is closely related to fluid intelligence [3], there is no definite answer as to why manipulation and retention of information are responsible for variations in individual fluid intelligence (henceforth referred to as Gf). The main account describing this relationship reported so far is based on memory capacity [4]. For instance, the role of capacity increases with items' difficulty of Raven's Progressive Matrices [5]. Although WM is known to play a role in intelligent behaviors, our goal is to go further to shed light on the process of information compression in WM as a determining factor of intelligence. Compression of information, that is, the capacity to recode information into a compact representation may be a key mechanism in accounting for the dual impact of the manipulation and retention of information on

The material in this section was adapted from **Structure emerges faster during cultural transmission in children than in adults**, Authors: Vera Kempe, Nicolas Gauvrit and Douglas Forsyth, Published in *Cognition*, Pages 247–254, Copyright 2015, with permission from Elsevier.

H. Zenil et al., *Methods and Applications of Algorithmic Complexity*, Emergence, Complexity and Computation 44,
https://doi.org/10.1007/978-3-662-64985-5_9

intelligence, particularly when individuals deal with new problems in the typical tests used to measure intelligence (such as Raven's Matrices).

Compression can potentially account for intelligence because many complex mental activities still fit well into rather low WM storage capacity—for similar ideas, see [6, 7] in artificial intelligence, and [8] in behavioral science. Our goal here is to show that the ability to compress information is a good predictor of WM performance and is, thereby, a good candidate for predicting intelligence: more compression simply means more available resources for processing other information, which can potentially have an impact on reasoning. For instance, a better prediction of fluid intelligence after increasing the length of the memoranda was shown in [9]. The authors observed that when list-length reached 5 items, simple spans became as good as complex spans in terms of predicting fluid intelligence. One possible explanation is that long-length lists need to be reorganized by individuals to be stored.

The memory span is about 4 items when the task is complex (i.e., rapid or dual [10]), whereas the span revolves around 7 items when the span task is simple (e.g., when the participants is only required to repeat back a sequence of simple items such as letters or digits, [11]). Therefore, there is a paradox in the usual STM/WM concepts, because it is difficult to conceive that the STM estimate is the highest if processing is not included in the concept of STM. Effectively, STM is conceptually thought to be primarily involved in storage; however, items can be manipulated and potentially compressed when the span task is simple. This may explain why the span can go as high as 7 items on the digit-span task. Conversely, while WM is primarily involved in both the manipulation and storage of information (conceptually), the capacity to manipulate the memorandum is generally limited in complex span tasks and the measure of interest is usually storage. The present study proposes and develops the idea that the concept of information compression can help eliminate the opposition between the STM and WM concepts and may be useful in accounting for intelligence in a comprehensive way. Effectively, one observation is that compression processes play a role in variations of the span [12] around the average 4 ± 1 WM capacity limit. In this study, the compression process allowed the formation of chunks that simplified participants' memorization. A second observation is that the previous 7 ± 2 estimation of short-term memory (STM) capacity may correspond to an overestimation due to compression of several items into larger chunks [13].

From this standpoint, our hypothesis is that the capability of simple spans (when using long-length lists and particularly when they can be compressed) to predict fluid intelligence could be due to information reorganization. More specifically, we hypothesize that information reorganization and intelligence should both rely on individuals' storage capacity, which hierarchically depends on a compression process that can help optimize the available storage. Storage capacity is supposed to be fixed for a given individual; however, any form of structuration of the stimulus sequence can contribute to extending the number of stored items. This optimization process could be helpful to make sense of the regularities that can be found in the items of the Raven or in the to-be-recalled sequences of repeated colors. The rationale is that humans use storage and compression in conjunction to solve the problems of the Raven and to reorganize (for instance, by chunking) the colors of the to-be-remembered sequences of repeated colors.

To study the ability to compress information, we developed a span task based on the SIMON©, a classic memory game from the 80s that can be played on a device with four colored buttons (red, green, yellow, and blue). The game consists of reproducing a sequence of colors by pressing the corresponding buttons. Our SIMON span task was developed on the reasoning that sequences of colors contain regularities that can be quantified to approximate compression opportunity. To estimate the compressibility our participants could be offered within thousands of different color sequences in the SIMON task, we used the ACSS defined in the first part of this book. To anticipate, the present results showed that this metric can indeed be used to predict WM performance and that intelligence is well predicted by the ability to compress information. We conclude that the ability to compress information in WM is the reason why both the manipulation and the retention of information are linked to intelligence.

9.1.1 Participants

One hundred and eighty-three students enrolled at the University of Franche-Comté in Besancon, France ($M_{age} = 21$; $SD = 2.8$) volunteered to participate in the experiment and received course credit in exchange for their participation.

9.1.2 Task

Each trial began with a fixation cross in the center of the screen (1000 ms). The to-be-memorized sequence, consisting of a series of colored squares appearing one after the other, was then displayed. Next, in the recall phase, four colored buttons were displayed, and participants could click on them to recall the whole sequence they had memorized and then validate their response. After each recall phase, feedback ("perfect" or "not exactly") was displayed according to the accuracy of the response.

Participants were administered either a spatial version or a nonspatial version of the task. In the spatial version ($N = 106$), the stimulus colored squares were displayed briefly lit up one at a time, in different locations on the screen, to show the to-be-remembered sequence, as in the original game. The spatial version was primarily run to follow the original game first; however, we also added a stringent condition (called the nonspatial version) to better control spatial encoding. In the nonspatial version ($N = 77$), the stimulus colors were displayed one after another in the center of the screen to avoid any visuo-spatial encoding strategy. To further discourage spatial strategies in both versions, the colors were randomly assigned to the buttons on the response screen for each trial, which resulted in the colors never being in the same locations across trials. The reason for avoiding spatial encoding was that our metric was developed for detecting regularities in one-dimensional sequences of symbols, not two-dimensional patterns. Given that our preliminary analysis indicated no significant difference between these two conditions, the two data sets were combined in the subsequent analyses.

Each session consisted of a single block of 50 sequences varying in length (from one to ten) and in the number of possible colors (from one to four). New sequences were generated for each participant, with random colors and orders, so as to avoid the presentation of the items in ascending length (two items, then three, then four, etc.). We chose to generate random sequences and measure their complexity a posteriori (as described below). A total of 9150 sequences (183 subjects $\times$ 50 sequences) were presented (average length = 6.24), each session lasted for 25 min on average.

Raven's Advanced Progressive Matrices (Gf). After a practice session using the 12 matrices in Set 1, the participants were tested on the matrices in Set 2 (36 matrices in all), during a timed session averaging 40 min. They were instructed to select the correct missing cell of a matrix from a set of eight choices. Correct completion of a matrix was scored one point; hence, the range of possible raw scores was 0–36.

9.1.3 Results

The effects of sequence length and the number of colors per sequence are indicative of memory capacity in terms of the number of items recalled, and provide an initial idea as to whether the repetition of colors generated interference during the recall process (which would predict lower performance) or permitted recoding of the sequences (which would predict better performance). First, we conducted a repeated-measures analysis of variance (ANOVA) with sequence length as a repeated measure, and the mean proportion of perfectly recalled sequences for each participant (i.e., proportion of trials in which all items in a sequence were correctly recalled6) as the dependent variable. Performance varied across lengths, $(9, 177) = 234.2$, $p < 0.001$, decreasing regularly as a function of length.

To confirm the reliability of our compressibility measure (inversely related to algorithmic complexity), we split each participant's sequences into two groups of equal complexity. This split-half method simulated a situation in which the participants were taking two equivalent tests. We obtained adequate evidence of reliability between the two groups of sequences ($r = 0.63$, $p < 0.001$; Spearman–Brown coefficient = 0.77).

We computed the overall link between compressibility and the correctness of responses. The Point-Biserial correlation between algorithmic complexity and correct recall of a series was −0.63, a number that is comparable to the correlation between the length of a series and its correct recall, −0.62. These correlations largely overtake the correlation between correct response and other index of complexity such as entropy (−0.43),the number of colors, (−0.50). This is a further justification of algorithmic complexity in our context.

A simple initial analysis targeting both the effect of the number of colors and the effect of complexity on accuracy (i.e., the sequence is perfectly recalled) showed that complexity ($\beta = -0.62$) took precedence over the number of colors ($\beta = -0.04$) in a multiple linear regression analysis when all 9150 trials were included. To further investigate the combined effects of complexity and list-length on recall in greater detail, we used a logistic regression approach to predict correct performance.

One interesting result of our complexity measure was revealed in the comparison between complexity of the stimulus and response sequences. When the participants failed to recall the right sequence, they tended to produce a simpler string. The mean complexity of the stimulus sequence was 25.3; however, the mean complexity of the responses was only 23.3 across all trials ($t(9149) = 26.24$, $p < 0.001$, Cohen's paired $d = 0.27$).

Following the regression on the data shown in Fig. 9.1, we computed a logistic regression for each subject to find the critical decrease in performance that occurred half-way down the logistic curve (i.e., the inflection point). The reason for this was that the participants were not administered the same sequences. This inflection point is thus based on complexity and simply indicates that the participants failed on sequences more than 50% of the time when the complexity level was above the inflection point. To estimate the relationship between each participant's inflection

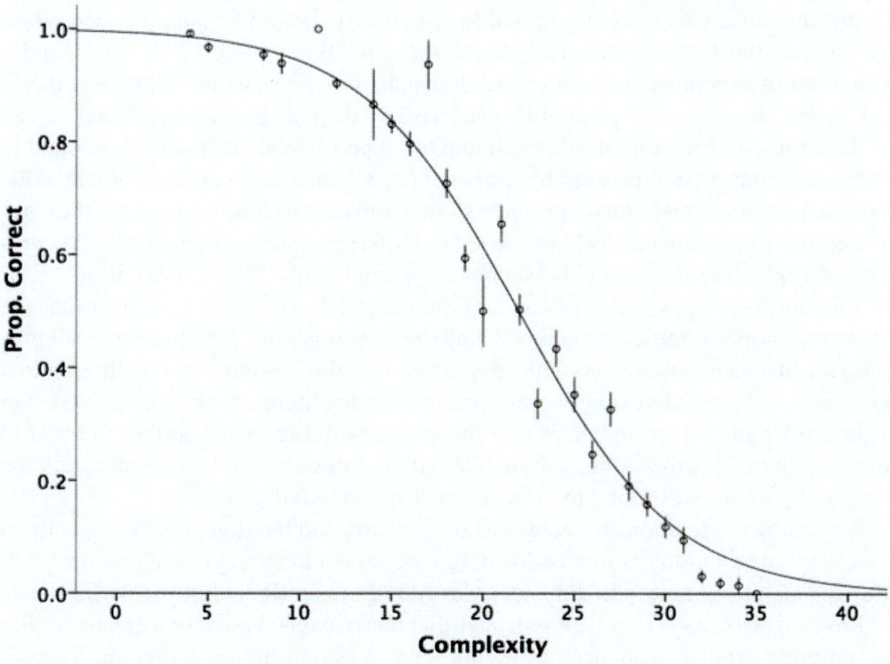

Fig. 9.1 Proportion of perfectly recalled sequences of colors as a function of complexity. *Note* Error bars show ± one standard error

point and his/her IQ, the data were input into confirmatory factor analysis (CFA) using IBM SPSS AMOS 21. A latent variable representing a construct in which storage and processing were separated during the task, and another latent variable representing a construct in which the two processes functioned together, were sufficient to describe performance [14]. The results suggest that Raven's matrices are better predicted by the construct in which storage and processing are combined ($r = 0.64$, corresponding to 41% of the shared variance, instead of $r = 0.36$ when separated). The combined construct can be assessed in the present study by our SIMON span task, a memory updating task, and a simple span task.

9.1.4 Discussion

Many complex mental activities still fit our rather low memory capacity. One observation is that challenging concepts (those of experts or those transmitted by culture) have been developed in the simplest possible way. Pedagogy is probably a quest for simplicity (for instance, the division algorithms first taught in universities are now taught in elementary school), and language could be under pressure to be compressible too (see Sect. 9.7). Short term memory capacity has been constant for a hundred years, and quite low, and maybe more elaborated reasoning can only occur when the underlying concepts are com- pressed to fit capacity. However, not all concepts are pre-compressed by previous generations, and sometimes intelligence may require compression to achieve lower memory demands to solve new problems. We, therefore, tested a new concept of intelligence based on the idea that optimal behavior can be linked to compression of information. Our experimental setup was developed to study this compression process by allowing participants to mentally reorganize the to-be-remembered material to optimize storage in WM. To follow up on our previous conceptualization wherein WM is viewed as producing, on the fly, a maximally compressed representation of regular sequences using chunks [12, 13], we used ACSS to estimate the compressibility factor. Taken together, our results suggest that the existence of opportunities for compressing the memorandum enhance the recall process (for instance, when fewer colors are used to build a sequence, resulting in low complexity). More interestingly, we found that having more structure interacted with sequence length, which indicates that the compressibility factor applies the best to long sequences. Furthermore, although length had a detrimental effect on recall, the complexity effect was found to be stronger than the length effect.

Regarding the relationship between compression and intelligence, the capacity to compress information (estimated by an individual's inflection point along the complexity axis of memory performance) was found to correlate with the performance on Raven's matrices. This result indicates that participants who have a greater ability to compress regular sequences of colors tend to obtain higher scores on Raven's matrices. This correlation is comparable to the one obtained from a composite measure of WM capacity. However, an exploratory analysis showed that our span task saturated two principal factors in a way very similar to that found for other tasks where processing is also devoted to storage (involving updating or free recoding of spatial information in the absence of a concurrent task), unlike the complex span tasks generally used to estimate Gf (remember that in complex span tasks, processing of the concurrent task is separate from storage in the main task, so it cannot support storage). A confirmatory factor analysis indicated greater predictability for the Raven test with a latent variable in which the storage and processing components were combined, in opposition to a latent variable representing complex span tasks in which storage and processing were separated. This latent variable shared 41% of the variance with Raven's matrices, which seems quite good.

These findings are largely consistent with previous studies suggesting that prediction of Gf by simple spans can reach that of complex spans by increasing list-lengths. Increasing digit list-lengths, for instance, might require the participant to further process the sequence to optimize capacity. If the digit storage capacity is 4 independent digits, the only way to increase capacity is to recode or group together a few items. Furthermore, the participant's linguistic experience with digits can presumably account for some additional form of compression due to a greater linguistic experience, and this observation plausibly supports our conclusion since digit span has strong links to intelligence.

The present study allows us to conclude that processing and storage should be examined together whenever processing is fully devoted to the stored items, and we believe that storage and processing must function together whenever there is information to be compressed in the memorandum. Thus, the ability to compress information in span tasks seems to be a good candidate for accounting for intelligence, along with other WM tasks (i.e., simple updating and short-term memory-span tasks) in which storage and processing also function together.

9.2 Randomness

The production of randomness by humans requires high-level cognitive abilities such as sustained attention and inhibition, and is impaired by poor working memory. Unlike other neuropsychological tests, random generation tasks possess specific features of interest: their demand on executive functions, especially inhibition processes, is high; and more importantly, training does not reduce this demand through automatization [15]. On the contrary, generating a random-like sequence requires continuous avoidance of any routine, thus preempting any automatized success.

Random generation tasks have been widely used in the last few decades to assess working memory, especially (sustained) inhibitive abilities [16], in normal subjects as well as in patients suffering from a wide variety of pathologies. In normal subjects, random generation varies with personal characteristics or states, such as belief in the paranormal [17] or cultural background [18, 19]. It affords insight into the cognitive effects of aging [20], hemispheric neglect [21], schizophrenia [22], aphasia [23], and Down syndrome [24].

As a rule, random generation tasks involve generating a random-like sequence of digits [25], nouns [20], words [26] or heads-or-tails [27]. Some authors have also offered a choice of more neutral items, such as dots, e.g., in the classical Mittenecker test [28]. Formally, however, these cases all amount to producing sequences of bits, that is 0 or 1 digits, since any object can be coded in this way.

In most studies, the sequence length lies between 5 and 50 items. Measuring the formal "randomness" of a given short sequence (say of length 5–50) is thus a crucial challenge. Apart from any objective and formal definition of randomness, researchers regularly use a variety of indices, none of which is sufficient by itself because of the profound limitations they all exhibit. Recently for instance, Schulter, Mittenecker and Papousek [29] provided software calculating the most widely used of such measures applied to the case of the Mittenecker Pointing Test, together with a comprehensive overview of the usual coefficients of randomness in behavioral and cognitive research. These tools provide a new way to describe how a given sequence differs from a truly random one. However, it is not fully satisfactory: multiple unsatisfactory measures do not result in a satisfactory description.

9.2.1 *The Usual Measures of Randomness*

The most common coefficients used to assess the quality of a pseudo-random production may be classified into three large varieties according to their main focus.

9.2.1.1 Departure from Uniformity

The simplest coefficients—even though they may rely upon sophisticated theories—are based on the mere distribution of outcomes, and are therefore independent of the order of the outcomes. In brief, they amount to the calculation of a distance between the observed distribution and the theoretical flat distribution, just as a chi-square would do.

Information theory [30] is often used as a basis for assessing randomness. Given a finite sequence s of N symbols repeatedly chosen from a set of n elements, the average symbol information, also called entropy, is given by $H(s) = -\sum_i p_i \log_2(p_i)$, where p_i is the relative frequency of an item i in the finite sequence. This entropy is maximal when the relative frequencies are all equal, when it amounts to $H_{\max} = \log_2(n)$.

Symbol redundancy (SR; see [29]) is an example of a coefficient arising from information theory. It is defined to be $SR = 1 - H/H_{\max}$ where H is the entropy. SR is no more than a measure of departure from uniformity. A sequence's SR does not depend on all aspects of the sequence, but only on the relative frequencies of each item comprising it. According to SR, a sequence comprised of 0s and 1s such as 000001 is weakly random as expected, but 010101, 000111 or 100101 are exactly equivalent to each other, since they all minimize SR to 0. This is the most obvious limitation of SR as a global measure of randomness, as well as of any measure relying on the mere distribution of symbols.

9.2.1.2 Normality

Beyond values depending on the distribution of individual symbols, one may consider pairs (dyads) or sequences of three (triads) adjacent outcomes. In a truly (infinite) random sequence, any dyad should appear with the same probability. Any distance between the distribution of dyads or triads (and so on) and uniformity may therefore be thought of as a measure of randomness. This is precisely what context redundancy CR_1 and coefficient of constraint CC_1 assess [29].

One may also consider dyads of outcomes separated by 1, 2 or k elements in the sequence, which is done, for instance, through CC_k and CR_k coefficients, a generalization of CC_1 and CR_1. Here we group methods of this kind under the rubric "normality assessment" for a reason that will soon become clear.

In mathematics, a sequence of digits $d_1, \ldots, d_n, \ldots$, with d_n lying within $[0, b]$—or equivalently, the real number written $0, d_1d_2d_3 \ldots$ in base $(b+1)$—is said to be normal in base b if the asymptotic distribution of any dyad, triad, or of finite sequences of k consecutive digits, is uniform. Any such series also manifests what may seem a stronger property: dyads of outcomes separated by a certain fixed number of other outcomes show, in the long run, a uniform distribution. Therefore, a normal sequence will be considered random by any normality assessment method.

Eventually, there will exist sequences produced by simple rules that are normal. The Coperland–Erdös sequence, arising from the concatenation of prime numbers (235711131719...) is an example. The Champernowne sequence [31], a concatenation of all integers from one:

$$123456789101112131415\ldots$$

or in base 2,

$$11011100101110111\ldots$$

is an even more simple example. There even exist rules to generate absolutely normal sequences, i.e. numbers that are normal in any base b [32]. Thus, sequences exist that would pass any normality test of randomness, despite being produced by a simple rule—which contravenes the notion that randomness is defined by an absence of regularity.

9.2.1.3 Gaps

Another variety of randomness coefficient is worked out using the rank distances between two identical items. For instance, in the sequence 12311, the distances between occurrences of the symbol 1 are 3 and 1. The frequency distribution of repetition distances (gaps) and the median of repetition gap distribution (MdG) are based on the study of the distance between two identical outcomes. They have proved useful in detecting the so-called cycling bias: people tend to repeat an item only after they have used every other available item once [33].

Gap methods are as limited as normality assessment is: a normal sequence will pass these tests and be considered truly random, even if a naive rule produces it.

9.2.1.4 Cognitive Complexity

Apart from these attempts, a variety of measures to assess *cognitive* complexity or randomness were put forward during the 60s and 70s (for an overview, see [34]). These calibrations of cognitive complexity are built for the most part on the idea of algorithmic complexity, although in practice the link with Turing machines (Turing machine is at the root of Kolmogorov complexity theory) is evanescent. Usually, they consist of a series of rules or pattern descriptions that human subjects supposedly use and are aware of. The complexity of a string is defined as a function of the minimum number of rules one has to use to produce the string in question. These indices depend heavily on the rules and pattern descriptions chosen and lack mathematical rationale, but are psychologically sound. Their purpose is not to set a normative and formal measure of randomness through complexity, but to capture the nature of human complexity perception.

Very recently, a more sophisticated approach based on changes has been suggested [35]. Given a n-long string s, define the change function of s, $c(s)$, as the $(n-1)$-long string $f(s)$ whose ith term is 0 if $s(i) = s(i+1)$, and 1 if not. From a string s of length n, build a $n \times n$ matrix whose jth line L_j is $c(L_{j-1})$. Complexity is defined as a weighted sum of the matrix coefficients. Aksentijevic and Gibson [35] claim that this *change complexity* is an alternative to the unfortunately uncomputable algorithmic complexity.

However, there is no mathematical rationale for using change complexity as a normative measure of complexity or randomness. Rather than being a computable version of algorithmic complexity, it seems like a refinement of the usual normality assessment tools. It captures some local structural aspects of the string to be rated, and will consider every normal sequence as perfectly random, even though it is produced by a simple algorithm. Change complexity is also heavily reliant on psychological considerations and subjective choices. For these reasons, it should be classified with the previous attempts to capture *perceived* randomness, but not as a formal measure of complexity. However useful it may be in capturing human perceived complexity or the production of pseudo-random strings, it is a bad candidate for rating human pseudo-random productions in a normative and objective fashion.

9.2.1.5 Psychological Justifications and Limitations

Notwithstanding their potential limitations, all the above-mentioned coefficients have proved useful in detecting some common biases in random generation. SR-like values capture outcome biases—the overuse of certain symbols [36]. Normality assessment accurately spots alternation biases—the tendency to avoid using the same symbol twice, e.g., HH or TT—or the inverse *repetition bias*. Context redundancy has also been linked with cognitive flexibility [37] , of which it constitutes an estimate. Gaps and related methods would diagnose the cycling bias, a tendency to repeat the same pattern or to exhaust every available symbol before a repetition [33]. For instance, if the available symbols are 1, 2, 3, 4, a subject might choose 1, 3, 4, 2, 1, postponing the second appearance of "1" until after every symbol has occurred once. Repetition avoidance is known to affect outcomes as far as 6 ranks forward [38], a bias that gap methods shed light on.

It is unclear whether these measures happen to capture the basic biases in human random generation, or whether, unfortunately, researchers have focused on these biases simply because they have had tools at their disposal for diagnosing them. As we have seen in the previous sections, a normal sequence such as that suggested by Champernowne, which is highly non-random to the extent that it is generated by a simplistic rule, would meet all random criteria using symbol distribution, normality assessment, gap methods as well as change complexity.

At this point, we may list three senses in which the usual randomness estimates are unsatisfactory: First, they do not capture non-standard biases in randomness, such as the existence of a simple generation rule, when such a rule begins to produce sequences bearing some resemblance to a truly random one, e.g., a normal sequence. Only a few features of a random sequence are captured by these tailored measures. Second, they lack a theoretical basis. Despite being based on formal probabilistic properties, they are nevertheless not subsumed by a theory of randomness. In fact, they neither use nor provide any definition of a random sequence. Third, partly as an upshot of the first two points, several coefficients are needed to sketch an acceptable diagnosis of the quality of randomness, whereas a single measure would allow the comparison of sequences.

This is why we suggest that ACSS should be used whenever one wants to assess the complexity of a human-made pseudo-random sequence. We must, however, underline the fact that the algorithmic complexity of short strings alone will not yield a description of any discrepancy detected between a theoretical and an experimental string. Investigating the manner in which human pseudo-random productions differ from truly random sequences (which is of course of great importance for psychology) will remain the domain of more specific measures focusing on particular characteristics of interest, such as the indices mentioned above.

9.2.2 *Understanding Random "Biases"*

When human try to produce a random sequence, they tend to maximize complexity. This can be thought of as a rational behavior, although it does lead to systematic biases in randomness production. We here briefly discuss such biases in light of the ACSS. We will show that ACSS accounts for several such biases such as the avoidance of global regularity or the alternation bias.

9.2.2.1 Detection of Global Regularities

Some binary sequences may seem simple from a global viewpoint because they show symmetry (1011 1101) or repetition (1011 1011). Let us consider the string $s = 1011$ as an example. We have $D(5)(s) = 3.267414 \times 10^{-3}$. The repetitions $ss = 10111011$ have a much lower probability $D(5)(ss) = 4.645999 \times 10^{-7}$. This is not surprising considering the fact that ss is much longer than s, but we may then wish to consider other strings based on s. In what follows, we will consider three methods (repetition, symmetrization, 0-complementation). The repetition of s is $ss = 10111011$, the "symmetrized" $s\bar{s} = 10111101$, and the 0-complementation 10110000. These three strings of the same length have different probabilities ($4.645999.10^{-7}$, 5.335785×10^{-7} and 3.649934×10^{-7} respectively).

Let us now consider all strings of length 3–6, and their symmetrizations, 0-complementations and repetitions. Figure 9.2 provides a visual representation of the results. In each case, even the minimum mean between the means of symmetrized, complemented and repeated patterns (dotted horizontal line) lies in the upper tail of the $D(5)$ distribution for $2n$-length strings, and this is even more obvious with longer strings. Symmetry, complementation and repetition are, on average, recognized by $D(5)$.

9.2.2.2 Alternations

Human pseudo-random productions have been described as exhibiting too much alternation: when trying to behave randomly, humans have a proclivity to produce '1' after '0' or '0' after '1'. The mean frequency of alternation in a random binary string is 0.5, but slightly superior frequencies (typically around 0.6) have been reported in human pseudo-random generation (e.g. [39]).

It is now widely believed that when human subjects try to behave randomly, they actually try to maximize the complexity of their responses. However, due to cognitive limitations, we probably are unable to produce binary sequences of maximal complexity, because this would require a too complicated algorithm. This theoretical intuition recently received experimental support when researchers found that children were more attracted by mildly complex patterns [40].

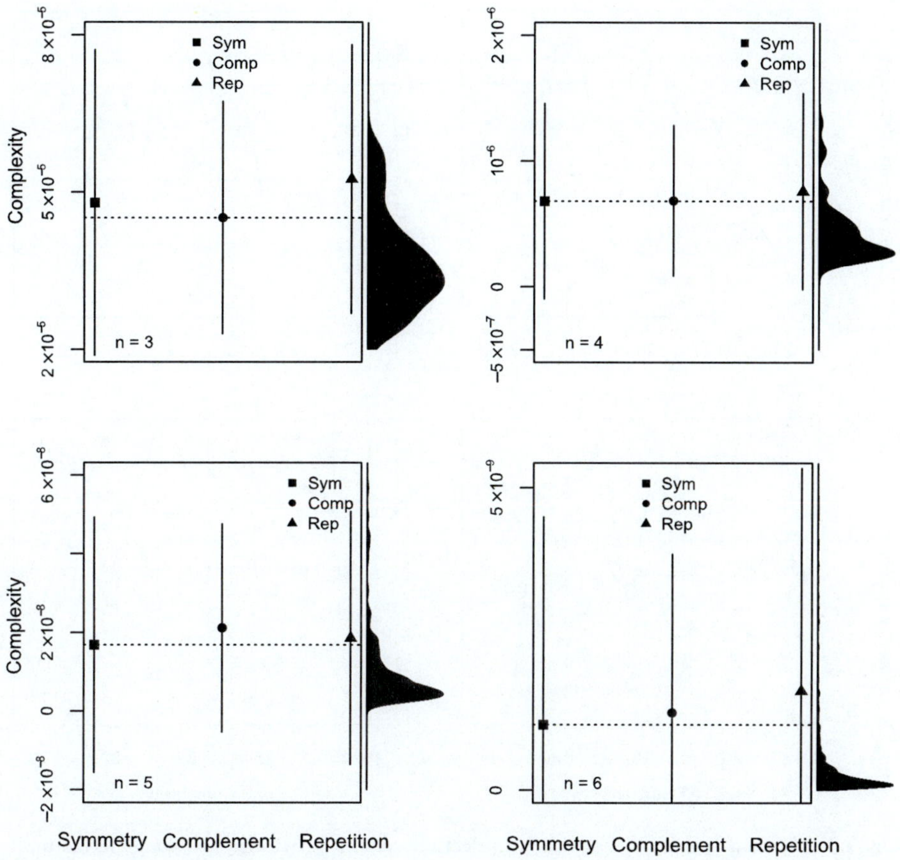

Fig. 9.2 Mean ± standard deviation of $D(5)$ of $2n$-long strings given by processes of symmetrization (Sym), 0-complementation (Comp) and repetition (Rep) of all n-long strings. The dotted horizontal line shows the minimum mean among Sym, Comp and Rep. The density of $D(5)$ for all $2n$-long strings is given in the right-margin

Figure 9.3 shows that binary strings of medium or mildly high complexity tend to exhibit an excess of alternations, whereas both simple and very complex strings tend to have an excess of repetitions. Complexity is a means to a deeper understanding of alternation bias. Formally, we suggest the following conjecture: When human subjects try to behave randomly and produce a binary string, they try to generate a complex sequence. Because their efforts are only partially rewarded, they produce strings of mild or just-above medium complexity, which usually have too much alternation.

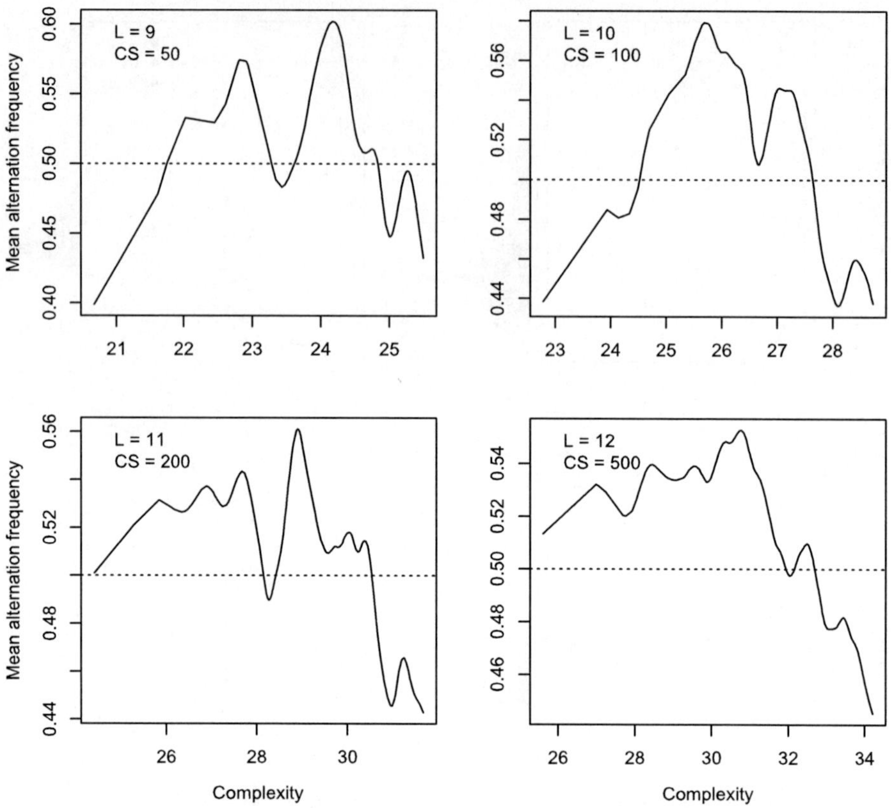

Fig. 9.3 For each length $l = 9$ to 12, we select clusters of CS strings and compute the mean frequency of alternations within these strings. The diagram displays the smooth-spline curve of the resulting function. The number CS of strings is chosen according to the total number of strings of length l, to ensure readability. The clusters are built thus: strings are sorted by increasing complexity. For a given complexity k, let j be the rank of the first string with complexity above k. Then the corresponding cluster comprises all strings with rank j to $j + CS - 1$

9.2.2.3 Bayesian Use of $D(5)$

Given a string s, we may now compute the probability of it being truly random (event R) against the hypothesis that it has been created by a Turing machine (event M). Let l be a length (for instance, $l = 12$). In this paragraph, we will denote the conditional probability of $D(5)$ when the length is l by P_l. Set the prior probability that the underlying process is random to $P(R) = 1/2$. We then have $P(s|R) = 2^{-l}$ and

$$P(R|s) = \frac{P(s|R)P(R)}{P(s)},$$

Table 9.1 A few examples of 12−long strings s with the associated probability of being random $P(R|s)$ and the number of 1s (6 meaning that s is perfectly balanced)

s	$P(R\|s)$	Ones
110100011100	0.975	6
101100100011	0.951	6
110001111001	0.942	7
101100111100	0.933	7
100000000011	0.519	3
110101001011	0.515	7
101110110111	0.180	9
101111101111	0.151	10
010101010101	0.097	6
010000000000	0.079	1
000000000000	0.017	0

with

$$P(s) = P(s|M)P(M) + P(s|R)P(R) = \frac{P_l(s)}{2} + 2^{-(l+1)}.$$

From this we derive

$$P(R|s) = \frac{1}{1 + 2^l P_l(s)}.$$

Table 9.1 gives a few examples of random binary strings of length 12 together with their probability of being random, and the number of 1s they include. As we can see from this table and as Fig. 9.4 visually confirms, the more complex strings (which also have a great probability of being random) are likely to be balanced, with approximately six 1 s. This again may help us understand the so-called "belief in the law of small numbers" [41] according to which subjects wrongly tend to generate equal numbers of each alternative while trying to produce random binary strings. Once again, this could be a result of an attempt to generate complex responses, that is, strings that are more likely to be random.

9.2.3 Applications

We provide now four illustrative applications of ACSS. The first two are short reports of new and illustrative experiments and the third one is a re-analysis of previously published data [42]. The last one deals with the dynamics of randomness production in children. Although these experiments are presented to illustrate the use of ACSS, they also provide new insights into subjective probability and the perception of randomness. Note that all the data and analysis scripts for these applications are also part of the `R` package `acss`.

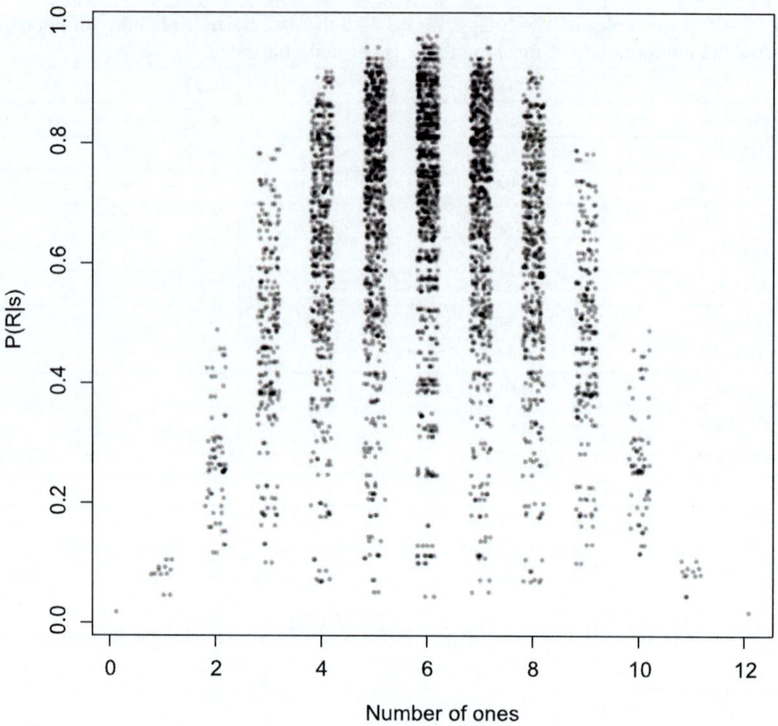

Fig. 9.4 Scatterplot of the probability of being random against the number of 1s in all 12−long strings. A jitter method has been applied on the number of 1s to increase readability

9.2.3.1 Humans are "Better Than Chance"

Human pseudo-random sequential binary productions have been reported to be overly complex, in the sense that they are more complex than the average complexity of truly random sequences [43]. Here, we test the same effect with non-binary sequences based on 4 symbols. To replicate the analysis, type `?exp1` at the `R` prompt after loading `acss` and execute the examples. A sample of 34 healthy adults participated in this experiment. Ages ranged from 20 to 55 (mean = 37.65, SD = 7.98). Participants were recruited via e-mail and did not receive any compensation for their participation. Participants were asked to produce at their own pace a series of 10 symbols using "A", "B", "C", and "D" that would "look as random as possible, so that if someone else saw the sequence, she would believe it to be a truly random one". Participants submitted their responses via e-mail. A one sample t-test showed that the mean complexity of the responses of participants is significantly larger that the mean complexity of all possible patterns of length 10 ($t(33) = 10.62$, $p < 0.0001$). The violin plot in Fig. 9.5 shows that human productions are more complex than

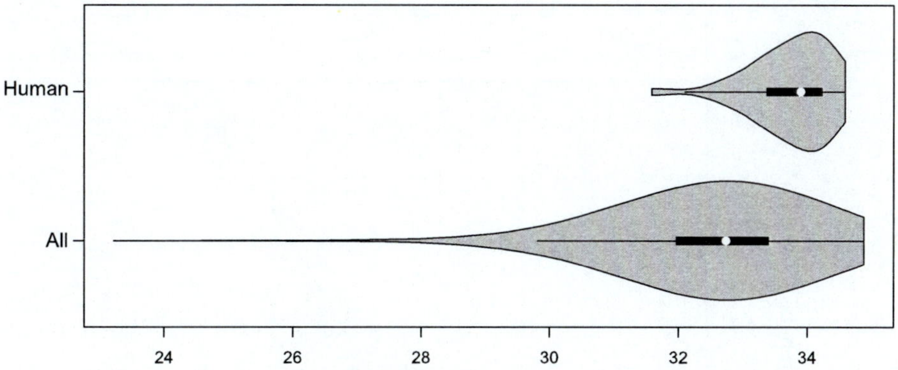

Fig. 9.5 Violin plot showing the distribution of complexity of human strings versus every possible pattern of strings, with 4-symbol alphabet and length 10

random patterns because humans avoid low-complexity strings. On the other hand, human productions did not reach the highest possible values of complexity.

These results are consistent with the hypothesis that when participants try to behave randomly, they in fact tend to maximize the complexity of their responses, leading to overly complex sequences. However, whereas they succeed in avoiding low-complexity patterns, they cannot build the most complex strings.

9.2.3.2 The Threshold of Complexity—A Case Study

Humans are sensitive to regularity and distinguish truly random series from deterministic ones [84]. More complex strings should be more likely to be considered random than simple ones. Here, we briefly address this question through a binary forced choice task. We assume that there exists an individual threshold of complexity for which the probability that the individual identifies a string as random is 0.5. We estimated that threshold for one participant. The participant was a healthy adult male, 42 years old. The data and code are available by calling `?exp2`.

A series of 200 random strings of length 10 from an alphabet of 6 symbols, such as "6154256554", were generated with the `R` function `sample()`. For each string, the participant had to decide whether or not the sequence appeared random.

A logistic regression of the actual complexities of the strings (K_6) on the responses is displayed in Fig. 9.6. The results showed that more complex sequences were more likely to be considered random (slope $= 1.9$, $p < 0.0001$, corresponding to an odds ratio of 6.69). Furthermore, a complexity of 36.74 corresponded to the subjective probability of randomness 0.5 (i.e., the individual threshold was 36.74).

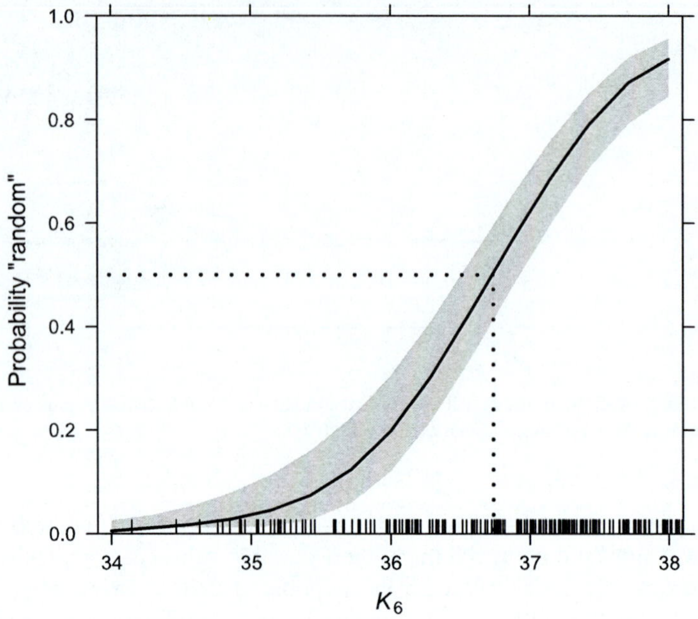

Fig. 9.6 Graphical display of the logistic regression with actual complexities (K_6) of 200 strings as independent variable and the observed responses (appears random or not) of one participant as dependent variable. The gray area depicts 95%-confidence bands, the black dots at the bottom the 200 complexities. The dotted lines show the threshold where the perceived probability of randomness is 0.5

9.2.3.3 The Span of Local Complexity

In a study of contextual effect in the perception of randomness, Matthews ([42], Experiment 1) showed participants series of binary strings of length 21. For each string, participants had to rate the sequence on a 6-point scale ranging from "definitely random" to "definitely not random". Results showed that participants were influenced by the context of presentation: sequences with medial alternation rate (AR) were considered highly random when they were intermixed with low AR, but as relatively non-random when intermixed with high AR sequences. In the following, we will analyze the data irrespective of the context or AR.

When individuals judge whether a short string of, for example, 3–6 characters, is random, they probably consider the complete sequence. For these cases, ACSS would be the right normative measure. When strings are longer, such as a length of 21, individuals probably cannot consider the complete sequence at once. Matthews [42] and others [85] have hypothesized that in these cases individuals rely on the local complexity of the string. If this were true, the question remains as to how local the analysis is. To answer this, we will reanalyze Matthews' data.

For each string and each span ranging from 3 to 11, we first computed the mean local complexity of the string. For instance, the string "XXXXXXXOOOOXXXOOOOOOO" with span 11 gives a mean local complexity of 29.53. The same string has a mean local complexity of 11.22 with span 5.

```
> sapply(local_complexity("XXXXXXXOOOOXXXOOOOOOO",
    11, 2), mean)
XXXXXXXOOOOXXXOOOOOOO
             29.52912
> sapply(local_complexity("XXXXXXXOOOOXXXOOOOOOO",
    5, 2), mean)
XXXXXXXOOOOXXXOOOOOOO
             11.21859
```

For each span, we then computed R^2 (the proportion of variance accounted for) between mean local complexity (a formal measure) and the mean randomness score given by the participants in Matthews' (2013) Experiment 1. Figure 9.7 shows that a span of 4 or 5 best describes the judgments with R^2 of 54% and 50%. Furthermore, R^2 decreases so fast that it amounts to less than 0.002% when the span is set to 10.

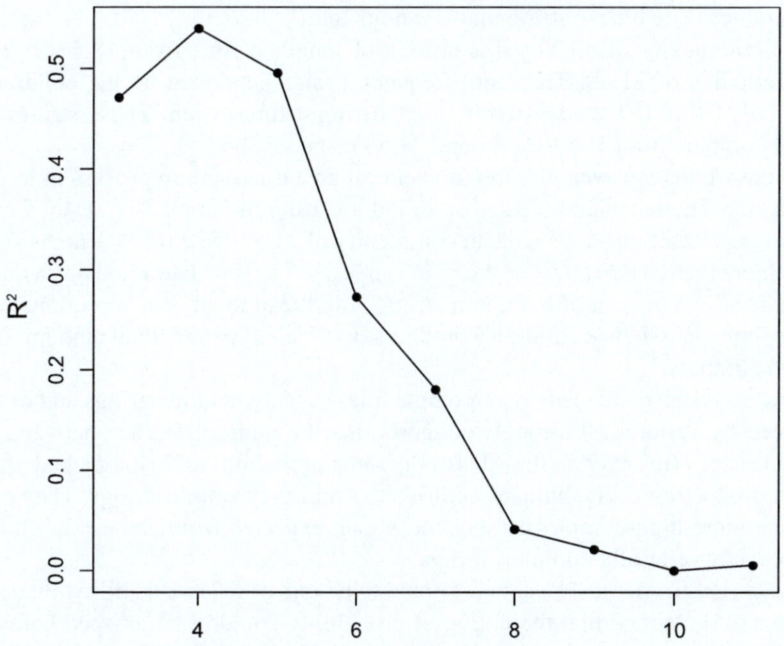

Fig. 9.7 R^2 between mean local complexity with span 3–11 and the subjective mean evaluation of randomness

These results suggest that when asked to judge if a string is random, individuals rely on very local structural features of the strings, only considering subsequences of 4–5 symbols. This is very near the suggested limit of the short term memory of 4 chunks [44]. Future researchers could build on this preliminary account to investigate the possible "span" of human observation in the face of possibly random serial data. The data and code for this application are available by calling `?matthews2013`.

9.2.3.4 Randomness in Children

68 children (35 male, 33 female) aged 7–11 participated in this experiment (mean age $\pm$ SD: $8, 7 \pm 1, 1$). All the children were pupils in 5 different classes from 2 public schools in France, grades 2–4. The youngest subjects (2d graders) were the first to be tested, followed 4 months later by the 3rd-graders, and 4 months after that by the 4th graders.

Each child was received individually in a room in the school, during class time. A token was presented to him or her. It was green on one side and red on the other. The experimenter explained what was meant by a "toss" and the token was then hidden so it would not distract the subject.

The instruction given to the child was to imagine that s/he tossed the token 8 times, and to say out loud which side, green or red, appeared each time. Each child thus produces one binary string that is 8 units long.

The complexity of all possible strings of length 8 runs from 18.53 to 22.68, with a median of 21.60. The more frequent strings generated by the children are 00101101, 01001101, and 11010010, appearing 4 times each. These strings have medial complexities (21.58, 21.37 and 21.58 respectively).

On the whole however, children do better than a true random process in terms of complexity. The mean complexity of a random string of length 8 is 21.46, whereas children's productions have a mean complexity of 21.74 (SD: 0.52). This is significantly more than 21.46 ($t(67) = 4.35$, $p < 5 \times 10^{-5}$). Of course, choosing a random Turing machine instead of a random string would lead to an over-representation of simple sequences, hence children's productions are also "better" than random Turing machine outputs.

Figure 9.8 shows the density of complexities of truly random strings and of those produced by children. It strongly suggests that the main difference between truly random strings (i.e. every string shares the same probability of being picked up) and human production is that humans contrive to avoid very simple strings. They do not generate more high-complexity sequences than expected from chance, but they do produce a lot of mildly complex strings.

The period from 7 to 11 years is known to be one of relative stability. Piaget and Inhelder [45] claimed that the notion of probability couldn't be grasped before the formal stage, at around age 11. In an experimental study of randomness perception in 7 to 16 year-olds, Green [46] finds no evolution between 7 and 11.

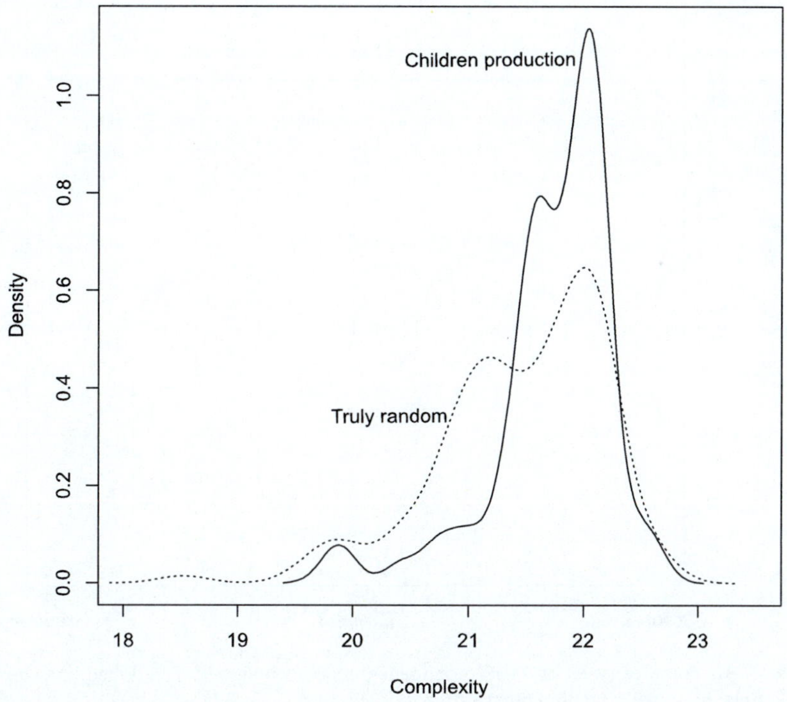

Fig. 9.8 Density of truly random strings' complexity (dotted) and of children's productions, as approximated by Gaussian kernel

One aim of the present experiment is to ascertain whether there is an evolution in subjective randomness at this particularly stable stage. Concerning complexity, we expected either no differences from grade 2 through grade 4, or an increase in complexity. Figure 9.9 displays the means and confidence intervals by grades. A one-way anova was applied to the data. Although there is no significant difference ($F(2, 65) = 0.68$, $p = 0.41$), a small effect [47] ($\eta^2 = 0.11$) appears in our sample and shows a slight progression of complexity with grade.

The main finding of this experiment is that even young children aged 7–11, from grade 2 to 4, can do better than picking a string from a uniform distribution to build up strings that "look random". This behavior is not an irrational bias: any string is certainly as likely to appear by chance, but not all of them show the same amount of evidence for a random underlying process. The possible slight evolution of complexity between grades 2 and 4 confirms that this period of childhood is relatively stable.

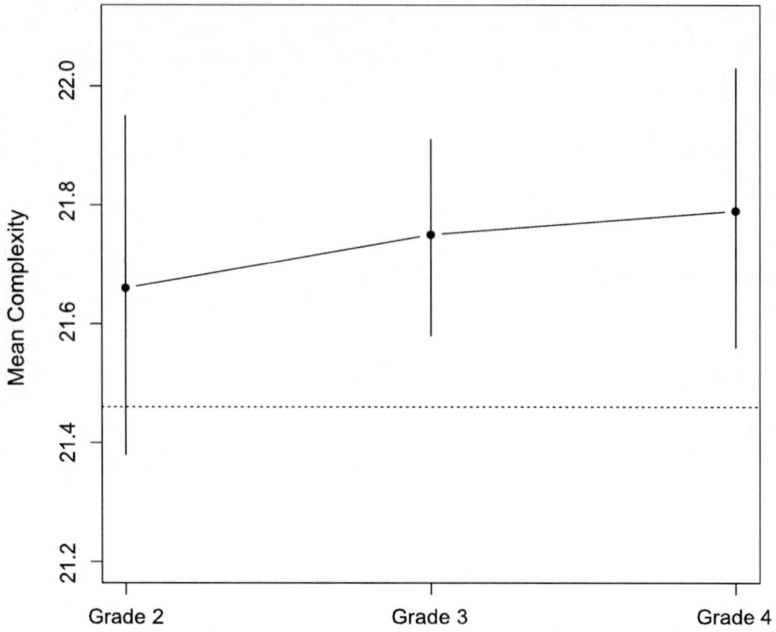

Fig. 9.9 Mean complexity K_5 and 95% confidence intervals by grade. The dotted line shows the mean complexity of truly random binary strings of length 8

9.2.4 Comments

For decades, researchers have relied upon algorithmic Kolmogorov–Chaitin complexity as a means to rate complexity and randomness in an objective way. Algorithmic complexity may seem to be a purely theoretical apparatus, useful only in abstract operations on infinite sequences. It has turned out to be useful in practice when applied to DNA sequences, or when long strings (of millions of digits, for instance) are involved. In such cases, compression methods seem to overcome the impossibility of an exact calculation of algorithmic complexity. Compression methods unfortunately don't apply well to short strings, which we encounter in the behavioral sciences. Since amending the theory to make it suitable for short strings was believed to be impossible, researchers have followed two routes:

1. Some renounced the ideal of a normative unique measure of complexity, and used a variety of indices focusing on special features of the strings they had to measure.
2. Others switched from universal Turing machines to tailor-made cognitive "machines" designed according to what they thought was important for humans. In doing so, they generated good estimates of *perceived* complexity, but lost the universality of algorithmic complexity.

CTM allows us to use algorithmic complexity even for very short strings. As we have shown, this may shed new light on some biases such as the alternation bias or the belief in the law of small numbers: they may be simple consequences of a tendency to be mildly complex. Because K_5 measures an objective complexity, we cannot expect human productions to be as complex as the more complex possible strings, though a mild and above average complexity may be expected.

9.3 Development

Knowledge forming the content of several academic fields, including mathematics, follows from precocious core knowledge [48], and then follows a specific developmental course along the lifespan [49].

Numerosity (our approximate implicit sense of quantity) has been a privileged target of recent research, because numbers form one of the main pillars of elementary mathematical knowledge, but the study of randomness perception and statistical reasoning has also yielded striking results in the field of probability: adults with no formal education [50] as well as 8–12 month-old children [51] have the wherewithal for simple implicit probabilistic reasoning. One of the toughest problems when it comes to Bayesian reasoning, however, is the detection of randomness, i.e., the ability to decide whether an observed sequence of events originates from a random source as opposed to produced by a deterministic origin [52]. Humans, adults and infants, have a keen ability to detect structure, both of statistic and algorithmic nature e.g. (0101... and 1234...) that only algorithmic complexity can detect (versus e.g. entropy).

Within the field of study devoted to our sense of complexity, the task of randomly arranging a set of alternatives is of special interest, as it poses almost insurmountable problems to any cognitive system. The complexity of a subject-produced pseudo-random sequence may serve as a direct measure of cognitive functioning, one that is surprisingly resistant to practice effects and largely independent of the kind of alternatives to be randomized. Although random item generation (RIG) tasks usually demand vocalization of selections, motor versions have comparable validity and reliability. RIG tasks are believed to tap our approximate sense of complexity (ASC), while also drawing heavily on focused attention, sustained attention, updating and inhibition. Indeed to produce a random sequence of symbols, one has to avoid any routine and inhibit prepotent responses. The ability to inhibit such responses is a sign of efficient cognitive processing, notably a flexibility assumed to be mediated by the prefrontal cortex.

Theoretical accounts of the reasons why RIG tasks are relevant tests of prefrontal functions are profuse, but pieces of experimental evidence are sparse. Sparse empirical factors indirectly validate the status of RIG tasks as measures of controlled processing, such as the detrimental effect of cognitive load or sleep deprivation [53] or the fact that they have proved useful in the monitoring of several neuropsychological disorders.

As a rule, the development of cognitive abilities across the lifespan follows an inverse U-shaped curve, with differences in the age at which the peak is reached [49]. The decrease rate following the peak also differs from one case to another, moving between two extremes. "Fluid" components tend to decrease at a steady pace until stabilization, while "crystalized" components tend to remain high after the peak, significantly decreasing only in or beyond the 60s [54]. Other evolutions may be thought of as a combination of these two extremes.

Two studies have addressed the evolution of complexity in adulthood, but with limited age ranges and, more importantly, limited 'complexity' measures. The first [55] compared young and older adults' responses and found a slight decrease in several indices of randomness. The second [20] found a detrimental effect of aging on inhibition processes, but also an increase of the cycling bias (a tendency to postpone the re-use of an item until all possible items have been used once), which tends to make the participants' productions more uniform.

In both studies, authors used controversial indices of complexity that only capture particular statistical aspects, such as repetition rate or first-order entropy. Such measures have proved some usefulness in gauging the diversity and type of long sequences (with e.g., thousands of data points) such as those appearing in the study of physiological complexity, but are inadequate when in comes to short strings (e.g., of less than a few tens of symbols), such as the strings typically examined in the study of behavioral complexity. Moreover, such indexes are only capable of detecting statistical properties. Authors have called upon algorithmic complexity to overcome these difficulties [56].

The main objective of the present study is to provide the first fine-grained description of the evolution over the lifespan of the (algorithmic) complexity of human pseudo-random productions. Secondary objectives are to demonstrate that, across a variety of different tasks of random generation, the novel measure of behavioral complexity does not rely on the collection of tediously long response sequences as hitherto required. The playful instructions to produce brief response sequences by randomizing a given set of alternatives are suitable for children and elderly people alike, can be applied in work with various patient groups and are convenient for individual testing as well as Internet-based data collection.

Participants with ages ranging from 4 to 91 performed a series of RIG tasks online. Completion time (CT) serves as an index of speed in a repeated multiple choice framework. An estimate of the algorithmic complexity of (normalized) responses was used to assess randomization performance (e.g., response quality). The testing hypothesis is that the different RIG tasks are correlated, since they all rely on similar core cognitive mechanisms, despite their differences. To ensure a broad range of RIG measurements, five different RIG tasks were selected from the most commonly used in psychology. The study was approved by the University of Zurich Institutional Review Board (Req00583).

9.3.1 Methods

The five tasks used, described in Table 9.2, are purposely different in ways that may affect the precise cognitive ability that they estimate. Figure 9.10 shows the screenshots of the tasks. For instance, some tasks draw on short-term memory because participants cannot see their previous choices (e.g., "pointing to circles"), whereas in other tasks memory requirements are negligible, because the participant's previous choices remain visible ("rolling a die"). Completion times across the various tasks showed a satisfactory correlation (Cohen's $\alpha = 0.79$), suggesting that participants did not systematically differ in the cognitive effort they devoted to the different tasks. Any difference between task-related complexities is thus unlikely to be attributable to differences in time-on-task.

Complexities were weakly to moderately positively correlated across the different tasks (Cohen's $\alpha = 0.45$), mostly as a consequence of the "filling the grid" task being almost uncorrelated with the other tasks. Despite this moderate link, however, all trajectories showed a similar pattern across the lifespan, with a peak around 25, a slow, steady decline between 25 and 60, followed by accelerated decline after 60 as shown in Fig. 9.11.

The hypothesis to test that the different tasks are positively related to each other was partially supported by the data, especially in view of the results obtained on the "filling the grid" task. At the same time, CTs showed correlations supporting the

Table 9.2 Description of the 5 RIG tasks used in the experiment. Order was fixed across participants

Task	Description
Tossing a coin	Participants had to create a series of 12 head-or-tails that would "look random to somebody else" by clicking on one of the two sides of a coin appearing on the screen. The resulting series was not visible on the screen (the participant could only see the last choice made)
Guessing a card	Participants had to select one of 5 types of cards (Zener cards), ten times. In contrast to the other tasks, they were not asked to make the result look random. Instead, they were asked to guess which card will appear after a random shuffle
Rolling a die	Participants had to generate a string of 10 numbers between 1 and 6, as random as possible ("the kind of sequence you'd get if you really rolled a die"). In contrast to the preceding cases, they could here see all previous choices, but could not change any of them
Pointing to circles	Participants had to point 10 times at one out of 9 circles displayed simultaneously on the screen. They could not see their previous choices. This task is an adaptation of the classical Mittenecker pointing test
Filling a grid	Participants had to blacken cells in a 3 × 3 grid such that the result would look randomly patterned, starting from a white grid. In contrast to the other tasks, they could see their choice and click as many times as they wished. Clicking on a white cell made it black, and vice versa

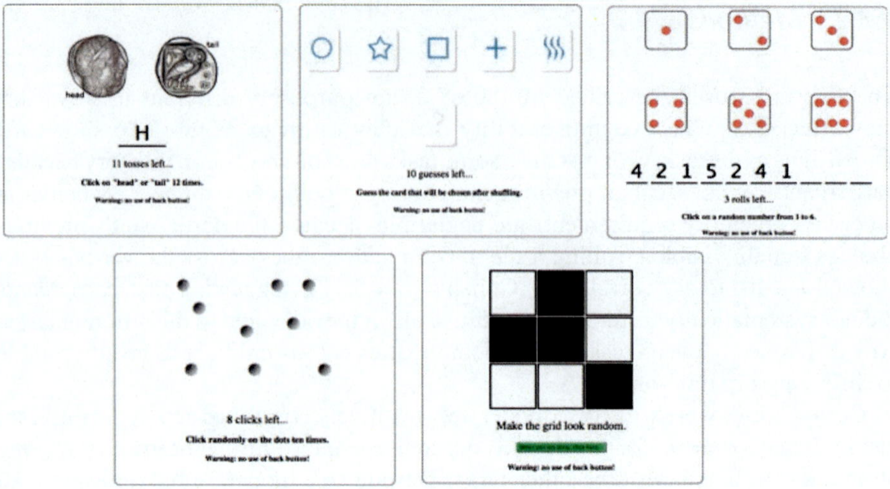

Fig. 9.10 Screenshots of the experimental implementation

testing hypothesis together with the developmental complexity curves in agreement pointing in the same direction. This suggests that all tasks do tap into our ASC as well as into other cognitive components with similar developmental trajectories, but that different tasks actually require different supplementary abilities or else they weight the components of these abilities differently.

The "filling the grid" task appeared unique in that it was loosely correlated with all the other tasks. The fact that it required binary responses cannot account for this lack of association, since the "tossing a coin" task yielded results uncorrelated with the "filling the grid" responses. Bi-dimensionality could possibly have had an effect, but the "pointing to circles" task was also unrelated to the grid task. On the other hand, one factor distinguished the grid task from all others in the set: the option offered to the participants to change their previous choices as many times as they wished. For that reason, the grid task may in fact have relied more on randomness perception, and less on inhibition and randomness production heuristics. Indeed, participants could change the grid until they felt it was random, relying more on their ASC than on any high order cognitive ability serving output structure. This hypothesis is supported by the fact that participants did indeed change their minds. There were only 9 cells (that could turn white or black) on the grids and participants' end responses had a mean of 4.08 ($SD = 1.8$) selected (black) cells thus generally favoring whiter configurations (possibly as a result of the all-white initial configuration). However, the number of clicks used by participants during this task was far larger ($M = 10.16$, $SD = 9.86$), with values ranging from 5 to 134 (this latter trying almost a fifth of all possible configurations). Thus the option to change previous choices in a RIG task may have been an important factor, and should accordingly be considered a novel variable in future explorations of randomization tasks (and balanced with an all-black configuration). In this view, the "filling the grid" task would reflect our ASC

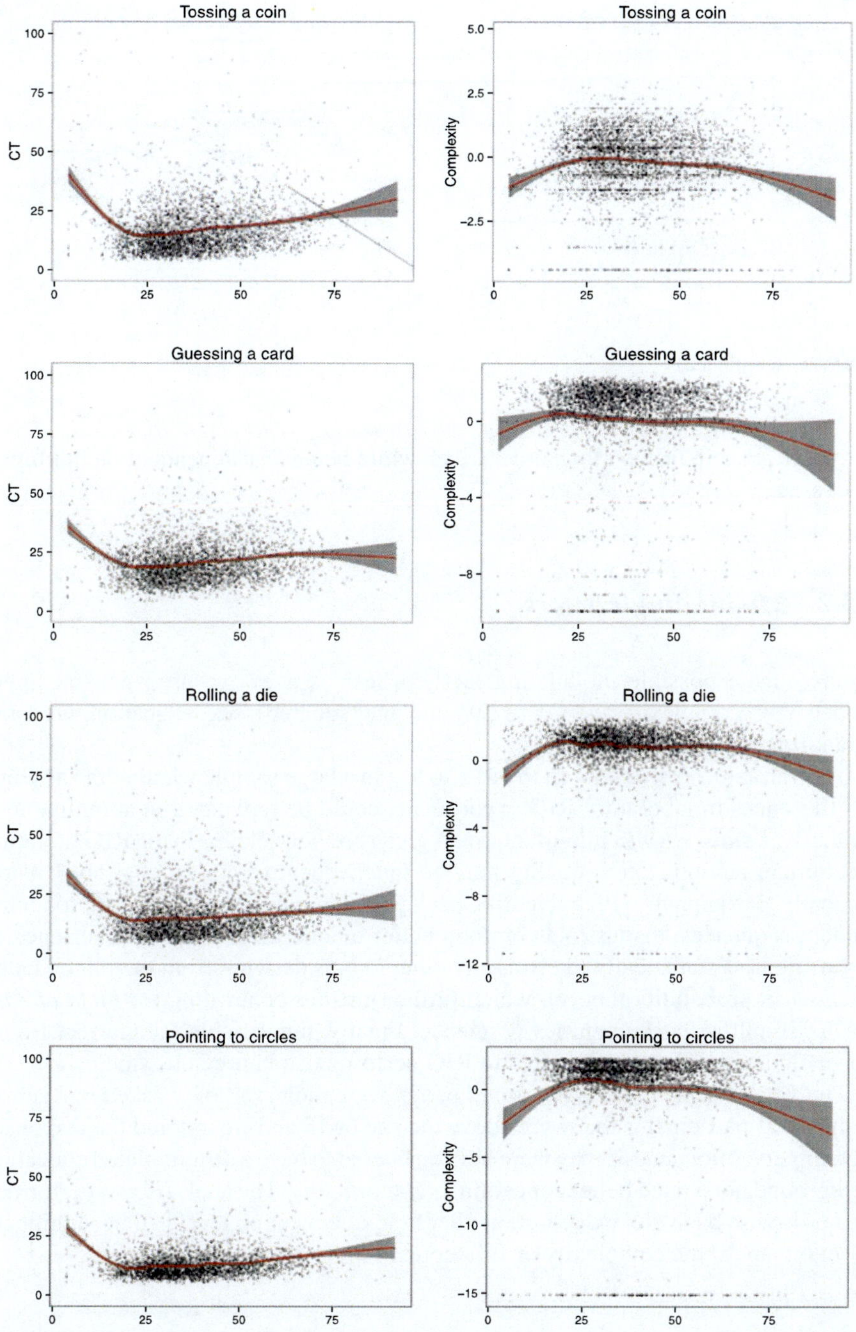

Fig. 9.11 Developmental curves of completion time and complexity, split by task, with trend curves and 95% confidence regions (shaded)

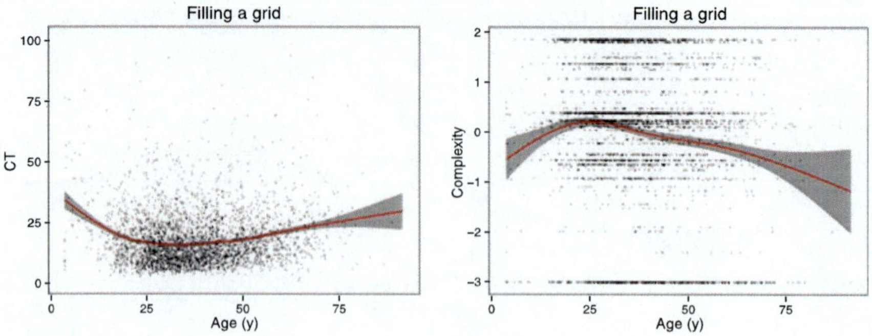

Fig. 9.11 (continued)

in a more reliable fashion than other tasks, while being less dependent on inhibition processes.

9.3.2 Modulating Factors

To investigate possible modulating factors (besides age), we used general linear models with complexity and CT as DV, and age (squared), sex, education, field and paranormal beliefs as IV.

The variable *Sex* was chosen in order to test in a large sample whether the absence of differences in laboratory RIG experiments could be replicated in an online test. Similarly, *Education* was important to test given previous claims in the RIG literature that human randomization quality may be independent of educational level. Paradoxically, participants with a scientific background may perform worse at producing random sequences, thanks to a common belief among them that the occurrence of any string is as statistically likely as any other (a bias deriving from their knowledge of classical probability theory), which further justifies controlling for *Field of education*, simplified as humanities v. science. Finally, the variable Paranormal Belief was included as it has been related to RIG performance in previous studies.

The variables *field* and *paranormal belief* were, however, only asked in a subset of the 3 313 participants that were above the age of 15 and we ignored the responses of younger participants as they were not considered to have a differentiated education background nor a fixed belief concerning paranormality. The analysis was performed on a task-wise basis. As we report, neither field or education level had no significant effect on any of the complexity or CT scores.

9.3.3 *Experiment*

A sample of 3 429 participants took part in the experiment (age range: 4–91y, M = 37.72, SD = 13.38). Participants were recruited through social networks, radio broadcasts and journal calls during a 10 months period.

9.3.3.1 Procedure

A specific web application was designed to implement the experiment online. Participants freely accessed to the website.[1] The experiment was available in English, French, German, and Spanish (all translated by native speakers for each language). In the case of young children as yet unable to read or use a computer, an adult was instructed to read the instructions out loud, make sure they were understood, and enter the child's responses without giving any advice or feedback. Participants were informed that they would be taking part in an experiment on randomness. They then performed a series of tasks before entering demographic information such as sex, age, mother tongue, educational level, and main field of education (science, humanities, or other) if relevant. Before each task, participants (or a parent, in the case of youngsters) read the specific instructions to the task and only then pressed a "start" key. This initiated the measurement of the completion time (CT) for a given task, which was recorded alongside the responses. Practice trials were not provided to maximize spontaneity of responding and to simultaneously avoid boredom effects and thus minimize drop-out rates.

One last item served as an index of paranormal beliefs and was included since probabilistic reasoning is among the factors associated with the formation of such beliefs. Participants had to rate on a 6-point Likert scale how well the following statement applied to them: "Some 'coincidences' I have experienced can best be explained by extrasensory perception or similar paranormal forces."

9.3.3.2 Measures

For each task, CT (in seconds) was recorded. The sum of CTs (total CT) was also used in the analyses. An estimate of the algorithmic complexity of participants' responses was computed using the `acss` function included in the freely publicly available acss R-package [57] that implements the complexity methods used in this project. Complexities were then normalized, using the mean and standard deviation of all possible strings with the given length and number of possible symbols, so that a complexity of 0 corresponds to the mean complexity of all possible strings. For each participant, the mean normalized complexity (averaged over the five tasks) was also computed, serving as a global measure of complexity.

---

[1] http://complexitycalculator.com/hrng/.

9.3.4 Results and Discussion

Sex had no effect on any of the complexity scores, but a significant one on two CT scores, with male participants performing faster in the first two tasks: "tossing a coin" ($p = 6.26 \times 10^{-10}$, $\eta_p^2 = 0.012$) and "guessing a card" ($p = 2.3 \times 10^{-10}$, $\eta_p^2 = 0.012$). A general linear model analysis of the total CT scores as a function of sex, age (squared), field, education level and paranormal belief was performed on the same subset and revealed a strongly significant effect of sex ($p = 9.41 \times 10^{-8}$, $\eta_p^2 = 0.009$), with male participants performing faster than female participants. A simpler model, including only sex and age (squared) as IV, still showed an effect of sex ($p = 6.35 \times 10^{-13}$, $\eta_p^2 = 0.016$), with male participants needing less time. The sex difference in CT, mostly appearing in adulthood (during the 60s) was in line with previous findings that in adults, choice CT is lower in men than in women.

Paranormal belief scores were unrelated to CTs for all tasks. However, they were negatively linked with the complexity score in the "filling a grid" task ($p = 0.0006$, $\eta_p^2 = 0.004$), though not with any other task.

Paranormal beliefs have been previously reported to be negatively linked to various RIG performances. Our results replicated these findings but only on the "filling a grid" task. One possible explanation is that the grid task actually measures randomness perception rather than randomness production, and that a paranormal bent is more strongly linked with a biased perception of randomness than with a set of biased procedures used by participants to mimic chance. This hypothesis is supported by the finding that believers in extrasensory perception are more prone to see "meaningful" patterns in random dot displays. Another complementary hypothesis is that the type of biases linked to beliefs in the paranormal only usually appear over the long haul, and are here preempted by the fact that we asked participants to produce short sequences (of 12 items at most). Indeed, when it comes to investigating the effects of paranormal belief on pure randomness production, rather long strings are needed, as the critical measure is the number of repetitions produced.

To get a better sense of the effect of paranormal belief, we performed a general linear model analysis of the mean complexity score as a function of sex, age (squared), field, education level and paranormal belief. For this analysis, we again used the subset of 3 313 participants over the age of 15. Paranormal belief no longer had an effect.

9.3.4.1 Mean Complexity and Total CT Trajectories

Our main objective was to describe the evolution over the lifespan of mean complexity, which is achieved here using an approximation of algorithmic complexity for short strings. The developmental curve of complexity found in Fig. 9.12, suggests that RIG tasks measure a combination of fluid mechanics (reflected in a dramatic performance drop with aging) and more crystallized processes (represented by a stable performance from 25 to 65 years of age). This trajectory indirectly confirms

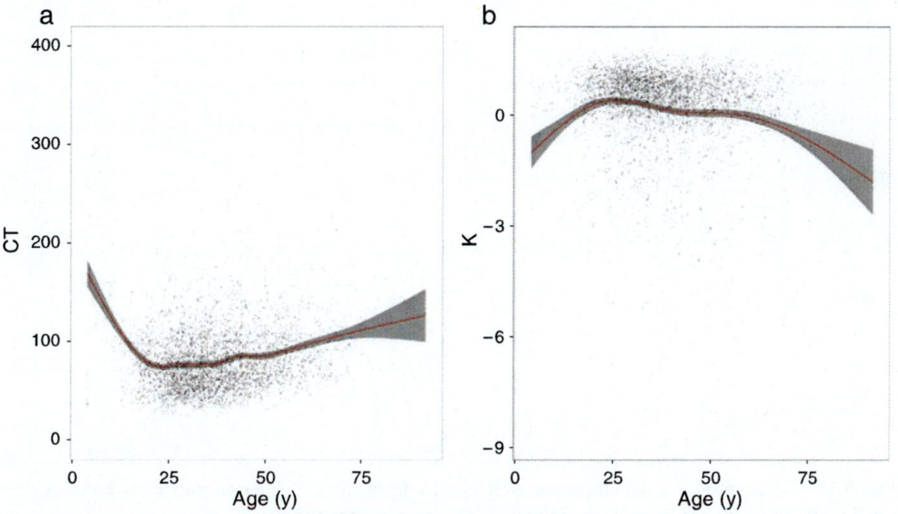

Fig. 9.12 **a** Total completion time (CT) and **b** mean complexity as a function of age, with trend curve and 95% confidence region (shaded area)

a previous hypothesis [20]: attention and inhibition decrease in adulthood, but an increased sense of complexity based on crystallized efficient heuristics counters the overall decline in performance.

Plotting complexity and CT trends on a single two-dimensional diagram allowed a finer representation of these developmental changes (Fig. 9.13). It confirmed the entanglement of complexity (accuracy) and CT (speed). In the first period of life (<25), accuracy and speed increased together in a linear manner. The adult years were remarkable in that complexity remained at a high level for a protracted period, in spite of a slow decrease of speed during the same period. This suggest that during the adult period, people tend to invest more and more computational time to achieve a stable level of output complexity. Later in life (>70), however, speed stabilizes, while complexity drops in a dramatic way.

These speed-accuracy trade-offs were evident in the adult years, including the turn toward old age. During childhood, however, no similar pattern is discernible. This suggests that aging cannot simply be considered a "regression", and that CT and complexity provide different complementary information. This is again supported by the fact that in the 25–60 year range, where the effect of age is reduced, CT and complexity are uncorrelated ($r = -0.012$, $p = 0.53$). These findings add to a rapidly growing literature that views RIG tasks as good measures of complex cognitive abilities.

We have gone further here in several respects than any previous literature. First, we present a set of data collected in RIG tasks with a broad variety of instructions as to what and how to randomize: our participants playfully solved binary randomization tasks along with multiple-alternative variants; they explicitly attempted to

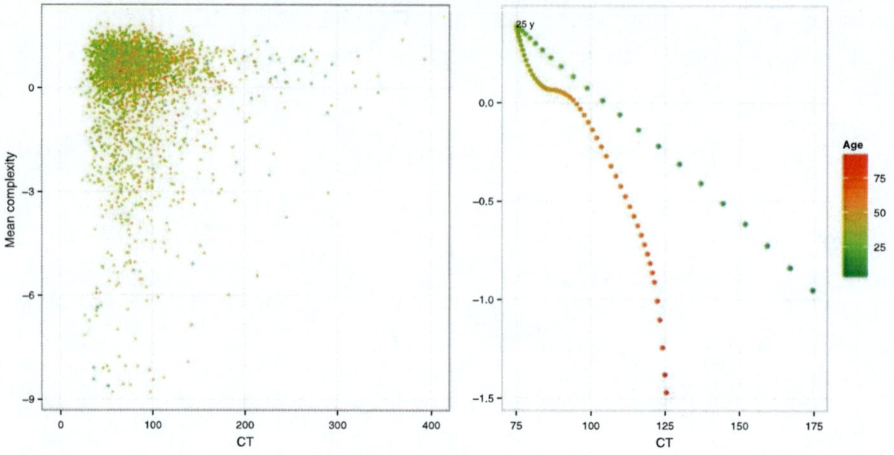

Fig. 9.13 Scatterplot and developmental change trend of the CT and complexity combined. The trend is obtained by use of smooth splines of CT and complexity (df = 7)

generate random sequences, but also distributed their responses in a guessing task, typically considered "implicit random generation". The expected outcome was unidimensional in some tasks and two-dimensional in others; constraints imposed by working memory capacity were high in some tasks, but almost absent in others. In the cognitive science literature, such diverse tasks have never been compared directly. We do not deny that the various tasks we used may tap into slightly different subcategories of prefrontal functioning, with some relying more on working memory and others on inhibitory control. Yet, we set out to illustrate the commonalities among the different tasks leaving a more fine-grained analysis to future studies.

Cross-sectional studies should try to relate behavioural complexity to the degree of maturation or degeneration of specific prefrontal cortex regions. Neuropsychological investigations could use the tasks and measures employed here with selected patient groups to perform lesion-symptom mappings, as has been done recently [57], but preferably in patients with focal vascular lesions. In parallel with such investigations, Internet-based work such as the project presented here may still play a powerful role. They may complement RIG tasks with brief behavioural tasks having a known neurocognitive basis and well-studied developmental trajectories. Thus, laboratory testing and web-based approaches may conjointly help pinpoint the cognitive mechanisms underlying the age-related modulation of behavioural complexity.

A second extension of the existing literature on subject-generated random sequences is the number of participants tested and their age-range. To date, only two studies have investigated age-related changes in RIG tasks with a range comparable to the one investigated here. They both compared groups of young adults and older adults and were thus unable to describe the continuous evolution of complexity across the lifespan.

Finally, one of the most exciting novel aspects of this research is that we have presented an estimate of algorithmic complexity that relies on sequences shorter than any that research on RIG reported in the psychological literature would have dared to use because of the limitations of other complexity indexes.

9.3.4.2 Conclusion

RIG tasks require a sense of randomness or complexity, as well as cognitive functions such as attention, inhibition and working memory. The evolution of algorithmic complexity over the lifespan is compatible with the idea that RIG tasks, even in a computerized and shortened format, reflect such abilities. The developmental curve reveals an evolution compatible with the concept of a combination of fluid and crystallized components in cognition, with the latter predominating.

Beyond the similarity of complexity trajectories, we found that the variety of RIG tasks offered different and probably complementary information about a participant's cognitive abilities. The exact component of cognition that is assessed by RIG tasks, and which factors differentiate the tasks, are still open questions.

Our findings shed light on the developmental change in ASC, on which inductive reasoning about randomness is built. They will hopefully further our understanding of human probabilistic core knowledge. Like other complex cognitive abilities, the trend in evidence here must not occlude important intra- and inter-subject variations. Age (squared) explains about 2% of the variance in mean complexity, and 4% of the variance in CT. Although age is the predominant variable, CT and complexity are also affected, in the case of some tasks, by sex, statistical intuition and paranormal belief.

9.4 Aesthetic Preferences

Within the long-standing line of research investigating the human judgment of beauty, complexity has been a prominent issue with contradictory conclusions (for a brief review see [58]). We believe that two factors contribute to this heterogeneity (for a similar view see [59]). Firstly, many studies on the perception of beauty have used pictures, paintings, portraits or natural world objects to increase ecological validity. By doing so, they have introduced unwanted confounding factors. Studies based on more abstract stimuli are probably easier to interpret. In this respect, experiments based on non figurative material, such as two-dimensional binary grids (i.e., grids with black and white cells, e.g., [60, 61]), are of special interest as they are easily modeled in a mathematical sound way as binary matrices, which gives access to well-defined measures of such as ACSS. Secondly, what is exactly meant by "complexity" is variable from one study to another, as many definitions of complexity exist. This is problematic as different types of complexity might result in different outcomes. In a recent paper, Friedenberg and Liby [62] investigated how participants rate the beauty of artificial 2-dimensional binary grids for various levels of complexity. The authors used different indices of complexity, listed below:

- The number of parts or blocks in the pattern (a block is a maximal subset of adjacent black cells; see Fig. 9.14 for examples).
- The total edge length, which is the perimeter of the figure defined by the black cells (see Fig. 9.15 for a examples).
- The GIF compression metrics, defined as GIF/BMP where "GIF" is the size of the image file in GIF format and "BMP" the size of the BMP image file of a given grid. BMP is a non-compressed format, and GIF is a lossless compressed format

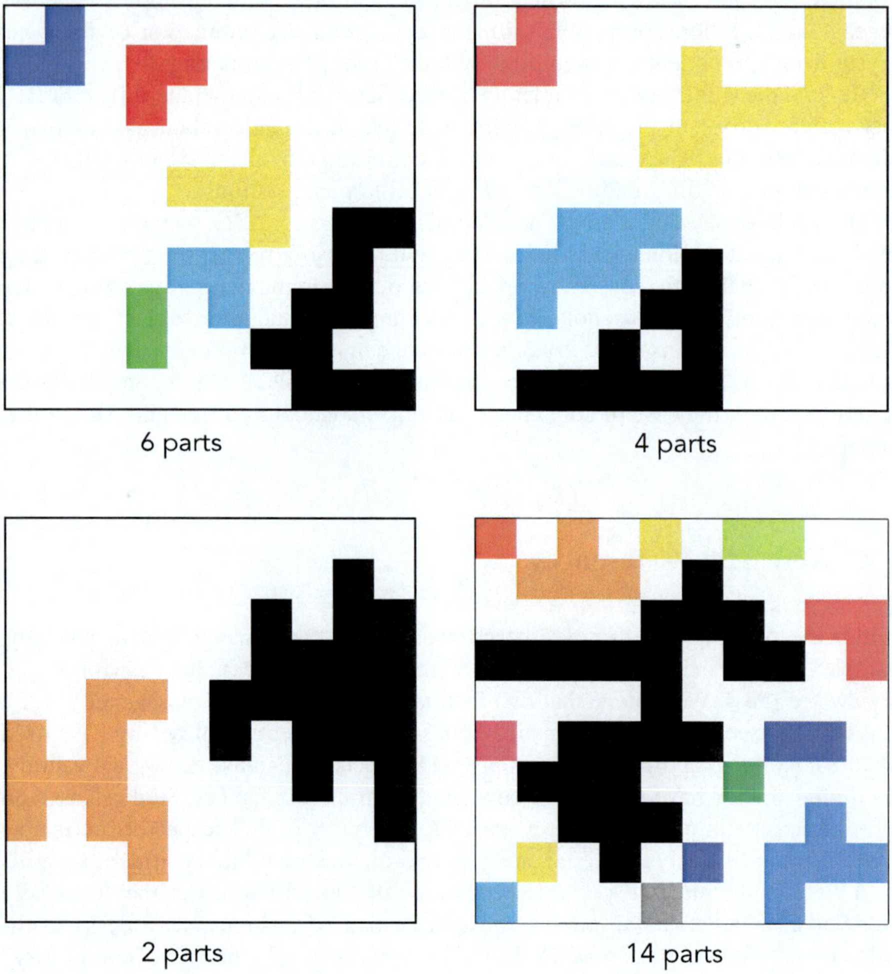

Fig. 9.14 Four examples of patterns with the corresponding number of parts. For better readability, parts have been colored

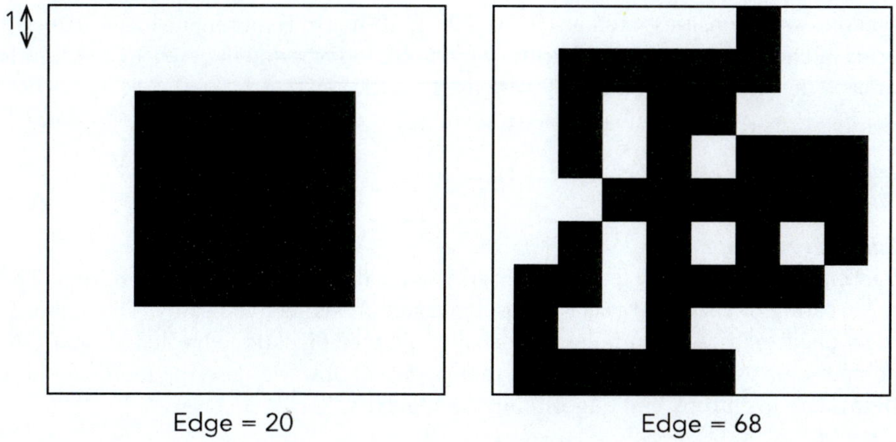

Fig. 9.15 Two patterns and the corresponding edge length, that is, the length of the black surface perimeter

based on the Lempel–Ziv–Welch (LZW) compression algorithm. Thus, the GIF compression metrics measures the non-redundancy of sub-patterns.

- The density, defined as the proportion of black cells in the grid.

Results showed that participants had a preference for evenly balanced patterns with large total edge length and high GIF compression metrics.

Crucially, in their introduction, Friedenberg and Liby [62] contrasted two fundamental notions of complexity, that is, entropy (to wit, how unbalanced a pattern is), and algorithmic complexity (to wit, how compressible a pattern is), but did not use them subsequently. Here, we reanalyze the data of both experiments using these two more conventional measures of complexity. We show that the general notion that people have a preference for complexity must be specified: they do show a preference for high entropy and more crooked figures, but there is no evidence that they prefer high algorithmic complexity per se. These results may help to understand why contradictory results have been found in previous research, and call for a cautious definition of complexity metrics.

9.4.1 General Method

9.4.1.1 Procedure

In their two experiments, Friedenberg and Liby used a series of square grids filled with white and black cells. Participants were asked to consider the white cells as the background, and the black cells as the figure. The grids (patterns) were pseudo-random, with densities fixed at 10, 20, 30,..., 100%. In Experiment 1, five 10×10 grids of each density (total: 50 grids) were used. In Experiment 2, ten 15×15 grids of each density (total: 100 grids) were used.

9.4.1.2 Measures

Subjective Beauty
Participants were asked to rate each grid in terms of their subjective beauty. The mean rating of each grid was used as a measure of subjective beauty. The 4 above-mentioned metrics of complexity (density, number of parts, edge length and GIF compression) were also used in our re-analysis. Below we describe them and how they relate to entropy and (algorithmic) complexity.

Edge Length and Number of Parts
As the definition of edge length was not fully detailed in the original paper by Friedenberg and Liby, we re-computed it with our own algorithm using `R` (Version 2.14; CRAN project; R Development Core Team, 2012). We also re-computed the number of parts. Here is the code we used to return the edge length of a matrix:

```
edge <- function(m){
  n <- nrow(m)
  res <- sum(m[1,])+sum(m[n,])+sum(m[,1])+sum(m[,n])
  for (i in 1:(n-1)){
    for (j in 1:(n-1)){
      if (m[i,j]+m[i,j+1]==1){res <- res+1}
      if (m[i,j]+m[i+1,j]==1){res <- res+1}}}
  return(res)}
```

The number of parts is a natural measure of "complexity". The more parts in the pattern, the more complex it might look on average. Nevertheless, it would be more accurately described as a measure of scatter, that is, how evenly spread the black cells are. For instance, a chessboard pattern achieves maximum possible number of parts, whereas it is usually not considered complex.

Edge length may look similar to the number of parts, but a single part with a given number of black cells can actually have very different edge lengths, depending on how crooked it is. Regular parts such as squares have small edge length, while more chaotic parts may have very large edge length (see Fig. 9.14). Edge length could be described as a measure of how crooked a figure is, which is of course linked to its algorithmic complexity as well as its entropy.

Here is the R-code we used to compute the number of parts:

```
nPart <- function(m){
  c <- ncol(m)
  l <- nrow(m)
  mcount <- matrix(c(1:(c*l)),nrow=l)*m
  ##
  change = T
  while (change == T){
    change = F;
    for (i in 1:l){
      for(j in 1:c){
        if (mcount[i,j]==0) {next;}
        if ((mcount[min(i+1,l),j] < mcount[i,j]) &
              mcount[min(i+1,l),j] != 0)
{mcount[i,j]=mcount[min(i+1,l),j]; change=T;}
        if ((mcount[i,min(j+1,c)] < mcount[i,j]) &
              mcount[i,min(j+1,c)] != 0)
{mcount[i,j]=mcount[i,min(j+1,c)]; change=T;}
        if (mcount[min(i+1,l),j] > mcount[i,j])
          {mcount[min(i+1,l),j]=mcount[i,j]; change=T;}
        if (mcount[i,min(j+1,c)] > mcount[i,j])
          {mcount[i,min(j+1,c)]=mcount[i,j]; change=T;}
      }
    }
  }
  return(length(table(mcount))-1)}
```

Entropy
Density can be related to Shannon's (first order) entropy in an exact mathematical way. In the case of a binary source such as the one used here, the entropy is a simple function of density. Indeed, the entropy H of a binary source corresponding to a density d is given by

$$H = -d \times \log_2(d) - (1 - d) \times \log_2(1 - d).$$

This function of d is an inverse U-shaped curve with a maximum at $1/2$. Friedenberg and Liby found an inverse U-shaped curve with a maximum at $1/2$ when plotting participants' ratings against density. As we will see below, this non-linear relation translates in a linear positive relation between aesthetic judgments and entropy. From here on, we will use entropy instead of density, as it carries the same information as density for our purpose.

Entropy has often been used as a measure of complexity, but as the above formula shows, it is a function of density. As a consequence, it depends on how far from a 50% density pattern the grid is, but it does not take into account how the black cells are spread across the grid. For instance, any grid with a density of 50%, including seemingly random patterns, but also regular patterns such as a chessboard, achieves maximum entropy. Although widely used as a complexity measure, entropy has been criticized for being too narrow in scope and equivalent to a χ^2 [63].

Algorithmic Complexity
GIF compression metrics is another natural means to assess algorithmic complexity. By using GIF compression, we get an estimate of the best-compressed version of the grid. However, the LZW algorithm on which the GIF compressor is based cannot detect all types of regularity. Actually, no compression algorithm can, since complexity is incomputable. In the case of the LZW algorithm, only sub-patterns repetitions can be detected. Thus, the GIF metrics would be more accurately defined as a measure of redundancy (of sub-patterns) than as a reliable estimate of the algorithmic complexity. Here, we used two-dimensional BDM (Chap. 6).

In all subsequent analyses, we set a significance threshold (alpha) at 0.01 to take multiple testing into account. Here is a list of measures of complexity used in this study and a description of the type of complexity they capture:

Entropy. Measures how unbalanced towards black or white the grid is. Does not take into account the organization of the black cells, only their density.
Number of parts. Measures how scattered the black cells are.
Edge length. Measures how crooked the black parts are. Captures the complexity of the contour.
GIF. Measures redundancy, i.e. how much sub-patterns are repeated in the grid.
BDM. Among the measures used here, this is arguably the best estimate of algorithmic complexity.

9.4.2 Results

Table 9.3 (above the diagonal) displays the correlational matrix corresponding to Experiment 1. All the measures of complexity, except the number of parts, are highly significantly correlated, illustrating the idea that they all capture some aspect of complexity. However while participants' responses are positively correlated to GIF compression metrics and edge length, they are uncorrelated to algorithmic complexity and entropy. Table 9.3 (below the diagonal) displays the correlational matrix corresponding to Experiment 2. All variables are now significantly positively linked to each other, with all p-values below 0.001, with the only exception of edge length and number of parts, which are less correlated, $r = 0.30$, $p < 0.01$.

The second experiment suggests that participants have a general preference for complexity, whatever particular aspect of complexity is taken into account, whereas the first experiment suggests that participants are sensitive to some aspects of complexity but not to algorithmic complexity per se. This apparent contradiction may come from the choice of stimuli, which increases the correlation between entropy

Table 9.3 Correlation matrix for experiments 1 and 2. BDM = Algorithmic Complexity as estimated using the Block Decomposition Method. Values above the diagonal correspond to Experiment 1, values below the diagonal refer to experiment 2. *Note* * $p < 0.01$, ** $p < 0.001$

	1	2	3	4	5	6
1. Response		0.68**	0.38*	0.20	0.35*	−0.32
2. Edge	0.88**		0.86**	0.74**	0.86**	0.27
3. GIF	0.89**	0.94**		0.91**	0.93**	0.52**
4. BDM	0.85**	0.92**	0.97**		0.95**	0.55**
5. Entropy	0.88**	0.93**	0.99**	0.99**		0.54**
6. Parts	0.49**	0.30*	0.52**	0.50**	0.49**	

and algorithmic complexity, leading to a spurious correlation between responses and algorithmic complexity.

Indeed, the authors have chosen to use an equal number of grids for each entropy level (0.1, 0.2, ..., 1). By comparison to a set of truly random grids, this artificially increases the variance of entropy: random grids are mostly of high entropy, with a low variance. For instance, in the set of stimuli used in Experiment 2, 10% of the grids have a 0 entropy (the all-black pattern with density 1, appearing 10 times). A true random matrix only has a probability of $1/2^{224}$, which is less than 4×10^{-68} of being chosen. As a consequence, entropy and algorithmic complexity are highly correlated when computed from the set of Experiment 2 stimuli ($r = 0.99$, $p < 0.0001$), but not with random 15×15 matrices. We randomly selected 500 15×15 matrices and computed entropy and algorithmic complexity. The resulting correlation is $r = 0.15$ (95% Confidence Interval: 0.06–0.24).

This suggests using partial correlations instead of correlations in order to estimate how complexity impacts aesthetic preferences independently of entropy.

In experiment 1, the correlation between responses and entropy, partialling out complexity, is still positive, $r = 0.49$, $p < 10^{-5}$, but the correlation between responses and algorithmic complexity partialling out entropy is significantly negative, $r = -0.50$, $p < 0.0001$. In experiment 2, the correlation between responses and entropy, partialling out complexity, is still positive, $r = 0.49$, $p < 10^{-6}$, on the other hand, the correlation between responses and complexity partialling out entropy is negative $r = -0.26$, $p < 0.01$.

9.4.3 Conclusive Discussion

The data from the two experiments by Friedenberg and Liby do support the notion that participants have a preference for some types of complexity—e.g., entropy—, but this is not true for all types of complexity—e.g, algorithmic complexity. We focused on 2 classical notions of complexity mentioned in the introduction, entropy and algorithmic complexity. Entropy is a function of density, with a maximum at 50%. Algorithmic complexity on the other hand is defined as the length of the shortest program that, running on a universal Turing machine (an abstract computer) produces the grid and halts. Although related, the two measures do not capture the same characteristics.

Participants did show a preference for high entropy across the two experiments. But, when controlling for entropy, they did not show any preference for high algorithmic complexity. On the contrary, participants seem to have a preference for a low algorithmic complexity when controlling for entropy.

In their introduction, Friedenberg and Liby first noticed that entropy and algorithmic complexity are the two prominent formal theories of complexity. But in the sequel, they used a series of metrics whose relations with either entropy or algorithmic complexity remain unclear. Here, we focused on the two principal mathematical concepts that have been related to complexity. Entropy is a measure of the overall unbalance between black and white cells, whereas algorithmic complexity measures a lack of structure, two different notions. The authors diverging results can be summarized in the light of entropy and algorithmic complexity: people show a preference for high entropy but a low algorithmic complexity (partialling out entropy).

The fuzzy notion of complexity comes in different flavors, and different definitions of complexity might yield dissimilar results. This might explain why contradictory conclusions have been drawn on how complexity impacts aesthetic preferences. These findings may help understand divergent results in the study of perceived beauty and complexity, but also call for the need to go beyond the unspecified term of complexity. The field would certainly benefit from a precise taxonomy of complexity.

9.5 Visual Cognition

Humans are extremely sensitive to patterns and regularities [84]. Our brains detect slight departures from randomness. Someone throwing 3 dice and getting three '6s' or the pattern '1, 2, 3' is likely to be stunned by the fact that these combinations are regular, to be incredulous in the face of this meager evidence that the dice are fair [64]. This general human feature (the discernment of rules governing the world) may be thought of as the cognitive basis of science, but also as an adaptive ability shaped by natural evolution to avoid predictable dangers.

In light of algorithmic complexity, we undertook an investigation into how humans perceive randomness, and how we learn (if we do) to perceive complexity. Hsu, Griffiths and Schreiber [65] advanced an interesting hypothesis: the frequency with which a pattern appears in real world scenes could explain how we perceive randomness, permitting us to infer complexity from the world we see. Hsu et al. [65] scanned a set of photographs of real world natural scenes and extracted every possible 4×4 array from these images. Then they computed the resulting probability distribution, and derived the randomness of each array x, defined as $random(x) = \log(P(x|r)/P(x|n))$, with $P(x|n)$ the relative frequency of the array

x in the natural scene database, and $P(x|r)$ the probability that this array appears by chance if every cell in the array is selected at random (either white or black). They chose 100 balanced arrays with probabilities of occurrence in real scenes ranging from low to high. They then had 77 subjects decide whether these arrays looked random or not. This led to a measure of subjective randomness (the proportion of participants declaring the array random) for each array.

They found that subjective probability and natural randomness were positively correlated on these particular 100 arrays ($r = 0.75,\ p < 0.0001$). We computed that Kolmogorov–Chaitin complexity is also significantly linked to the subjective perception of randomness. With the arrays published in Hsu et al. [65], we found a correlation of $r = 0.52$ ($p < 0.0001$) between two-dimensional algorithmic complexity and subjective probability. This pattern of correlations is not surprising. Because the world can be thought of as a generator of patterns, like a random computer program, the probability that an array will occur in the world is then linked to its algorithmic complexity. Indeed, as computed with the 100 arrays of Hsu et al. [65], we also found a positive correlation between algorithmic complexity and natural scene statistics ($r = 0.50,\ p < 0.0001$).

Could natural scene statistics account for human perception of algorithmic complexity? This would be in line with recent results in neuroscience reported by Berkes, Orban, Lengyel and Friser [66]. They analyzed cortical activity in ferrets, and compiled evidence in favor of the hypothesis that our brain learns an optimal internal probabilistic model of the environment, based on natural world frequencies—see also [51] for examples of children's rapid adaptation to natural frequencies. But how much of our perception of randomness is attributable to learning through the natural world? To answer this question, we performed a mediation analysis using scaled data. A regression of subjective randomness on both algorithmic complexity and natural scenes statistics gives an adjusted R-squared equal to 0.58, ($p < 0.0001$). Figure 9.16 displays the coefficients linking complexity to subjective randomness ($0.19,\ p = 0.013$) and natural scenes statistics to subjective randomness, controlling for algorithmic complexity ($0.66,\ p < 0.0001$). In this figure, "Algorithmic complexity" stands for the Kolmogorov–Chaitin complexity of the arrays, as approximated by the method described in Sect. 5.1. "Natural statistics" refers to the random function defined above, in which $P(x|n)$ stands for the frequency of array x in the natural scenes dataset. Last, "subjective randomness" is a shorthand for $\log(p(x))$, where $p(x)$ designates the proportion of participants who indicate that x is seemingly random. A Sobel test confirms the mediational role of natural scene statistics ($z = 4.78,\ p < 0.0001$).

The result suggests that our perception of complexity is partially driven by the perception of natural scenes. However, it is fair to underscore two points that may prejudice the values found here. First, we cannot control the link between "natural scene statistics" (i.e. the "random" function) and the choice of the set of pictures. Second, because the 100 arrays chosen for use here fall within certain parameters (they are all balanced, and have been chosen in such a way that they are evenly distributed on the natural scene statistics scale), variance of natural scene statistics could be artificially high. In the following experiment, we overcome these two possi-

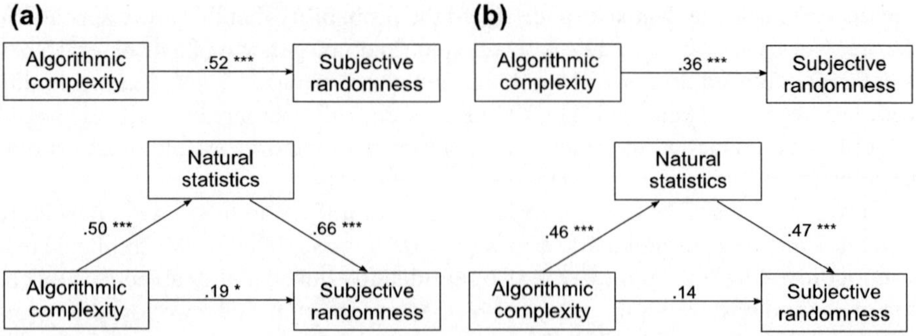

Fig. 9.16 Mediation analysis, performed with scaled data as computed from the dataset of Hsu et al. [subplot A] and with our experimental data [subplot B]. In each subplot, the top graph displays the standardized correlation coefficient. The bottom graph displays (1) the standardized regression coefficient between natural statistics and algorithmic complexity, (2) the standardized regression coefficient between complexity and subjective randomness, and (3) the partial standardized regression coefficient between natural scenes statistics and subjective randomness, controlling for algorithmic complexity. *Note* * $p < 0.05$, *** $p < 0.001$

ble drawbacks in order to get a clear view of the possible mediational role of natural scene statistics in the perception of complexity.

9.5.1 *Method*

We perform an experiment similar to the one presented above, but releasing some constraints that could affect the results. We do not impose that every pattern is balanced in terms of white and black cells. We do not choose still nature shots only. Our hypothesis is that even when these constraints are relieved, natural scene statistics will play a mediational role.

9.5.1.1 Participants

A sample of 100 participants (59 male, 41 female) was recruited via the Amazon Mechanical Turk. Hired "workers" from the Mechanical Turk were required to have a 90% approval rating on previous Mechanical Turk tasks (HITs) and at least 50 previous HITs approved. Ages in years ranged between 19 and 55 ($mean \pm sd = 30.6 \pm 8$). Participants were paid 0.30 USD for their participation. The experiment duration ranged from 84s to 289s ($mean \pm sd = 212.2 \pm 45$). Older participants in this sample showed a slight tendency to need more time ($r = 0.15$).

9.5.1.2 Stimuli

Hsu et al. [65] used a set of 62 pictures previously used by Doi, Inui, Lee, Wachtler and Sejnowski [67] to compute the natural scenes statistics. All pictures were still nature shots, including no faces, urban scenes or artificial objects. Therefore, the random function may vary if computed with other sets of pictures. To test this hypothesis, we applied the method used by Hsu et al. [65] to a new set of 100 random pictures, taken from the Wikimedia Commons database.[2] The sample included natural scenes but also animals and non-natural objects such as buildings. Then we binarized the pictures to black and white pixels using the median as the threshold. We then divided each image into 4×4 adjacent binary square arrays and calculated (for the whole set of 100 images) the probability of each square.

The resulting "random" function is strongly correlated to the data obtained by Hsu et al. [65] when computed on their choice of 100 arrays ($r = 0.91$, $p < 0.0001$), which validates the method. The correlation between natural scenes statistics (function random) and algorithmic complexity was 0.50 when computed on the 100 arrays chosen by Hsu et al. [65]. However, because the choice of arrays was not random, the correlation could well be overestimated. When computed on every possible 4×4 array found while scanning the 100 random pictures, the correlation remains highly significant, although slightly lower ($r = 0.42$, $p < 0.0001$), confirming the previous result.

We then picked at random a sample of 100 arrays from among all the arrays found in our set of 100 images, using the sample function in R. We did not contrive to obtain balanced arrays, a departure from the design of Hsu et al. [65]. Figure 9.17 displays the 100 arrays obtained by random selection.

9.5.1.3 Procedure

The procedure mirrored the one used in Hsu et al. [65], although our experiment took place online. Participants filled out a questionnaire similar to that used by Hsu et al. They were informed that a series of arrays would appear on the screen, and that their task was to decide whether the arrays were produced by a random process or by a nonrandom process. For each array, they were asked to press a button, either "random" or "not random" according to their perception.

9.5.2 Results and Discussion

The data were analyzed with the same method as Hsu et al. [65]. Algorithmic complexity is positively correlated with natural statistics ($r = 0.46$, $p < 0.0001$) and

[2] http://commons.wikimedia.org/wiki/Special:Random/Image.

Fig. 9.17 The 100 arrays used in our experiment together with their algorithmic complexity (above each array). Arrays are ordered according to their natural scene frequency (from more to less frequent arrays)

subjective randomness ($r = 0.36$, $p < 0.0001$), as are subjective randomness and natural statistics ($r = 0.56$, $p < 0.0001$).

A multiple regression of subjective randomness on algorithmic complexity and natural scene statistics yields an adjusted R-squared of 0.31 ($p < 0.0001$). Figure 9.16B displays the coefficients linking complexity to subjective randomness (0.14, $p = 0.15$) and natural scenes statistics to subjective randomness, controlling for algorithmic complexity (0.47, $p < 0.0001$). A Sobel test confirms the mediational role of natural scene statistics ($z = 3.66$, $p < 0.001$).

Perhaps as an upshot of the eschewal of constraints as compared with the Hsu et al. [65] study (balanced arrays scattered along the natural probability range), the coefficients are now smaller. However, the patterns of correlations are remarkably similar. These results suggest that natural scene statistics are indeed an important element in the perception of complexity. The correspondence between the reanalysis and the subsequent experiment also suggests that there is some objective natural probability of arrays, linked both to our perception of complexity and to the formal definition of complexity arising from Kolmogorov–Chaitin theory. Although our perception of complexity may be largely explained by natural scene statistics, this does not preempt the possibility of a complementary means of perception, which could eventually turn out to be innate.

9.6 Conspiracy Theories

Immediately after the *Charlie Hebdo* murders on January 7, 2015, in Paris, amateur theories purportedly explaining the events began to flourish online. Such fast and widespread hypotheses after major events are now familiar and are known as conspiracy theories, defined as unverified claims of conspiracy (i.e., a secret agreement between powerful individuals to enforce and hide a malevolent agenda) with sensationalistic implications.

Conspiracy theories typically involve interpreting errant data, that is, unaccounted-for elements in the official narrative, as evidence of a conspiracy rather than mere anomalies. In the *Charlie Hebdo* case, the errant data were, for example, discrepancies in different pictures of the offenders' car or the fact that one perpetrator left his ID card in the car. This highlighting of alleged incongruities is sometimes summarized by the expression "nothing happens by accident" and has been outlined as a critical mechanism of conspiracist ideation. In a similar vein, McCauley and Jacques [68], as well as Leman and Cinnirella [69], suggested that the common heuristic that big events have big causes may lead to conspiracy thinking, which indicates that major events that are targeted by conspiracy theories are not intuitively attributed to minor causes, such as chance.

Early social and psychological mechanisms for conspiracist ideation involved the role of nonclinical paranoia, but more recently, a number of factors have been found to correlate with adherence to conspiracy theories. For instance, an association with anomia (distrust toward authorities, feelings of powerlessness, and feelings of dissatisfaction about one's life) has been established across a number of studies. The emotional state of anxiety has also been linked to a higher belief in conspiracy theories, as has the political attitude of right-wing authoritarianism. People believing conspiracy theories were also prone to the conjunction fallacy (an error of probabilistic reasoning in which people overestimate the likelihood of co-occurring events). Although such findings suggest a polarized or irrational mindset, Franks, Bangerter, and Bauer [70] argued that conspiracist ideation could serve important social and cultural functions, such as making sense of ambiguous, threatening events and seeking to restore order in an uncertain world.

In this view, conspiracist ideation has been associated with a tendency to downplay or deny coincidences and a need to perceive structure [71]. According to this approach, the conspiracist mind sees specific events as too complex or convenient to have arisen by chance alone—as illustrated by the expression "nothing happens by accident"—and this leads to a preference for sophisticated plots.

Whitson and Galinsky [71] investigated a link between the rejection of randomness and conspiracist ideation. In their framework, lack of experienced control over events increases the (illusory) perception of patterns in noisy backgrounds. Belief in conspiracy theories would thus be akin to "the identification of a coherent and meaningful interrelationship among a set of random or unrelated stimuli" (p. 115). They found that participants who lacked a sense of control tended to perceive a greater likelihood of conspiracy than did participants who had a sense of control. However, the scenario used in this study to measure conspiracy belief was an interpersonal conspiracy (involving colleagues and a boss), not a large-scale conspiracy as commonly understood. Moreover, Van Prooijen and Jostmann [72] noted that the evidence for a direct relationship between uncertainty and conspiracy beliefs is mixed. Indeed, people sometimes preserve a sense of order by increasing the faith they have in governmental institutions.

Further evidence suggesting a link, albeit indirect, between conspiracist ideation and the rejection of randomness comes from the correlation found between endorsement of conspiracy theories and paranormal beliefs. Such beliefs have repeatedly been found to be associated with the misperception of random noise as ordered patterns [17].

Thus, although there seems to be wide agreement about the importance of randomness perception for the belief in conspiracy theories, direct evidence to support this claim is lacking. Here, we set out to explore the relationship between the ability to judge the randomness of binary strings and conspiracist ideation. We predicted that participants who tend to judge binary strings to be less random overall would be more prone to conspiracist ideation. This is a very straightforward and conservative test of the hypothesis that conspiracist ideation is linked to a low prior for randomness (i.e., a general tendency to dismiss randomness as a possible cause for an event). If this is the case, then subjective ratings of randomness for binary strings among believers in conspiracy theories should be lower than average, which would provide strong evidence that, for these people, nothing indeed happens by accident.

9.6.1 General Method

In all three experiments reported in this section, we tested whether increased belief in conspiracy theories correlates with a lower prior for randomness, following the experimental design used by Matthews [42]. Participants were presented with strings of 12 Os and Xs that represented binary outcomes and were asked to rate the randomness of each string. This rating was our measure of subjective randomness. The current theory of randomness perception is based on the idea that when people face a given stimulus s, they estimate the probability that s is random (R), as opposed to the hypothesis that it was created by an unknown deterministic process (D), in a way that is similar to applying Bayes's rule:

$$p(R|s) = \frac{p(s|R)p(R)}{p(s|R)p(R) + p(s|D)p(D)}$$

On the right-hand side of this formula, $p(s|D)$ refers to the probability that s occurs from an unknown nonrandom process. $p(R)$ is the prior for randomness (i.e., the subjective probability assigned by the participant to the a priori hypothesis that the process leading to s is chance). Because $p(R|s)$ decreases with $p(R)$, the average subjective randomness score for a given participant is an indication of his prior for randomness.

In addition, to ascertain that participants did in fact behave in a way that is correctly modeled by this framework, we compared subjective randomness with the normative value arising from algorithmic information theory. In this formal approach, $p(s|D)$ is the algorithmic probability of s (approximated by $D(5)$ in Chap. 3). Setting a prior for randomness (usually at 0.5, following the conventions of Bayesian methods), we could compute $p(R|s)$ using Bayes's formula. In what follows, we refer to the normative value of $p(R|s)$ as *true randomness*. We computed the true randomness of binary strings using function `prob_random` included in the `acss` R-package. This approach allowed determination of the extent to which participants' judgments of the randomness of meaningless binary strings approached a robust algorithmic approximation of the actual content of those strings, in terms of complexity. In other words, it measures whether subjective (psychological) assessments of randomness matched a purely mathematical construct of randomness that is unlikely to be consciously accessible.

9.6.2 Experiment 1

9.6.2.1 Method

Participants ($N = 107$) first read the following instructions:

> In this questionnaire, you will have to judge sequences composed of the two symbols X and O. They may correspond to heads or tails, to even or odd digits, or to series of successes and failures. For instance, the series "tails, heads, heads, tails" could be represented by OXXO. The series "success, failure, success" could be represented by OXO, etc. Some of these sequences were created using a real fair coin, which means that they are random series. Other sequences, however, correspond to nonrandom processes: They may come from computer calculations or from sport performances (for example, series of success and losses of a basketball team). For each sequence, your task is to decide whether the sequence was produced by a truly random process or could be the outcome of nonrandom processes; you will do this by selecting the response that seems more appropriate to you, from "certainly random" to "certainly not random" or any intermediate response. Please answer as fast and accurately as possible.

Forty 12-character strings of Xs and Os were displayed in the questionnaire. The strings were semi-randomly chosen in such a way that they would cover the whole range of possible complexity. Participants were instructed to rate the randomness of each string on a 6-point Likert scale with anchors of certainly random and certainly not random. To increase the readability of the statistics, we recoded this rating so that 0 corresponded to certainly not random, 1 corresponded to certainly random, and 0.2, 0.4, 0.6, and 0.8 corresponded to intermediate responses.

In a second part of the study, ostensibly an unrelated project aimed at validating a French translation of a questionnaire on political opinions, participants filled out the 15-item Generic Conspiracist Beliefs Scale [73], which captures generic conspiracist ideation without alluding to specific conspiracies. They also completed four items, taken from Wagner-Egger and Bangerter [74], measuring adherence to classical conspiracy theories (the September 11 attacks, the assassination of John F. Kennedy, Princess Diana's car crash, and the Apollo 11 moon landing). Next, participants read two scenarios describing situations of potential conspiracies, one interpersonal and the other political, and were asked to rate the perceived probability of conspiracy in these situations. These four conspiracy scales (general, classical, interpersonal, and political) were counterbalanced in four orders. Finally, participants were asked to provide some demographic information.

9.6.2.2 Results and Discussion

No order effect was found for the measures of adherence to conspiracy theories. We thus aggregated the data across the four orders. The reliability of the scales used was satisfactory. All measures of belief in conspiracy theories were significantly associated with each other. However, subjective randomness was not significantly correlated with any measures of belief in conspiracy theory.

The mean subjective randomness of strings was linked to their true randomness, $r(38) = 0.52$, $p < 0.001$. Figure 9.18 displays a scatterplot showing mean subjective randomness of strings as a function of true randomness.

The subjective-randomness measure showed that there are indeed individual differences in global perception of randomness. However, the data did not support our initial hypothesis that conspiracist ideation is linked to a low prior for randomness, because no conspiracy theory subscale correlated with subjective randomness.

The mean subjective randomness of a binary sequence was (moderately) correlated with its true randomness (Fig. 9.18). This finding confirmed that our participants had a sound (albeit approximate) perception of randomness or complexity (i.e., an implicit sense of what true randomness looks like), and thus that they genuinely attempted to perform the task. All measures of belief in conspiracy theories were correlated with each other, thus replicating the now standard finding that conspiracy-theory beliefs involve a monological belief system.

Thus, although the data replicated two robust findings from the literature on randomness perception and conspiracist ideation, Experiment 1 failed to provide evidence of a link between the two domains. In this experiment, randomness was

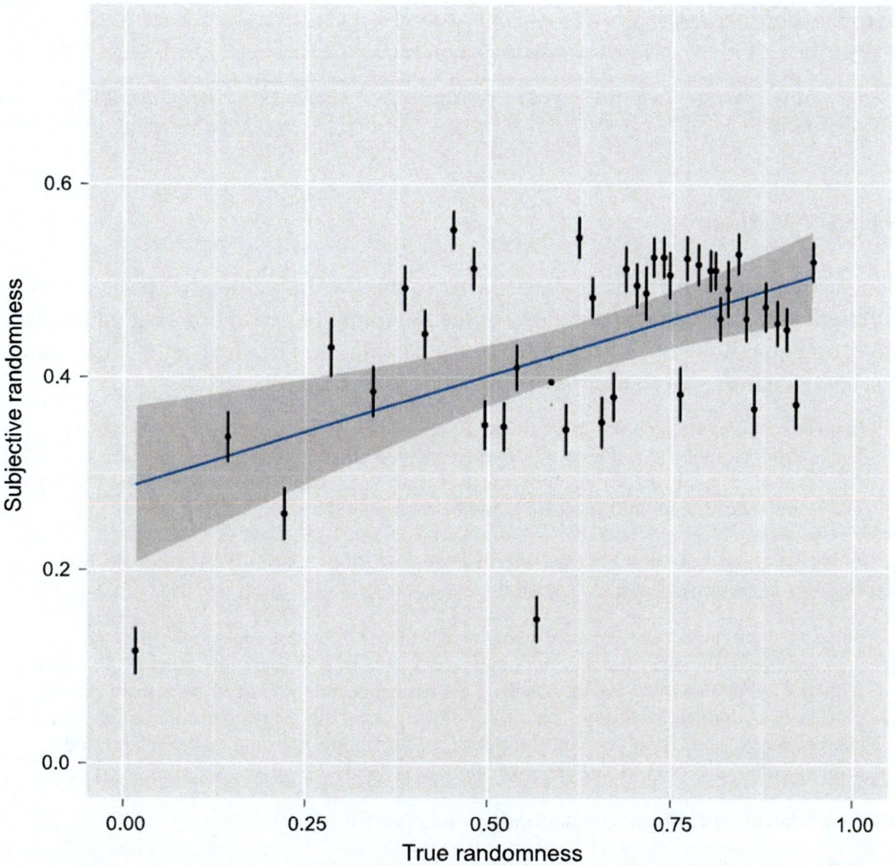

Fig. 9.18 Results from Experiment 1. Scatterplot (with best-fitting regression line) of mean subjective randomness as a function of true randomness (with a prior set at 0.5). Error bars represent ±1 SEM. The shaded area represents the 95% confidence interval around the regression line

contrasted with an unspecified deterministic process. However, previous research has shown that judgments of randomness can be modulated by contextual factors, potentially leading to differential assessments of chance, mechanistic, or intentional processes. Therefore, it may be that conspiracist ideation is linked to a low prior for randomness but only when people are resolving a choice between randomness and human agency, and especially a malevolent agency. Experiment 2 was designed to address this possibility.

9.6.3 Experiment 2

Participants were 123 first-year psychology students. None took part in Experiment 1.

9.6.3.1 Method

The method for Experiment 2 was similar to that for Experiment 1, but with a key difference: There were two versions of the randomness-perception task, one for the inventing condition and one for the cheating condition. The first set of instructions for both versions was similar to the instructions in Experiment 1:

> In this questionnaire, you will have to judge series composed of the two symbols X and O. They may correspond to heads or tails, to even and odd numbers, or to series of successes and failures. For instance, the series "tails, heads, heads, tails" could be represented by OXXO. The series "success, failure, success" could be represented by OXO, etc.

After this, there was a second set of instructions. In the inventing condition, the second set read as follows:

> Some of these series were created using a real fair coin, which means that they are random series. On the other hand, some other series have been invented by humans. For each sequence, your task is to decide whether the sequence was produced by a truly random process or by human invention; you will do this by selecting the response that seems more appropriate to you, from "certainly random" to "certainly invented" or any intermediate response. Please answer as fast and accurately as possible.

In the cheating condition, the second set read as follows:

> In the context of a game, two players have to isolate themselves in a room, toss a coin a couple of times, and write down the obtained series. One of the players correctly follows the rule and tosses the coin, but the other intentionally cheats: He makes up the series without tossing the coin. For each sequence, your task is to decide whether the sequence was produced by the honest player (randomly produced by tossing a coin) or by the cheater (intentionally produced); you will do this by selecting the response that seems more appropriate to you, from "certainly random" to "certainly invented" or any intermediate response. Please answer as fast and accurately as possible.

Participants then had to rate the randomness of each 40 strings of Xs and Os (the same strings as in Experiment 1) on a 6-point Likert scale with anchors of certainly random and certainly not random. The rationale for these two conditions was to allow investigation of the possible role of nefarious intent (i.e., cheating) compared with a more neutral intent (i.e., inventing) because conspiracy theories are usually defined as involving deliberate mischief rather than mere hoaxing. Participants were randomly assigned to one condition or the other. We deliberately did not specify how an actual cheater would behave in such a situation, leaving it to the participants to resolve its paradoxical nature: Indeed, a perfect cheater would create highly complex series, and a very clumsy (or maximally mischievous) cheater would come up with highly ordered series. This paradox is inherent in conspiracy theorizing: The plotting agents can be considered either inordinately efficient in hiding their tracks (leaving an event to be "too good to be true") or transparently flawed (leaving "too many coincidences" behind them).

The second part of the study was similar to the second part of Experiment 1, except that the order of the four conspiracy-theory scales was kept constant because no order effects had been observed in Experiment 1.

9.6.3.2 Results and Discussion

The reliability of the scales used was satisfactory overall. There were no significant differences in subjective randomness between conditions, Welch two-sample t test, $t(118.1) = 0.76$, $p = 0.45$, or between Experiment 1 and Experiment 2 results, Welch two-sample t test, $t(202.1) = 1.49$, $p = 0.14$. The mean subjective randomness of strings was highly correlated in the two conditions, $r(38) = 0.95$, $p < 0.001$, and moderately correlated with true randomness-inventing condition: $r(38) = 0.40$, $p < 0.05$; cheating condition: $r(38) = 0.51$, $p < 0.001$. As in Experiment 1, subjective randomness was not significantly correlated with any of our four measures of belief in conspiracy theories. This was true in both conditions.

Whereas Experiment 1 contrasted randomness with an unspecified deterministic process, Experiment 2 used a more specific, nonrandom anchor, especially relevant to conspiracy theories: a deliberate human intervention that, in the cheating condition, could be interpreted as an attempt to induce the appearance of randomness. We found no significant link between subjective randomness and any of the conspiracy theory scales. Again, the data did not support our primary hypothesis, but they did confirm that participants were able to distinguish randomness from structure. In light of the cheater's paradox highlighted earlier, the results suggest that both people who believe in conspiracy theories and those who do not use similar criteria to discern random from created sequences.

These negative findings could result from a lack of heterogeneity in our sample of psychology students and perhaps an excess of binary strings for them to evaluate, which could have induced tiredness and stereotypical responses. Experiment 3 was designed to improve on these possible issues and thus reinvestigated the cheating condition with increased statistical power, a more diverse population than in the previous experiments, and a shortened paradigm.

9.6.4 Experiment 3

Participants were recruited via social networks and e-mails and redirected to a URL for an online experiment; the URL remained available for 1 week after the initial announcement. After a week, 217 French-speaking participants had freely taken part in the experiment.

9.6.4.1 Method

The method in Experiment 3 was similar to that in the two previous experiments. In this experiment, however, we contrasted randomness only with cheating (or intentional deceit). The experiment was run online to obtain a larger and more varied sample of participants. Finally, we included additional items related to political ideology and optimism to control for other possible variables related to conspiracist ideation. Moreover, in Experiment 2, the instructions mentioned two players, one of whom cheated. This may have triggered a 0.5 prior that would not have come into play otherwise. Thus, in Experiment 3, we provided no hint about the prior probability, to avoid this possible shortcoming.

In the first part of the study, participants completed the randomness-perception task. They were asked to read the following instructions:

> In the context of a game of chance, participants used a coin bearing an X symbol on one side and an O on the other. They had to throw the coin 12 times and write down the obtained sequence of Xs and Os. In what follows, you will see some of these sequences. Some of them have been correctly produced by the throwing of the coin. However, we know that others have been made up by cheating participants; instead of throwing the coin, they directly wrote down a sequence of Xs and Os. For each sequence, your task is to decide whether the sequence was produced by a truly random process or by a cheater; you will do this by clicking the response that seems most appropriate to you, from "certainly random" to "certainly cheating" or any intermediate response. Please answer as fast and as accurately as possible.

Fifteen 12-character strings of Xs and Os were then displayed on the screen, one after the other. The strings were chosen in such a way that they covered the whole range of possible complexity. Participants were asked to rate the randomness of each string on a 6-point Likert scale with anchors of certainly random to certainly cheating.

The second part of the study was similar to the second part of Experiment 1. Finally, they rated their own political general orientation on a 6-point Likert scale with anchors of far left (liberal) and far right (conservative), and they rated their pessimism and optimism on a 6-point Likert scale with anchors of extremely pessimistic and extremely optimistic.

9.6.4.2 Results and Discussion

The reliability of the scales used was satisfactory. Contrary to our initial hypothesis, the correlation between conspiracist ideation and subjective randomness was again found to be nonsignificant, $r(217) = -0.10$, $p = 0.13$, 95% confidence interval, or $CI = [-0.23, 0.03]$. (A power analysis indicated that our sample size was sufficient for the experiment to have 80% power to detect any correlation greater than 0.17 at a p level of 0.05.) Likewise, we found no correlation between self-reported optimism and general conspiracist ideation. However, scores for optimism and randomness were positively correlated: More optimistic participants were more prone to perceive a string to be randomly produced; correspondingly, pessimistic participants were more

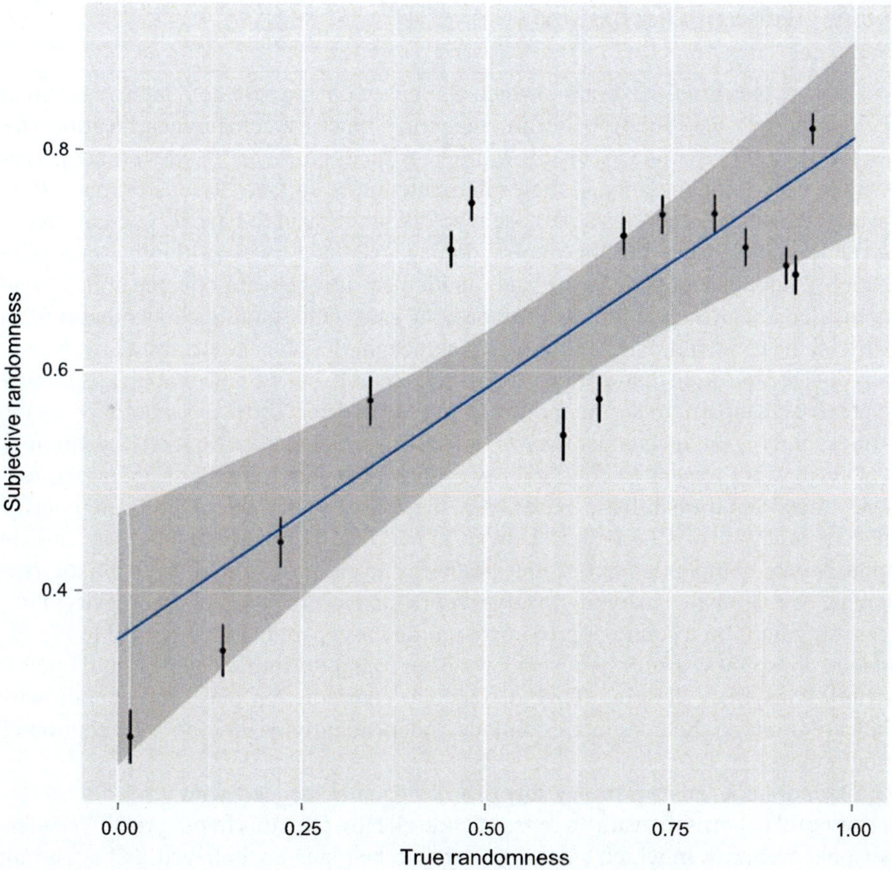

Fig. 9.19 Results from Experiment 3. Scatterplot (with best-fitting regression line) of mean subjective randomness as a function of true randomness. Error bars represent ± 1 SEM. The shaded area represents the 95% confidence interval around the regression line

prone to perceive a string to have been produced by a cheater. Finally, a weak but significant positive correlation between conspiracy ideation and political background was found: Participants with a liberal ideology were slightly less prone to conspiracist ideation than participants with a conservative ideology.

There was a strong correlation between subjective and true randomness, $r(13) = 0.84$, $p < 0.001$: Participants' intuitions about the probability of randomness conformed to the normative view given by algorithmic probability (Fig. 9.19).

In Experiment 3, we specifically contrasted randomness with cheating (i.e., malevolent deception) and used a large and diverse sample, which maximized the likelihood of confirming our initial hypothesis that people with conspiracy-theory beliefs have a lower prior for randomness. However, even in this particular condition and despite a high power, no significant correlation appeared.

9.6.5 General Discussion

Contrary to our initial prediction, which was based on a widely held interpretation of the psychology underlying belief in conspiracy theories, conspiracist ideation was not found to be associated with a low prior for randomness, even when randomness was opposed to deliberate and malevolent intentions. In fact, both participants who favored conspiracy theories and those who did not proved to have an accurate sense of randomness, attuned to the true randomness defined by algorithmic complexity theory. Although the concept of true randomness might seem counterintuitive and its mathematical foundations are not easy to grasp, this finding shows that human participants can tap into it while looking at meaningless binary strings. How exactly they do it is a question that we defer to further research, but we note that (a) this finding guarantees that participants genuinely attempted to detect order or randomness in the strings, and (b) the mechanisms involved seem unrelated to conspiracist ideation.

Our findings appear to contradict two lines of research. First, an increased tendency to exhibit probabilistic biases (e.g., the conjunction fallacy) has been reported in conspiracy-theory believers. However, our paradigm investigated the human approximate-complexity sense, not the ability to perform sound probabilistic reasoning. Probabilistic reasoning involves more than a perception of complexity: Poor reasoning built on a sound approximate randomness perception may lead to biases. In fact, conspiracy-theory believers have been found to simultaneously hold contradictory conspiracy theories. This tolerance to contradiction may well explain why they are prone to the conjunction fallacy, independently of their core perception of randomness.

Second, belief in conspiracy theories has been associated with a need to detect meaningful patterns to restore a sense of control. However, this finding relied on interpersonal scenarios in which a conspiracy could be plausibly believed. In the current study, we used a more conservative approach to investigate randomness perception, aiming at a core randomness sense and focusing mostly on large-scale conspiracy theories.

It thus seems that the nothing-happens-by-accident heuristic, if it applies at all in the case of conspiracist ideation, would involve high levels of information processing and cognitive integration rather than a core, low-level, and quasisensory disposition. This does not preclude the need for further research on deep individual idiosyncrasies that could shed light on conspiracist ideation, but the present negative finding certainly puts the focus on higher-level approaches to understand, and possibly remediate, this increasingly widespread phenomenon. Climate-change skepticism is an analogous phenomenon: Research has shown that it is unrelated to scientific literacy but is strongly dependent on more general ideological and cultural factors. Negative results, such as the one we obtained in the current study, can thus help narrow the psychological and social mechanisms involved in specific belief structures.

In conclusion, one cannot take the expression "nothing happens by accident" at face value as an explanation for conspiracist ideation. Nevertheless, the lack of a direct link between a general low prior for randomness and conspiracy-theory beliefs does not rule out the possibility that a low prior for randomness could in fact be a consequence, rather than a cause, of conspiracist ideation for general or specific conspiracy-theory beliefs. Consequently, people who believe or even develop conspiracy theories themselves may dismiss any intervention of chance for events that they perceive to be best explained by their imagined conspiracy.

9.7 The Emergence of Grammar

Cultural traits and social conventions owe not just their survival but also their structural characteristics to the process of cultural transmission. Starting with experimental studies of iterated reproduction of narratives and drawings [75], a growing line of research investigating technological and communicative systems demonstrates that cultural transmission of information across generations of learners reduces entropy, thereby rendering cultural traits functionally relevant and suitable to serve as social markers. This issue has received particular attention in research on the evolution of language, where it has been suggested that structure emerges because language is gradually shaped by the information processing capacity of successive generations of learners. Computational and experimental studies of iterated language learning show that languages with inconsistent mappings between forms and meanings become progressively more learnable and more compositional, i.e. they acquire a more systematic and predictable relationship between form features and meaning dimensions. Iterated learning amplifies the biases that affect learners' inductive inferences leading to gradual emergence of structure. Moreover, cultural transmission reveals not just inductive biases that operate during iterated learning of artificial languages or even meaningless colour sequences but also general memory biases that operate during serial reproduction of information. However, experimental studies of cultural transmission have so far only focused on adult learners who generally are at peak cognitive ability, yet many cultural traits and social conventions are learned and transmitted by children whose information processing capacity is more limited. To fully understand how processing constraints shape the biases that lead to the emergence of structure it is important to compare cultural transmission between cognitively and socially mature and immature learners. This is the aim of the study presented in this chapter.

Inspiration for this study comes from the domain of language learning, where the existence of regularizing biases has been attributed to either language-specific or domain-general mechanisms. Language-specific biases imply access to an innate Universal Grammar [76], often conceived of as an evolved adaptation to the requirement of constraining the hypotheses of learners faced with limited input [77]. Domain-general biases, on the other hand, arise from information processing constraints, which (according to the Less-Is-More Hypothesis [78]) can facilitate language learning. Several mechanisms have been proposed to account for beneficial effects of processing limitations. On the one hand, limited working memory capacity may make it easier for children to attend to smaller units, e.g. grammatical morphemes, in input sequences, which can go unnoticed by adults processing larger input segments. Connectionist simulations using recurrent networks [79] and experimental evidence from adult learners have demonstrated that imposing limitations on working memory capacity, or on input complexity, leads to superior outcomes in terms of mastery and generalisation of the underlying linguistic structure. On the other hand, regularisation and emergence of structure have been linked to immature executive control. [80] showed that when inputs contained inconsistencies in the use of grammatical forms, adults tried to match the probabilities of occurrence of each form, while 5–7-year old children regularised the inconsistencies by using predominantly one or the other form. Thus, compared to adults, children are less able to inhibit pre-potent responses, although adults have been found to regularise as well if increased input complexity taxes their processing resources [81]. Both of these domain-general mechanisms lead to the prediction that structure should emerge more rapidly during cultural transmission of information in chains of children than of adults, a prediction that is in line with the notion that children play a crucial role in creolisation, e.g. as described for the emergence of Nicaraguan Sign Language [82].

One of the reasons direct experimental evidence comparing iterated language learning between adults and children has not yet been provided may be due to the difficulty of administering artificial language learning regimens to large numbers of children. During artificial language learning, learners are exposed to multiple exemplars one at a time, and rely on frequency of occurrence to make inductive probabilistic inferences about the underlying rules that generate these examples. This is a time-consuming process difficult to implement in a sufficiently controlled manner on a scale large enough to allow for statistical comparison between chains of children and of adults. In an attempt to provide proof of concept, the present study therefore investigates serial reproduction of visuo-spatial patterns, assuming that similar regularising biases operate during transmission of information dispersed not in time but in space. We asked learners to recreate patterns of dots distributed randomly on a grid by placing the same number of dots into the appropriate cells of an empty grid. To ensure task compliance, we used a sticker task (a familiar and highly enjoyable activity for children). As with sequence learning, learning of spatial configurations also requires learners to infer rules that can be used to reconstruct patterns for recall.

We take this procedural component as being loosely analogous to learning grammar. For sequences and for spatial configurations, emergence of structure can be conceptualised as reduction in Kolmogorov–Chaitin (or algorithmic) complexity. It captures the notion of structure in terms of the length of the procedure required to generate it (the shorter the algorithm required to generate a pattern, the more structured and less complex the pattern is). This approximation is superior to alternative attempts to capture degree of non-randomness such as measures based on the mere distribution of outcomes, as it does not focus on intuitively plausible biases that may affect randomness generation but rather constitutes an objective measure that captures all potential biases. Crucially, if processing constraints shape the biases that lead to the emergence of structure, we expect that algorithmic complexity will be reduced faster in transmission chains of children than of adults.

9.7.1 Method

Participants: Ninety adults, all undergraduate students, participated in the study, comprising nine diffusion chains of 10 generations each. Three randomly generated seed patterns were used to start three diffusion chains per seed. Ninety primary school pupils aged 5;1 to 8;2 years also participated in the study. 30 children were recruited from grade 1, 30 from grade 2, and 30 from grade 3. There were three diffusion chains per grade level, each starting with a different one of the three random seeds.

Materials: We prepared 180 grids of size 10×10 comprising an area of 15×15 cm., printed on A4 paper, and slips containing 12 black round stickers with a diameter of 12 mm each. To control for effects of starting conditions across age groups, three seed patterns were generated by randomly selecting 12 numbers out of 100 using the Excel random number generator, and equally distributed across the adult and the child chains, matching for primary school grade level.

Procedure: Participants were given the 12 stickers, and seated at a table with a sheet containing the empty 10×10 grid in front of them. They were instructed to look at a pattern of 12 black dots distributed on the grid for 10 s, and to recreate this pattern by placing the provided stickers inside the cells of the empty grid. The experimenter then placed the random dot pattern (at the beginning of each chain), or the dot pattern produced by the previous participant, on top of the empty grid for 10 s, which were timed using a stopwatch. After removal of the target pattern, participants placed all 12 stickers onto the empty grid.

9.7.2 Results

Below we report the analyses of four dependent variables that allowed us to quantify learnability, structure and identifiability of the patterns. Each variable was analysed using Page's L trend test for $m = 9$ chains and $n = 10$ generations, to identify continuous trends over the course of transmission. This test was conducted separately for children and for adults. Each variable was also analysed using a $10 \times 2 \times 3$ mixed-type ANOVA with Generations (1–10) as repeated measures, and Age (children vs. adults) and Random Seed (1–3) as between-subjects factors. This analysis provides information about whether there are differences between children and adults, whether age differences depend on position in the transmission chain, and whether the initial conditions influence the outcome. A second $2 \times 2 \times 3$ mixed-type ANOVA combined values from early (1–3) and late (4–10) generations, to ascertain whether trends that occur over the course of transmission interacted with age or initial conditions.

Transmission Accuracy: Transmission accuracy was calculated as the percentage of dots correctly placed on cells in the grid, and showed continuous improvement in children and in adults (see Fig. 9.20). The ANOVA revealed a significant effect of Generations but no effects of Age, Seed or any of the interactions. Combining data for the three early and the seven late generations confirmed that transmission accuracy

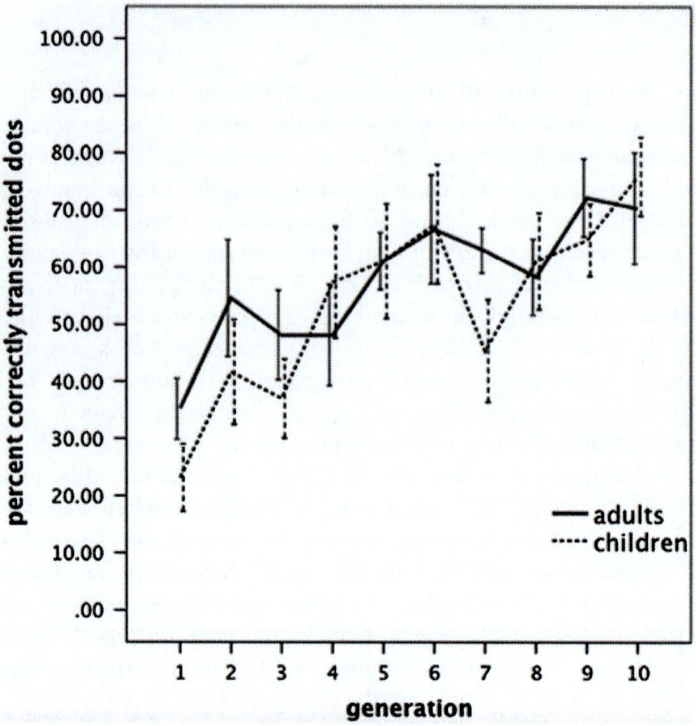

Fig. 9.20 Mean transmission accuracy per generation for adults (solid line) and for children (dotted line). Error bars indicate ± 1 S.E.M

increased, an effect that did not interact with Age or Random Seed. Thus, for children and adults, transmission accuracy increased at the same rate, independently of the specific random pattern presented at the outset.

Clustering: One way in which structure can be imposed on random dot patterns is by combining them into clusters of connected dots, a process akin to chunking (a mechanism that can reduce memory load, in this case for the spatial positioning of the dot clusters). Here, a cluster was defined as any group of dots placed into cells sharing either a side or a corner. Page's L trend test showed continuous reduction of the number of clusters in children and in adults (see Fig. 9.21). The ANOVA revealed significant main effects of Generation and Age, but no effect of Seed or of any of the interactions. Combining data for the three early and the seven late generations confirmed these main effects. Thus, numbers of cluster diminished over the course of transmission with children producing fewer cluster than adults from quite early on.

Algorithmic Complexity: Algorithmic complexity of the two-dimensional patterns was computed using BDM (Chap. 6). In this study, each 10×10 grid thus yielded sixteen 4×4 tiles with a maximum overlap of 2 columns or 2 rows. Given

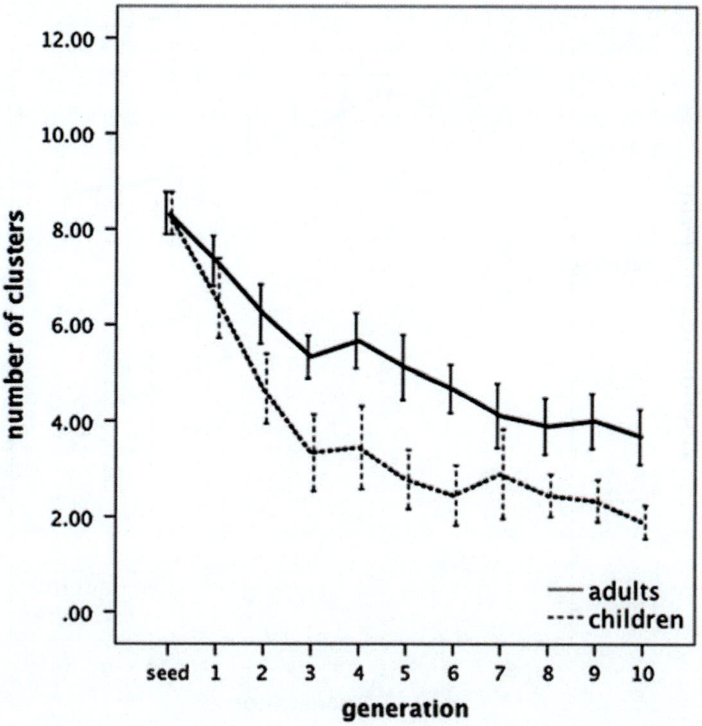

Fig. 9.21 Mean number of clusters per generation for adults (solid line) and for children (dotted line). Error bars indicate ± 1 S.E.M

the two-dimensional complexity of each tile, we computed an estimation of the pattern complexity using the formula:

$$\sum_{p}(\log_2(n_p) + K(p)),$$

where p denotes the different types of 4 × 4 patterns, n_p the number of occurrences of each type of pattern p, and $K(p)$ the complexity of p as computed using CTM (Sect. 5.1). For this measure, Page's L trend test showed a continuous reduction in children and in adults (see Fig. 9.22). The ANOVA revealed significant main effects of Generation and Age, but no effect of Random Seed or of any of the interactions. Combining the data for the three early and the seven late generations confirmed the decrease in complexity, as well as the effect of Age, yet showed no interactions. These results indicate that complexity decreased over the course of transmission, and that it was lower overall in the children.

Identifiability: To maintain consistency in starting conditions across children and adults, we had presented identical seeds to both groups. This allowed us to test whether initial conditions determined the particular attractor states a given transmis-

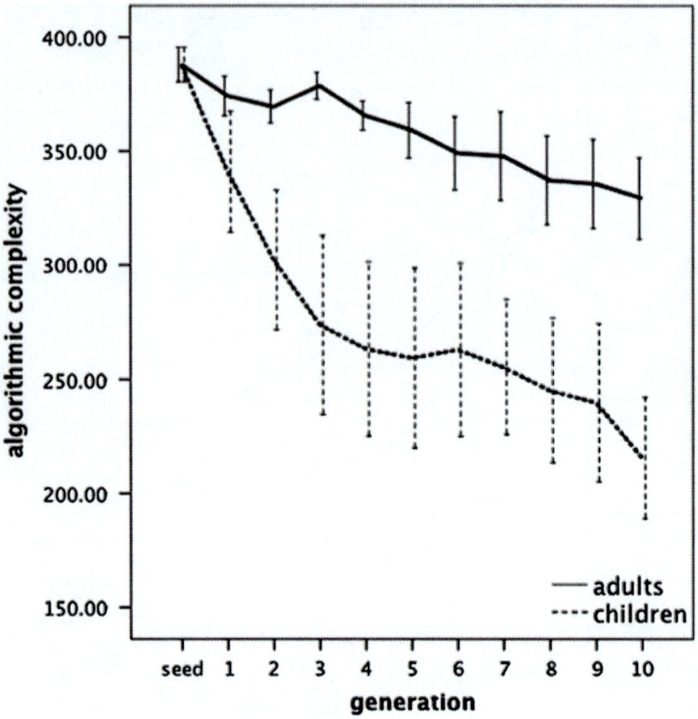

Fig. 9.22 Mean estimated algorithmic complexity per generation for adults (solid line) and for children (dotted line). Error bars indicate ± 1 S.E.M

sion chain had settled into, or whether transmission gave rise to uniquely identifiable lineages despite identical initial conditions. For each pattern, we quantified identifiability by computing the proportion of within-chain similarity divided by the sum of within-chain similarity and across-chain similarity, where across chain similarity was defined as similarity of patterns originating from the same seed in a given generation. Values greater than 0.5 indicate greater within-chain similarity; values below 0.5 indicate greater similarity with patterns of the same generation in the different chains originating from the same seed.

As Fig. 9.23 illustrates, from the first chain onwards, each reproduced pattern resembled other patterns within the chain more than other patterns of the same generation in other chains originating from the same seed. A one-sample Wilcoxon's signed ranks test for each generation showed that for the adults, identifiability was significantly greater than the expected chance level of 0.5 for all but generation 1 (all $Zs > 2.42$, $p < 0.05$), and for the children, all generations exceeded the chance level (all $Z > 2.10$. $p < 0.05$). Thus, even when originating from the same random seed, by generation 2 patterns already started forming distinct lineages. Page's L trend test showed that for adults and children, resemblance to other patterns within a chain increased over the course of transmission. The lack of effects of Random Seed

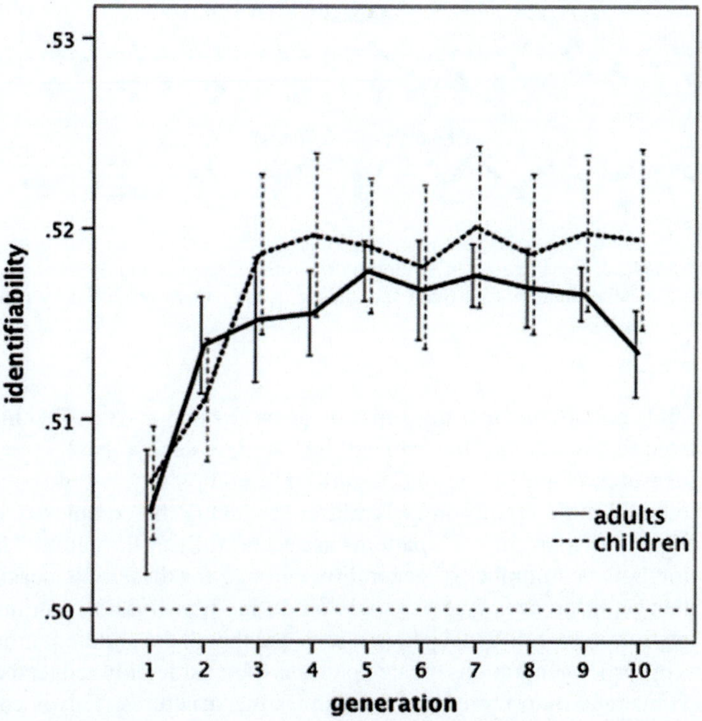

Fig. 9.23 Mean identifiability per generation for adults (solid line) and for children (dotted line). The dashed line represents chance level. Error bars indicate ± 1 S.E.M

and Age in both ANOVAs indicates that formation of lineages was not contingent upon initial conditions, and that children did not differ from adults in terms of how much each pattern resembled other patterns within its lineage.

9.7.3 Discussion

Our study tested whether children and adults differed in terms of how fast learnability improved and structure emerged during transmission of random dot patterns. Consistent with findings on iterated learning of artificial languages and random sequences in adults as well as recent findings on cultural transmission of visuo-spatial patterns in non-human primates [83] we found that transmission accuracy of visuo-spatial patterns increased in children as well, and to a similar extent as in adults. However, compared to adults, visual working memory capacity of children is limited, and children's memory representations of visuo-spatial pattern are less stable, reaching adult levels of performance in visual working memory tasks only around age 7. Given these

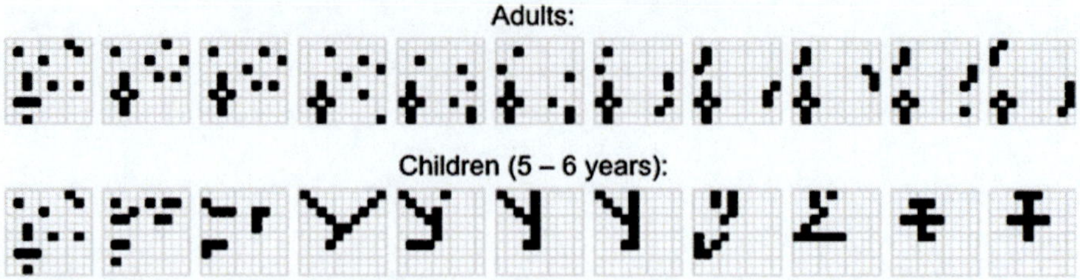

Fig. 9.24 Example of transmission of random dot patterns in adults and 5–6-year-old children, starting from the same random seed (left) and ending in the pattern produced by generation 10 (right)

limitations, it is remarkable that transmission accuracy in 5–8-year-old children did not differ from that of adults. How can children achieve such a feat?

Using measures of clustering and algorithmic complexity, we discovered that structure emerged more readily in the children, reducing the complexity to a level that allowed them to reproduce the patterns as successfully as the adults. The lack of an interaction between number of generation and age for these two measures indicates that the age differences emerged early on during transmission, as illustrated in Fig. 9.24. Inspection of individual patterns revealed that in the children, more radical innovations in earlier generations created patterns of considerably reduced complexity, which converged onto more easily transmissible structures. This is compatible with findings that show an inverse relationship between amount of information transmitted and rate of change during transmission. How can children's processing limitations encourage such structural innovations? One possible explanation is related to the reconstructive nature of memory: When recalling complex visual patterns, learners may be biased by acquired prototypical representations of shapes, and in children, these prototypes may be simpler and less numerous. Another possibility is that limited executive control makes it more difficult for children to inhibit pre-potent action routines. To illustrate this in simple words: having just placed two stickers onto adjacent cells of the grid, it may be very tempting to just continue placing stickers on the entire row or column of cells as this appears to be an intuitively plausible and well-practiced routine to follow. Adults are likely to be more capable of inhibiting such pre-potent action routines. Thus, both of these explanations support the notion that the cognitive demands of generating and reproducing complexity are higher for children than for adults. Crucially, we have demonstrated that because reproducing complexity is less feasible for children it makes them more prone to 'injecting' structure, which then is amplified over the course of cultural transmission. Importantly, the emergence of unique patterns in each chain shows that distinct lineages develop because structuring is not contingent upon initial conditions but shaped by learner biases.

It certainly is a far stretch from simple random dot patterns to complex cultural traits such as language. Still, our findings draw attention to the possibility that whenever children act as agents of cultural transmission, the structure of the resulting traits can be enhanced due to children's specific processing limitations. Future research will have to explore to what extent, at what ages, and in what ways the innovative potential that arises from children's processing limitations affects the transmission of the various complex traits and conventions that make up our surrounding human culture.

References

1. Kempe, V.: Gauvrit, Nicolas, Forsyth, Douglas: structure emerges faster during cultural transmission in children than in adults. Cognition **136**, 247–254 (2015)
2. Dieguez, S.: Wagner-Egger, Pascal, Gauvrit, Nicolas: nothing happens by accident, or does it? a low prior for randomness does not explain belief in conspiracy theories. Psychol. Sci. **26**(11), 1762–1770 (2015)
3. Kane, M.J., Hambrick, D.Z., Conway, A.R.A.: Working memory capacity and fluid intelligence are strongly related constructs: comment on Ackerman, Beier, and Boyle (2005). Psychol. Bull. **131**(1), 66–71 (2005)
4. Oberauer, K., et al.: Individual differences in working memory capacity and reasoning ability. Variation in Working Memory, pp. 49–75 (2007)
5. Little, D.R., Lewandowsky, S., Craig, S.: Working memory capacity and fluid abilities: the more difficult the item, the more more is better. Front. Psychol. **5**, 239 (2014)
6. Baum, E.B.: What is Thought? MIT Press, Cambridge (2004)
7. Hutter, M.:On the existence and convergence of computable universal priors. In: Proceedings of the 14th International Conference on Algorithmic Learning Theory (ALT-2003). Ed. by Springer. Vol. 2842, Sapporo, Lecture Notes on Artificial Intelligence, Vol. 2842, Lecture Notes on Artificial Intelligence (2003), pp. 298–312
8. Ferrer-i-Cancho, R., et al.: Compression as a universal principle of animal behavior. Cognit. Sci. **37**(8), 1565–1578 (2013)
9. Unsworth, N., Engle, R.W.: On the division of shortterm and working memory: an examination of simple and complex span and their relation to higher order abilities. Psychol. Bull. **133**(6), 1038–1066 (2007)
10. Cowan, N.: Metatheory of storage capacity limits. Behav. Brain Sci. **24**(1), 154–176 (2001)
11. Miller, G.A.: The magical number seven, plus or minus two: some limits on our capacity for processing information. Psychol. Rev. **63**(2), 81 (1956)
12. Chekaf, M.: Cowan, Nelson, Mathy, Fabien: Chunk formation in immediate memory and how it relates to data compression. Cognition **155**, 96–107 (2016)
13. Mathy, F., Feldman, J.: Whats magic about magic numbers? Chunking and data compression in short-term memory. Cognition **122**(3), 346–362 (2012)
14. Chekaf, M., et al.: Compression in working memory and its relationship with fluid intelligence. Cognit. Sci. (2018)
15. Towse, J.N., Cheshire, A.: Random number generation and working memory. Eur. J. Cognit. Psychol. **19**(3), 374–394 (2007)
16. Miyake, A., et al.: The unity and diversity of executive functions and their contributions to frontal lobe tasks: a latent variable analysis. Cognit. Psychol. **41**, 49–100 (2000)
17. Brugger, P., Landis, T., Regard, M.: A sheep-goat effectin repetition avoidance: extra-sensory perception as an effect of subjective probability? Br. J. Psychol. **81**(4), 455–468 (1990)
18. Vandewiele, M., et al.: Number and color preferences in four countries. Percept. Motor Skills **63**(2), Pt 2, 945–946 (1986). ISSN: 1558-688X. http://search.ebscohost.com.gate3.inist.fr/login.aspx?direct=true&db=psyh&AN=1988-13529-001&lang=fr&site=ehost-live
19. Strenge, H., Lesmana, C.B., Suryani, L.K.: Random number generation in bilingual Balinese and German students: preliminary findings from an exploratory cross-cultural study. Percept. Motor Skills **109**, 61–75 (2009)

20. Heuer, H., Janczyk, M., Kunde, W.: Random noun generation in younger and older adults. Q. J. Exp. Psychol. **63**(3), 465–478 (2010)
21. Loetscher, T., et al.: Eye position predicts what number you have in mind. Curr. Biol. **20**(6), R264–R265 (2010)
22. Ka-Shing Chan, K., et al.: Random number generation deficit in early schizophrenia. Percept. Motor Skills **112**(1), 91–103 (2011)
23. Proios, H., Asaridou, S.S., Brugger, P.: Random number generation in patients with aphasia: a test of executive functions. Acta Neuropsychol. **6**, 157–168 (2008)
24. Rinehart, N.J., et al.: Pseudo-random number generation in children with high-functioning autism and Asperger's disorder. Autism **10**(1), 70–85 (2006)
25. Loetscher, T., Bockisch, C., Brugger, P.: Eye position predicts what number you have in mind. Curr. Biol. **20**(6), 264–265 (2009)
26. Taylor, K.I., et al.: Semantic and phonemic sequence effects in random word generation: a dissociation between Alzheimer's and Huntington's disease patients. J. Int. Neuropsychol. Soc. **11**(3), 303–310 (2005). ISSN: 1469-7661. http://search.ebscohost.com.gate3.inist.fr/login.aspx?direct=true&db=psyh&AN=2005-05122-009&lang=fr&site=ehost-live
27. Hahn, U., Warren, P.: Perceptions of randomness: why three heads are better than four. Psychol. Rev. **116**(2), 454–461 (2009)
28. Mittenecker, E.: Die Analyse zuflliger Reaktionsfolgen. Zeitschrift fr Experimentelle und Angewandte Psychologie **5**, 45–60 (1958)
29. Schulter, G., Mittenecker, E., Papousek, I.: A computer program for testing and analyzing random generation behavior in normal and clinical samples: the Mittenecker Pointing Test. Behav. Res. Methods **42**(1), 333–341 (2010)
30. Shannon, C.E.: A mathematical theory of communication. Bell Syst. Techn. J. **27**, pp. 379–423, 623–656 (1948)
31. Champernowne, D.G.: The construction of decimals normal in the scale of ten. J. Lond. Math. Soc. **8**, 254–260 (1933)
32. Becher, V., Figueira, S.: An example of a computable absolutely normal number. Theor. Comput. Sci. **270**, 947–958 (2002)
33. Ginsburg, N., Karpiuk, P.: Random generation: analysis of the responses. Percept. Motor Skills **79**, 1059–1067 (1994)
34. Simon, H.A.: Complexity and the representation of patterned sequences of symbols. Psychol. Rev. **79**(5), 369 (1972)
35. Aksentijevic, A., Gibson, K.: Complexity equals change. Cognit. Syst. Res. **15-16**, 1–16 (2012). ISSN: 1389-0417. http://search.ebscohost.com.gate3.inist.fr/login.aspx?direct=true&db=psyh&AN=2012-24458-002&lang=fr&site=ehost-live
36. Nickerson, R.S.: The production and perception of randomness. Psychol. Rev. **109**(2), 330 (2002)
37. Stoffers, D., et al.: Motor perseveration is an early sign of Parkinsons disease. Neurology **57**, 2111–2113 (2001)
38. Chapanis, A.: Human production of "random" numbers. Percept. Motor Skills **81**, 1347–1363 (1995)
39. Tubau, E., Lpez-Moliner, J.: Knowing what to respond in the future does not cancel the influence of past events. PLoS ONE **4**(5), e5607 (2009)
40. Kidd, C., Piantadosi, S.T., Aslin, R.N.: The goldilocks effect: human infants allocate attention to visual sequences that are neither too simple nor too complex. PLoS **7**(1), e36399 (2012)
41. Tversky, A., Kahneman, D.: Belief in the "Law of small numbers". Psychol. Bull. **76**, 105–110 (1971)
42. Matthews, W.J.: Relatively random: context effects on perceived randomness and predicted outcomes. J. Exp. Psychol.: Learn. Mem. Cognit. **39**(5), 1642 (2013)
43. Gauvrit, N., et al.: Algorithmic complexity for short binary strings applied to psychology: a primer. Behav. Res. Methods **46**(3), 732–744 (2014)
44. Cowan, N.: The magical number 4 in short-term memory: a reconsideration of mental storage capacity. Behav. Brain Sci. **24** (01 Feb. 2001), 87–114. ISSN: 1469-1825. https://doi.org/10.1017/S0140525X01003922. http://journals.cambridge.org/article_S0140525X01003922

45. Piaget, J., Inhelder, B.: La gense de l'ide de hasard chez l'enfant. Presses Universitaires de France (1951)
46. Green, D.R.: Children's understanding of randomness: report of a survey of 1600 children aged 7-11 years. In: Proceedings of ICOTS 2, pp. 287–291 (1986)
47. Cohen, J.: A power primer. Psychol. Bull. **112**(1), 155–159 (1992)
48. Spelke, E.S., Kinzler, K.D.: Core knowledge. Dev. Sci. **10**(1), 89–96 (2007)
49. Halberda, J., et al.: Number sense across the lifespan as revealed by a massive Internet-based sample. Proc. Natl. Acad. Sci. **109**(28), 11116–11120 (2012)
50. Fontanari, L., et al.: Probabilistic cognition in two indigenous Mayan groups. Proc. Natl. Acad. Sci. **111**(48), 17075–17080 (2014)
51. Tgls, E., et al.: Pure reasoning in 12-month-old infants as probabilistic inference. Science **332**(6033), 1054–1059 (2011)
52. Williams, J.J., Griffiths, T.L.: Why are people bad at detecting randomness? A statistical argument. J. Exp. Psychol.: Learn. Memory Cognit. **39**(5), 1473 (2013)
53. Sagaspe, P., et al.: Inhibition and working memory: effect of acute sleep deprivation on a random letter generation task. Can. J. Exp. Psychol. (Revue canadienne de psychologie experimentale) **57**(4), 265–273 (2003)
54. Craik, F.I.M., Bialystok, E.: Cognition through the lifespan: mechanisms of change. Trends Cognit. Sci. **10**(3), 131–138 (2006)
55. Van der Linden, M., Beerten, A., Pesenti, M.: Age-related differences in random generation. Brain Cognit. **38**(1), 1–16 (1998)
56. Chater, N., Vitnyi, P.: Simplicity: a unifying principle in cognitive science? Trends Cogni. Sci. **7**(1), 19–22 (2003)
57. Gauvrit, N., et al.: Package acss. Compr. R Arch. Netw. (2014). https://cran.r-project.org/web/packages/acss/
58. Forsythe, A., et al.: Predicting beauty: fractal dimension and visual complexity in art. Br. J. Psychol. **102**(1), 49–70 (2011)
59. Nadal, M., et al.: Visual complexity and beauty appreciation: explaining the divergence of results. Empir. Stud. Arts **28**(2), 173–191 (2010)
60. Palumbo, L., et al.: Examining visual complexity and its influence on perceived duration. J. Vis. **14**(14), 3–3 (2014)
61. Spehar, B., et al.: Universal aesthetic of fractals. Comput. & Graph. **27**(5), 813–820 (2003)
62. Friedenberg, J., Liby, B.: Perceived beauty of random texture patterns: a preference for complexity. Acta Psychol. **168**, 41–49 (2016)
63. Gauvrit, N., Morsanyi, K.: The equiprobability bias from a mathematical and psychological perspective. Adv. Cognit. Psychol. **10**(4), 119 (2014)
64. Falk, R., Konold, C.: Making sense of randomness: implicit encoding as a basis for judgment. Psychol. Rev. **104**(2), 301 (1997)
65. Hsu, A.S., Griffiths, T.L., Schreiber, E.: Subjective randomness and natural scene statistics. Psychon. Bull. & Rev. **17**(5), 624–629 (2010)
66. Berkes, P., et al.: Spontaneous cortical activity reveals hallmarks of an optimal internal model of the environment. Science **331**(6013), 83–87 (2011)
67. Inui, T., et al.: Spatiochromatic receptive field properties derived from information-theoretic analyses of cone mosaic responses to natural scenes. Neural Comput. **15**(2), 397–417 (2003)
68. McCauley, C., Jacques, S.: The popularity of conspiracy theories of presidential assassination: a Bayesian analysis. J. Pers. Soc. Psychol. **37**(5), 637 (1979)
69. Leman, P.J., Cinnirella, M.: A major event has a major cause: evidence for the role of heuristics in reasoning about conspiracy theories. Soc. Psychol. Rev **9**, 18–28 (2007)
70. Franks, B., Bangerter, A., Bauer, M.: Conspiracy theories as quasi-religious mentality: an integrated account from cognitive science, social representations theory, and frame theory. Front. Psychol. **4**, 424 (2013)
71. Whitson, J.A., Galinsky, A.D.: Lacking control increases illusory pattern perception. Science **322**(5898), 115–117 (2008)

72. Prooijen, J.-W., Jostmann, N.B.: Belief in conspiracy theories: the influence of uncertainty and perceived morality. Eur. J. Soc. Psychol. **43**(1), 109–115 (2013)
73. Brotherton, R., French, C.C., Pickering, A.D.: Measuring belief in conspiracy theories: the generic conspiracist beliefs scale. Front. Psychol. **4**, 279 (2013)
74. Wagner-Egger, P., Bangerter, A.: The truth lies elsewhere: correlates of belief in conspiracy theories. Revue Internationale de Psychologie Sociale **20**(4), 31–61 (2007)
75. Bartlett, F.C., et al.: Remembering: A Study in Experimental and Social Psychology. Cambridge Books Online. University Press, Cambridge (1995). ISBN: 9780521483568. https://books.google.es/books?id=WG5ZcHGTrm4C
76. Bickerton, D.: Roots of language. Karoma. Ann. Arbor. **10**, 268–284 (1981)
77. Pinker, S., Bloom, P.: Natural language and natural selection. Behav. Brain Sci. **13**(4), 707–727 (1990)
78. Newport, E.L.: Maturational constraints on language learning. Cognit. Sci. **14**(1), 11–28 (1990)
79. Elman, J.L.: Learning and development in neural networks: the importance of starting small. Cognition **48**(1), 71–99 (1993)
80. Kam, C.L.H., Newport, E.L.: Regularizing unpredictable variation: the roles of adult and child learners in language formation and change. Lang. Learn. Dev. **1**(2), 151–195 (2005)
81. Kam, C.L.H., Newport, E.L.: Getting it right by getting it wrong: when learners change languages. Cognit. Psychol. **59**(1), 30–66 (2009)
82. Senghas, A., Coppola, M.: Children creating language: how Nicaraguan sign language acquired a spatial grammar. Psychol. Sci. **12**(4), 323–328 (2001)
83. Claidiere, N., et al.: Cultural evolution of systematically structured behaviour in a non-human primate. Proc. R. Soc. Lond. B: Biol. Sci. **281**(1797), 20141541 (2014)
84. Yamada, Y., Kawabe, T., Miyazaki, M.: Pattern randomness aftereffect. Sci. Rep. **3** (2013)
85. Hahn, U.: Experiential limitation in judgment and decision. Topics Cognit. Sci. **6**(2), 229–244 (2014)

Appendix
Source Code of the Turing Machine Simulator

This appendix presents the complete source code of the Turing machine simulator used to run all machines in $Red(s, k)$. This is a C++ program that can be compiled with GCC using GMP:

```
g++ -o TMsimulator TMsimulator.cpp -lgmpxx -lgmp
```

The execution of the simulator follows the following format:

```
TMsimulator <s> <k> <r> <first> <last>
```

Where `<s>` is the number of states, `<k>` the number of symbols, `<r>` the maximum runtime and `<first>` and `<last>` are integers representing machines in the reduced enumeration for (s, k) (Chap. 2). The simulator will run all machines from `<first>` to `<last>` and will return a table with the output strings of all halting machines and non-halting codes of the rest. See more details on Sect. 2.5.

```
// File: TMsimulator.cpp
#include <stdlib.h>
#include <gmp.h>
#include <iostream>
#include <sstream>
#include <map>
using namespace std;

typedef int symbol;
enum direction {DIR_LEFT, STOP, DIR_RIGHT};
```


H. Zenil et al., *Methods and Applications of Algorithmic Complexity*,
Emergence, Complexity and Computation 44,
https://doi.org/10.1007/978-3-662-64985-5

```
struct instruction {    // Entries in transition table
  int head_state;       // – new state
  symbol write_symbol;  // – written symbol
  enum direction dir;   // – head direction
};
typedef struct instruction instruction;

struct turing_machine_state { // The state of a TM
  int head_state;           // – current state
  int head_position;        // – tape position of the head
  int max_head_position;    // – most–right visited cell
  int min_head_position;    // – most–left visited cell
  int escapees;             // – to detect escapees
  map<int, symbol> tape;    // – the tape
};
typedef struct turing_machine_state
        turing_machine_state;

struct turing_machine {          // Data of the TM
  int number_states;             // – number of states
  int number_colors;             // – number of symbols
  int running_state;             // – running codes
  turing_machine_state state;// – state of the TM
  instruction** transition_table; // – transition table
};
typedef struct turing_machine turing_machine;

// Executes a step in the given TM
int run_step(turing_machine *m){

  int state1 = m->state.head_state;   // current state
  int cell = m->state.head_position; // current position
  symbol s1 = m->state.tape[cell];    // reading symbol

  // Gets the transition to apply
  instruction tr = m->transition_table[state1][s1];

  m->state.tape[cell] = tr.write_symbol; // write symbol
  m->state.head_state = tr.head_state; // change state

  // Stops (without moving the head) if halting state
  if(m->state.head_state == -1){
    m->running_state=0;
    return 0;
  }
```

```
    bool escap = false; // flag to detect escapees

    // Moving the head
    switch(tr.dir){
    case DIR_LEFT:                          // LEFT
      m->state.head_position--;             // move
      if(m->state.min_head_position >
         m->state.head_position){
        m->state.min_head_position--; // new cell visited
        escap = true;                       // activate flag
      };
      break;
    case DIR_RIGHT:                         // RIGHT
      m->state.head_position++;             // move
      if(m->state.max_head_position <
         m->state.head_position){
        m->state.max_head_position++; // new cell visited
        escap = true;                       // activate flag
      };
      break;
    }

    // Detecting escapees
    if(escap){  // only executed if 'escap' flag is true
      m->state.escapees++;          // increments escapees
      if(s1 == 0 && m->state.head_state == state1){
        m->running_state = -3;  // short escapee
        return 0;
      }
      if(m->state.escapees > m->number_states){
        m->running_state = -4;  // long escapee
        return 0;
      }
    } else{  // if no new cell was visited
      m->state.escapees = 0;        // set counter to 0
    }

    // Detecting cycles of order two
    if(tr.write_symbol == s1){
      symbol s2 =                // the following read symbol
        m->state.tape[m->state.head_position];
      instruction tr2 =          // the following transition
        m->transition_table[m->state.head_state][s2];
      if (tr2.head_state == state1 && // previous state
      tr2.write_symbol == s2 &&     // the same symbol
```

```
      tr2.dir != tr.dir){              // to previous cell
        m->running_state = -5;           // cycle detected
        return 0;
      }
    }
    return 1;
  }

  // Initialization of a Turing Machine
  // - states: number of states
  // - colors: number of symbols
  // - numberTM: big integer representing the TM number
  turing_machine init_turing_machine(int states,
        int colors, mpz_t numberTM){

    int i = 0;
    int j= 0;

    turing_machine m;
    m.number_colors = colors;        // set symbols
    m.number_states = states;        // set states

    // Create the transition table
    m.transition_table = new instruction*[states];
    for(i=0;i<states;i++)
      m.transition_table[i] = new instruction[colors];

    // Set the initial state
    m.state.head_state = 0;
    m.state.head_position = 0;
    m.state.max_head_position = 0;
    m.state.min_head_position = 0;
    m.state.escapees = 0;

    // Filling the transition table
    mpz_t gr; // rest
    mpz_t gc; // quotient
    mpz_init(gr);
    mpz_init(gc);

    int rest;
    int halts = -2; // to detect TMs without transitions
                    // to halting state
    // Filling the initial transition
    mpz_fdiv_qr_ui(gc, gr, numberTM, colors*(states-1));
```

```
        mpz_set(numberTM, gc);

        m.transition_table[0][0].dir = DIR_RIGHT;
        rest = mpz_get_ui(gr);
        m.transition_table[0][0].write_symbol =
              (rest % colors);
        rest = rest/colors;
        m.transition_table[0][0].head_state =
              1+(rest % (states-1));
        // Other transitions from the initial state
        for(j=1;j<colors;j++){
          mpz_fdiv_qr_ui(gc, gr, numberTM,
                         colors*((2*states)+1));
          mpz_set(numberTM, gc);
          rest = mpz_get_ui(gr);
          if(rest<colors){
            m.transition_table[0][j].head_state = -1;
            m.transition_table[0][j].write_symbol = rest;
            m.transition_table[0][j].dir = STOP;
            halts = -1; // transition to halting state
          } else {
            rest = rest - colors;
            m.transition_table[0][j].write_symbol =
                  (rest % colors);
            rest = rest/colors;
            m.transition_table[0][j].dir =
                  ((rest % 2)==0 ? DIR_RIGHT : DIR_LEFT);
            rest = rest / 2;
            m.transition_table[0][j].head_state =
                  (rest % states);
          }
        }

        // Transitions from other states
        for (i=1; i<states; i++) {
          for(j=0;j<colors;j++){
            mpz_fdiv_qr_ui(gc, gr, numberTM,
                           colors*((2*states)+1));
            mpz_set(numberTM, gc);
            rest = mpz_get_ui(gr);
            if(rest<colors){
          m.transition_table[i][j].head_state = -1;
          m.transition_table[i][j].write_symbol = rest;
          m.transition_table[i][j].dir = STOP;
          halts = -1; // transition to halting state
```

```
        } else {
          rest = rest - colors;
          m.transition_table[i][j].write_symbol =
                    (rest % colors);
          rest = rest/colors;
          m.transition_table[i][j].dir =
                    ((rest % 2)==0 ? DIR_RIGHT : DIR_LEFT);
          rest = rest / 2;
          m.transition_table[i][j].head_state =
                    (rest % states);
        }
      }
    }
    m.running_state = halts; // will indicate if there is
    mpz_clear(gr);           // a halting transition
    mpz_clear(gc);

    return m;
}

// Returns the output string of a Turing machine
string outputTM(turing_machine *m){
    string sout;
    stringstream strm;
    if(m->running_state == 0){
      int i;
      for (i=m->state.min_head_position;
           i<=m->state.max_head_position; i++)
        strm << (m->state.tape[i]);
    } else {
      strm << m->running_state;
    }
    strm >> sout;
    return sout;
}

// Deallocate the memory used by the machine
void deleteTT(turing_machine *m){
    int i;
    m->state.tape.clear();
    for(i=0; i<m->number_states;i++)
      delete(m->transition_table[i]);
    delete(m->transition_table);
}
```

```
// Main function
int main(int argn, char* argv[]) {

  turing_machine machine;

  // map to store output strings
  map<string, unsigned long long> results;
  string out;
  int i;

  int s = atoi(argv[1]);          // number of states
  int k = atoi(argv[2]);          // number of symbols
  int runtime =  atoi(argv[3]);   // maximum runtime

  mpz_t TM;                        // first TM to run
  mpz_init_set_str(TM,argv[4],10);

  mpz_t Last;                      // last TM to run
  mpz_init_set_str(Last,argv[5],10);

  mpz_t acc;
  mpz_init(acc);

  // Run all machines in the block
  while(mpz_cmp(TM,Last)<=0){
    mpz_set(acc,TM);
    machine = init_turing_machine(s,k,acc);  // init TM
    if(machine.running_state==-1){    // no -2 detected
      // Run the machine
      for(i=1; i<=runtime && run_step(&machine); i++){};
      out = outputTM(&machine);        // get output
      ++results[out];                  // store output
    } else
      ++results["-2"];          // no halting transition

    deleteTT(&machine);  // free used memory
    mpz_add_ui(TM,TM,1); // number of the next machine
  }
  mpz_clear(TM);
  mpz_clear(Last);
  mpz_clear(acc);

  // Print the results
  map<string,unsigned long long>::iterator it;
  for(it=results.begin(); it!=results.end(); ++it)
```

```
        {
          cout<<(*it).first <<","<< (*it).second <<'\n';
        }
    }
```

Index

A
Absolutely normal sequences, 199
Algorithmic complexity, 231
Algorithmic probability, 40, 81
Alternation bias, 202
Aphasia, 197
Apollo 11 moon landing, 238
Approximate Sense of Complexity, 213
11-9 attacks, 238

B
Bayes' theorem, 10
Binarized pictures, 233
Bioinformatics
  regulatory networks, 40
  sequence analysis, 40
Bit, 8
Busy Beaver, 4, 31, 39, 40, 43–48, 50, 55, 62, 65, 103

C
Cellular automata, 3–5
  compressibility, 12
  compression-based classification, 12
  Elementary Cellular Automata, 4, 5, 12
  gliders, 6
  Wolfram classification, 5, 12
Champernowne sequence, 199
Change complexity, 200
Channel capacity, 9
Charlie Hebdo, 235
Chi-square distribution, 198
Chunking, 210
Claude Shannon, 8
Climber, 106
Coding Theorem, 39, 40, 43, 104, 106
  validation by compressibility, 110
Cognitive complexity, 200
Cognitive effects of aging, 197
Cognitive flexibility, 201
Completion time, 219
Compression algorithms, 40, 110
  Lempel-Ziv algorithm, 11
  lossless compression, 10
  lossy compression, 10
Conditional probability, 8–10
Conspiracy theories, 235
Context redundancy, 201
Coperland–Erdös sequence, 199
Core knowledge, 213
Correlation, 11
  Pearson correlation coefficient, 12
  Spearman correlation coefficient, 12
Cycling bias, 200

D
Data compression, 10
Departure from uniformity, 198
Down syndrome, 197

E
Elementary Cellular Automata, 121
Entropy, 8, 9, 198

F
Filling the grid task, 216
Fluid intelligence, 191

H. Zenil et al., *Methods and Applications of Algorithmic Complexity*,
Emergence, Complexity and Computation 44,
https://doi.org/10.1007/978-3-662-64985-5

Frequency distribution, 40

G
Game of Life, 7, 112
Gap methods, 200
Generic Conspiracist Beliefs Scale, 238
Grammar, 245

H
Halting probability, 40
Human complexity perception, 200
Human production of randomness, 197, 201, 206

I
Image complexity, 230
Independent occurrences, 7
Indices of randomness, 198
Information Theory, 8, 198
Inhibition processes, 197
Inhibitive abilities, 197
Intelligence, 191
Iterated language learning, 245

J
John Conway, 7
John F. Kennedy, 238

K
Kolmogorov–Chaitin complexity, 231
Kolmogorov complexity
    complexity of a string, 200

L
Language evolution, 245
Lempel-Ziv algorithm, 39
Levin's semi-measure, 40, 43
Lifespan, 214
Local complexity of a string, 209
Logical Depth, 83, 87

M
Measures of randomness, 198
Mechanical Turk, 232
Memory span, 192
Mersenne Twister, 30
Mittenecker Pointing Test, 198
Mixed radix number, 23

N
Natural scene statistics, 230, 234, 235
Neuropsychological tests, 197
Normal distribution, 11
Normality, 199
Normal sequence, 200, 201

P
Paranormal beliefs, 220
Pattern recognition, 230
Political background, 243
Post tag systems, 3
Power-law distribution, 11
Princess Diana's car crash, 238
Probability theory, 7, 10
    axioms of probability, 7

R
Random generation tasks, 197
Random Item Generation, 213
Randomness coefficient, 199
Raven's Matrices, 194
Regularizing biase, 245
Runtime distribution, 31, 103
Runtime sample, 30

S
Sample
    runtime sample, 33
Schizophrenia, 197
Shannon's noisy-channel coding theorem, 10
Short-term memory, 191
SIMON game, 193
Sobel test, 234
Stephen Wolfram, 5
Symbol Redundancy, 198
Symmetry, 202

T
True randomness, 237
Turing machine, 200
    blank symbol, 20, 25, 26, 28
    complete enumeration, 18–19
    completions, 25–26
    cycle, 29, 37
    direction, 17, 27, 34, 37

escapee, 28, 37
execution, 20, 39
filter, 27–29
formalism, 3, 17
graphical representation, 20
halting state, 4, 17, 20–22, 27–29, 40, 62, 80
initial state, 21, 22, 27
initial transition, 21, 22, 24, 25, 27, 30
instruction, 17, 27
minimal number of instructions, 80
minimum runtime, 82
model of computation, 3
non-halting machine, 21, 25–29
number, 19, 23
output, 4, 20, 29, 37
random halting machine, 81, 103
random machine, 30
reduced enumeration, 20–24, 27
runtime, 27, 29
short escapee, 28
simulator, 27–38
space, 25, 31
state, 17, 20, 27–29
step, 20, 28
symbol, 17, 27, 29
symmetries, 20–22, 25
tape, 20, 28, 29, 34
transition, 27, 29, 102
transition table, 17, 27, 28
two-dimensional machine, 34–38, 101, 103

U
Uniformity, 198
Universal Distribution, 39, 102

W
Wikimedia Commons, 233
Working memory, 191, 197, 210

Z
Zener cards, 215